J. Fort Freidl

Studies in Surface Science and Catalysis 163

FISCHER-TROPSCH SYNTHESIS, CATALYSTS AND CATALYSIS

Studies in Surface Science and Catalysis

Advisory Editors: B. Delmon and J.T. Yates
Series Editor: G. Centi

Vol. 163

FISCHER-TROPSCH SYNTHESIS, CATALYSTS AND CATALYSIS

Edited by

B.H. Davis
Center for Applied Energy Research
University of Kentucky
Lexington, Kentucky, U.S.A.

M.L. Occelli
MLO Consultants
Atlanta, Georgia, U.S.A.

ELSEVIER

Amsterdam – Boston – Heidelberg – London – New York – Oxford – Paris
San Diego – San Francisco – Singapore – Sydney – Tokyo

Elsevier
Radarweg 29, PO Box 211, 1000 AE Amsterdam, The Netherlands
The Boulevard, Langford Lane, Kidlington, Oxford OX5 1GB, UK

First edition 2007

Library of Congress Cataloging-in-Publication Data
A catalog record for this book is available from the Library of Congress

British Library Cataloguing in Publication Data
A catalogue record for this book is available from the British Library

ISBN-13:   978-0-444-52221-4
ISBN-10:   0-444-52221-2
ISSN: 0167-2991

> For information on all Elsevier publications
> visit our website at books.elsevier.com

Printed and bound in The Netherlands

07 08 09 10 11   10 9 8 7 6 5 4 3 2 1

Foreword

In general, there are two approaches to the production of substitutes for crude petroleum. In one of these, the organic material is heated at high temperatures under a high pressure of hydrogen. In the other approach, the organic material is converted to a mixture of hydrogen and carbon monoxide (syngas) and this syngas is converted to hydrocarbons by conversion over suitable catalysts. The papers included in the present volume are concerned with the indirect liquefaction approach.

The introduction of the catalytic synthesis of ammonia was widely recognized. The Nobel Prize in 1918 for chemistry was awarded to Fritz Haber for his developments that led to the synthesis of ammonia from the elements. The development of the very high pressure ammonia synthesis and its commercial success gave Germany a decided leadership position in high pressure process during the early part of the twentieth century. Rapidly following the ammonia synthesis, the commercial production of methanol from synthesis gas was a commercial success. After much work, Bergius finally was able to show that heating coal at high temperatures under high pressures of hydrogen led to the production of liquid products. Fritz Fischer, director of the coal research laboratory, worked to develop a coal conversion process that could compete with the direct process developed by Bergius. During the 1920s, the work by Fischer and coworkers led to what is now known as the Fischer-Tropsch process. The advances in high pressure process technology led to the Nobel Prize being awarded in 1932 to Bergius and Carl Bosch; however, the Fischer-Tropsch scientific advances were not afforded this honor. The Fischer-Tropsch process also lost out to the direct coal liquefaction process in the production of synfuels in Germany during the 1935-1945 period, for both technological and political reasons.

During the energy crisis of the 1970s the direct and indirect coal liquefaction processes received much attention. During this period the direct coal liquefaction process received more attention in the U.S., with four large scale demonstration plants being operated. At that time, the major goal of producing synfuels was to provide a source of gasoline and the direct liquefaction process provided high octane gasoline due to its high aromatics content. Today the direct coal liquefaction process is out of favor, primarily because of the high aromatics content and the reduction of the high heteroatom content which greatly exceed today's environmental requirements. This, plus the advances in Fischer-Tropsch technology during the intervening thirty years, leads to the concentration of the effort to produce commercial quantities of synfuels upon the Fischer-Tropsch technology. In addition to the fifty year efforts by Sasol that now produces about 150,000 bbl/day, Shell Oil (15,000 bbl/d) and PetroSA (formerly Mossgas; 40,000 bbl/d) became commercial producers in the early 1990s. Sasol has brought on line a 35,000 bbl/d plant in Qatar in mid-2006.

The present book addresses four major areas of interest in Fischer-Tropsch synthesis (FTS). The first three contributions address the development of FTS during the early years in Germany and Japan and more recently by BP. The next section includes eight contributions that relate to the development of catalysts for FTS, their structure and changes that occur during use. The third section contains six contributions that relate to impact of various process conditions upon the productivity and selectivity of the FTS operation. The final section consists of six contributions relating to the FTS process and the conversion of the primary products to useful fuels. Most of these contributions are based on presentations at the 2005 Spring National Meeting of the American Chemical Society, held in San Diego in 2005.

Table of Contents

Fischer-Tropsch Synthesis, Catalysts and Catalysis
B.H. Davis and M.L. Occelli (Editors)

A History of the Fischer-Tropsch Synthesis in Germany 1926-45

Anthony N. Stranges

Department of History, Texas A&M University, College Station, TX 77843-4236

1. Introduction: twentieth-century synthetic fuels overview

The twentieth-century coal-to-petroleum, or synthetic fuel, industry evolved in three stages: (1) invention and early development of the Bergius coal liquefaction (hydrogenation) and Fischer-Tropsch (F-T) synthesis from 1910 to 1926; (2) Germany's industrialization of the Bergius and F-T processes from 1927 to 1945; and (3) global transfer of the German technology to Britain, France, Japan, Canada, the United States, South Africa, and other nations from the 1930s to the 1990s.

Petroleum had become essential to the economies of industrialized nations by the 1920s. The mass production of automobiles, the introduction of airplanes and petroleum-powered ships, and the recognition of petroleum's high energy content compared to wood and coal, required a shift from solid to liquid fuels as a major energy source. Industrialized nations responded in different ways. Germany, Britain, Canada, France, Japan, Italy, and other nations having little or no domestic petroleum continued to import petroleum. Germany, Japan, and Italy also acquired by force the petroleum resources of other nations during their 1930s-40s World War II occupations in Europe and the Far East. In addition to sources of naturally-occurring petroleum, Germany, Britain, France, and Canada in the 1920s-40s synthesized petroleum from their domestic coal or bitumen resources, and during the 1930s-40s war years Germany and Japan synthesized petroleum from the coal resources they seized from occupied nations. A much more favorable energy situation existed in the United States, and it experienced few problems in making an energy shift from solid to liquid fuels because it possessed large resources of both petroleum and coal.

Germany was the first of the industrialized nations to synthesize petroleum when Friedrich Bergius (1884-1949) in Rheinau-Mannheim in 1913 and Franz Fischer (1877-1947) and Hans Tropsch (1889-1935) at the Kaiser Wilhelm Institute for Coal Research (KWI) in Mülheim, Ruhr, in 1926 invented processes for converting coal to petroleum. Their pioneering researches enabled IG Farben, Ruhrchemie, and other German chemical companies to develop a technologically-successful synthetic fuel industry that grew from a single commercial-size coal liquefaction plant in 1927 to twelve coal liquefaction and nine F-T commercial-size plants that in 1944 reached a peak production of 23 million barrels of synthetic fuel.

Britain's synthetic fuel program evolved from post-World War I laboratory and pilot-plant studies that began at the University of Birmingham in 1920 on the F-T synthesis and in 1923 on coal liquefaction. The Fuel Research Station in East Greenwich also began research on coal liquefaction in 1923, and the program reached its zenith in 1935 when Imperial Chemical Industries (ICI) constructed a coal liquefaction plant at Billingham that had the capacity to synthesize annually 1.28 million barrels of petroleum. British research and development matched

Germany's, but because of liquefaction's high cost and the government's decision to rely on petroleum imports rather than price supports for an expanded domestic industry, Billingham remained the only British commercial-size synthetic fuel plant. F-T synthesis in the 1930s-40s never advanced beyond the construction of four small experimental plants: Birmingham, the Fuel Research Station's two plants that operated from 1935 to 1939, and Synthetic Oils Ltd. near Glasgow [1].

Britain and Germany had the most successful synthetic fuel programs. The others were either smaller-scale operations, such as France's three demonstration plants (two coal liquefaction and one F-T), Canada's bitumen liquefaction pilot plants, and Italy's two crude petroleum hydrogenating (refining) plants, or technological failures as were Japan's five commercial-size plants (two coal liquefaction and three F-T) that produced only about 360,000 barrels of liquid fuel during the World War II years [2].

The US Bureau of Mines had begun small-scale research on the F-T synthesis in 1927 and coal liquefaction in 1936, but did no serious work on them until the government expressed considerable concern about the country's rapidly increasing petroleum consumption in the immediate post-World War II years. At that time the Bureau began a demonstration program, and from 1949 to 1953 when government funding ended, it operated a small 200-300 barrel per day coal liquefaction plant and a smaller fifty barrel per day F-T plant at Louisiana, Missouri. In addition to the Bureau's program, American industrialists constructed four synthetic fuel plants in the late 1940s and mid-1950s, none of which achieved full capacity before shutdown in the 1950s for economic and technical reasons. Three were F-T plants located in Garden City, Kansas; Brownsville, Texas; and Liberty, Pennsylvania. The fourth plant was a coal liquefaction plant in Institute, West Virginia [3].

Following the plant shutdowns in the United States and until the global energy crises of 1973-74 and 1979-81, all major synthetic fuel research and development ceased except for the construction in 1955 of the South African Coal, Oil, and Gas Corporation's (SASOL) F-T plant in Sasolburg, south of Johannesburg. South Africa's desire for energy independence and the low quality of its coal dictated the choice of F-T synthesis rather than coal liquefaction. Its Johannesburg plant remained the only operational commercial-size synthetic fuel plant until the 1970s energy crises and South Africa's concern about hostile world reaction to its apartheid policy prompted SASOL to construct two more F-T plants in 1973 and 1976 in Secunda.

The 1970s energy crises also revitalized synthetic fuel research and development in the United States and Germany and led to joint government-industry programs that quickly disappeared once the crises had passed. Gulf Oil, Atlantic Richfield, and Exxon in the United States, Saarbergwerke AG in Saarbrüken, Ruhrkohle AG in Essen, and Veba Chemie in Gelsenkirchen, Germany, constructed F-T and coal liquefaction pilot plants in the 1970s and early 1980s only to end their operation with the collapse of petroleum prices a few years later [4].

In the mid-1990s two developments triggered another synthetic fuel revival in the United States: (1) petroleum imports again reached 50 percent of total consumption, or what they were during the 1973-1974 Arab petroleum embargo, and (2) an abundance of natural gas, equivalent to 800,000,000,000 barrels of petroleum, but largely inaccessible by pipeline, existed. Syntroleum in Tulsa, Oklahoma; Exxon in Baytown, Texas; and Atlantic Richfield in Plano, Texas, developed modified F-T syntheses that produced liquid fuels from natural gas and thereby offered a way of reducing the United States's dependence on petroleum imports. The Department of Energy (DOE) at its Pittsburgh Energy Technology Center through the 1980s-90s also continued small-scale research on improved versions of coal liquefaction. DOE pointed out that global coal reserves greatly exceeded petroleum reserves, anywhere from five to twenty-four times, and that it expected petroleum reserves to decline significantly in 2010-2030. Syntroleum, Shell in Malaysia, and SASOL and Chevron in Qatar have continued F-T research, whereas DOE switched its coal liquefaction research to *standby*. The only ongoing coal liquefaction research is a pilot plant study by Hydrocarbon Technologies Incorporated in Lawrenceville, New Jersey, now Headwaters Incorporated in Draper, Utah.

A combination of four factors, therefore, has led industrialized nations at various times during the twentieth century to conclude that synthetic fuel could contribute to their growing liquid fuel requirements: (1) the shift from solid to liquid fuel as a major energy source, (2) the invention of the Bergius and F-T coal-to-petroleum conversion or synthetic fuel processes, (3) recognition that global petroleum reserves were finite and much less than global coal reserves and that petroleum's days as a plentiful energy source were limited, and (4) the desire for energy independence.

With the exception of South Africa's three F-T plants the synthetic fuel industry, like most alternative energies, has endured a series of fits and starts that has plagued its history. The historical record has demonstrated that after nearly 90 years of research and development synthetic liquid fuel has not emerged as an important alternative energy source. Despite the technological success of synthesizing petroleum from coal, its lack of progress and cyclical history are the result of government and industry uninterest in making a firm and a long-term commitment to synthetic fuel research and development. The synthetic fuel industry experienced intermittent periods of intense activity internationally in times of crises, only to face quick dismissal as unnecessary or uneconomical upon disappearance of the crises. Even its argument that synthetic liquid fuels are much cleaner burning than coal, and if substituted for coal they would reduce the emissions that have contributed to acid rain formation, greenhouse effect, and to an overall deterioration of air quality has failed to silence its critics. The hope of transforming its accomplishments at the demonstration stage into commercial-size production has not yet materialized.

The history of the synthetic fuel industry's fits and starts remains only partially written, with much of the historical interest having focused on Germany's coal hydrogenation process because it was the more advanced and contributed much more significantly to Germany's liquid fuel supply than the F-T synthesis. Coal hydrogenation produced high quality aviation and motor gasoline, whereas the F-T synthesis gave high quality diesel and lubricating oil, waxes and some lower quality motor gasoline. The two processes actually were complementary rather than competitive, but because only coal hydrogenation produced high quality gasoline it experienced much greater expansion in the late 1930s and war years than the F-T synthesis, which hardly grew at all. F-T products were mainly the raw materials for further chemical syntheses with little upgrading of its low quality gasoline by cracking because of unfavorable economics. Hydrogenation also experienced greater development because brown coal (lignite), the only coal available in many parts of Germany, underwent hydrogenation more readily than a F-T synthesis. In addition, the more mature and better developed hydrogenation process had the support of IG Farben, Germany's chemical leader, which successfully industrialized coal hydrogenation beginning in 1927 [5].

Despite its smaller size and lower production, the 9 F-T plants contributed 455,000-576,000 metric tons of coal-derived oil per year during the war years 12-15 percent of Germany's total liquid fuel requirement. The historical analysis that follows examines the T-T's invention and industrial development during several decades of German social, political , and economic unrest and complements the historical literature on Germany's coal hydrogenation process. The historical examination of the two processes provides a more complete history of Germany's synthetic fuel industry.

2. Early development of the F-T synthesis: catalysts, conditions, and converters

Germany has virtually no petroleum deposits. Prior to the twentieth century this was not a serious problem because Germany possessed abundant coal reserves. Coal provided for commercial and home heating; it also fulfilled the needs of industry and the military, particularly the navy. In the opening decade of the twentieth century, Germany's fuel requirements began to change. Two reasons were especially important. First, Germany became increasingly dependent on gasoline and diesel oil engines. The appearance of automobiles, trucks, and then airplanes made a plentiful supply of gasoline essential. Moreover, ocean-going ships increasingly used diesel oil rather than coal as their energy source. Second, Germany's continuing industrialization and urbanization led to the replacement of coal with smokeless liquid fuels that not only had a higher energy content but were cleaner burning and more convenient to handle.

Petroleum was clearly the fuel of the future, and to insure that Germany would never lack a plentiful supply, German scientists and engineers invented and developed two processes that enabled them to synthesize petroleum from their country's abundant coal supplies and to establish the world's first technologically successful synthetic liquid fuel industry [6]. Bergius in Rheinau-Mannheim began the German drive for energy independence with his invention and early development of high-pressure coal hydrogenation in the years 1910-25. Bergius crushed and dissolved a coal containing less than 85 percent carbon in a heavy oil to form a paste. He reacted the coal-oil paste with hydrogen gas at high pressure (P = 200 atmospheres = 202.6 x 10^2 kPa) and high temperature (T = 400°Celsius) and obtained petroleum-like liquids. Bergius sold his patents to BASF in July 1925, and from 1925 to 1930 Matthias Pier (1882-1965) at BASF (IG Farben in December 1925) made major advancements that significantly improved product yield and quality. Pier developed sulfur-resistant catalysts, such as tungsten sulfide (WS_2), and separated the conversion into two stages, a liquid stage and a vapor stage [7].

Figure 1. Friedrich Bergius

A decade after Bergius began his work Fischer and Tropsch at the Kaiser-Wilhelm Institute invented a second process for the synthesis of liquid fuel from coal. Fischer and Tropsch reacted coal with steam to give a gaseous mixture of carbon monoxide and hydrogen and then converted the mixture at low pressure (P = 1-10 atmospheres = 1.013-10.013 x 10^2 kPa) and a temperature (T = 180-200° Celsius) to petroleum-like liquids. Fischer and his co-workers in the 1920s-30s developed the cobalt catalysts that were critical to the F-T's success, and in 1934 Ruhrchemie acquired the patent rights to the synthesis.

Fischer had received the PhD at Giessen in 1899, where he studied under Karl Elbs (1858-1933) and his research focused on the electrochemistry of the lead storage battery. He continued his electrochemical studies spending a semester with Henri Moissan (1852-1907) in Paris, the years 1901-2 in Freiburg's chemical industry and 1902-4 at the University of Freiburg's physiochemical institute. Upon leaving Freiburg Fischer Conducted additional research from 1904 to 1911 in the institutes of Wilhelm Ostwald (1853-1932) in Leipzig and Emil Fischer in Berlin and from 1911 to 1914 at the Technische Hochschule in Berlin-Charlottenburg.

Emil Fischer (1852-1919) had an interest in Fischer's electrochemical work, and as a leading figure in establishing the KWIs beginning in 1912 he invited Fischer to direct the new institute for coal research planned for Mülheim in the Ruhr valley. The institute, which opened on 27 July 1914 was the first KWI located outside of Berlin-Dahlem, and like the others the Imperial Ministry of Education provided funding for the operating and administration costs whereas private industrial firms paid for the building and equipment. The Ruhr industries, particularly Hugo Stinnes, supported the Mülheim institute.

Figure 2. Franz Fischer

Figure 3. Hans Tropsch

Fischer had planned to study a coal-to-electricity direct path conversion, but with the institute's opening four days before World War I began and Germany's lack of petroleum quickly becoming apparent, the institute's program shifted from basic research on coal to methods of converting coal to petroleum. This wartime work was the institute's first comprehensive research program. It involved the decomposition of coal and the production of tar from the low-temperature carbonization (LTC) of different coals, giving yields of 1-25 percent, and the extraction (solution) of a coal with different organic solvents such as alcohols, pyridine, benzene, and petroleum ether at various temperatures and pressures. The extraction studies showed that decreasing the coal's particle size by grinding increased tar yields. With benzene as the solvent at 270°C and 55 atm Fischer and W. Gluud in 1916 obtained tar yields many times the low yields obtained at atmospheric pressure. These early studies on coal also led Fischer and Hans Schroder in 1919 to propose their controversial lignin theory of coal's origin in which during the peat-bog stage of coal's formation the cellulose material in the original plant material decomposed leaving only the more resistant lignin that then changed into humus coal.

With the wartime coal investigations well underway, Fischer's interest shifted to a different hydrocarbon reaction.   In 1913 Badische Anilin-und Soda-Fabrik (BASF) in Ludwigshafen patented a process for the catalytic hydrogenation (reduction) of carbon monoxide to give hydrocarbons other than methane, alcohols, ketones, and acids. According to the patent, hydrocarbon synthesis occurred best with an excess of carbon monoxide (2:1 carbon monoxide, hydrogen volume mixture) at 300-400°C, 120 atm, and the metals cerium, cobalt, or molybdenum, or their alkali-containing (sodium hydroxide) metallic oxides as catalysts. Because of World War I and priority given to industrializing the ammonia and methanol syntheses, BASF never continued its hydrocarbon synthesis [8]. Upon learning of BASF's patent Fischer decided to test its claims. Working with Tropsch he began investigating the catalytic reduction of carbon monoxide at various temperatures and pressures but using excess hydrogen gas, a 2:1 hydrogen: carbon monoxide volume mixture they called synthesis gas. This avoided carbon monoxide decomposition ($2CO \rightarrow C + CO_2$) which deposited carbon (soot) on the catalyst and rendered it ineffective.

The experiments with synthesis gas continued into the 1920s, and in 1923 Fischer and Tropsch showed that reacting the gas in a tubular, electrically-heated converter at high temperature and pressure, 400-450°C and 100-150 atm, and with alkali-iron instead of metallic oxide catalysts, gave a mixture of oxygen-containing organic compounds, such as higher alcohols, aldehydes, ketones, and fatty acids, that they called synthol. The reaction produced no hydrocarbons [9]. Additional studies in 1925-1926 using small glass combustion tubes 495 millimeters (mm) long, a gas-heated horizontal aluminum block furnace, and different reaction conditions, cobalt-iron catalysts at 250-300°C and 1 atm eliminated completely the oxygenated compounds. The product contained only hydrocarbon gases (ethane, propane, butane) and liquids (octane, nonane, isononene) with a boiling point range of 60-185°C [10].

Fischer continued his investigations into the 1930s, constructing a small pilot plant in Mülheim in 1932. The plant contained a series of converters five meters (m) high, 1.2 m wide, 12 mm thick walls, immersed in an oil bath for cooling and operated at the same conditions he had used earlier (2:1 hydrogen : carbon monoxide volume mixture, 190-210°C, 1 atm) but with a catalyst having the weight ratio 100 nickel-25 manganese oxide-10 aluminum oxide-100 kieselguhr. The catalyst, containing previously untested nickel, which differed in atomic number from iron and cobalt only by one and two units, had a short four to six week lifetime because of sulfur poisoning. The total yield per cubic meter (m^3) of synthesis gas consumed was only 70 grams (g) of a 58-octane number gasoline and a diesel oil boiling above 220°C [11].

Two years later Fischer's decade-long research moved to the next level with the construction in 1934 of the first large pilot plant in which he planned to solve the synthesis' three main problems and synthesize hydrocarbons from carbon monoxide and hydrogen. Ruhrchemie AG, a company Ruhr coal industrialists founded, envisioned the F-T synthesis as an outlet for its surplus coke, and upon acquiring the patent rights to the synthesis in 1934, constructed the pilot plant in Oberhausen-Holten (Sterkrade-Holten), near Essen. The plant operated at the conditions used in Fischer's small pilot plant and had an annual capacity of 1,000 metric tons (7,240 barrels) of motor gasoline, diesel oil, and lubricating oil.

Although the larger pilot plant demonstrated the overall success of the F-T synthesis, its three main problems, removing the large amount of heat released in the gas stream during the reaction, the nickel catalyst's short lifetime, and the significant loss of catalytic metals (nickel, manganese, aluminum) during their recovery (regeneration) for reuse, persisted during the operation. The nickel catalyst's poor performance forced Fischer and Ruhrchemie to abandon its use for commercial development. At this time research resumed with the more active but expensive cobalt catalysts. Oberhausen-Holten subsequently became the production center for a standardized cobalt catalyst used in all the F-T plants constructed later in the 1930s, for all the development work on synthetic motor fuel and lubricating oil, and for the oxo process [12].

The successful pilot plant research and development at Oberhausen-Holten was the major turning point in the F-T synthesis. By November 1935, less than three years after Germany's Nazi government came to power and initiated the push for petroleum independence, four commercial-size Ruhrchemie licensed F-T plants were under construction. Their total annual capacity was 100,000-120,000 metric tons (724,000-868,000 barrels) of motor gasoline, diesel oil, lubricating oil, and other petroleum chemicals. The motor vehicle products comprised 72 percent of the total capacity. Petroleum chemicals made up the remaining 28 percent and included alcohols, aldehydes, soft waxes which when oxidized gave the fatty acids used to produce synthetic soap and edible fat (margarine), and heavy oil for conversion to the inexpensive detergent *Mersol*.

All the plants were atmospheric pressure (1 atm) or medium pressure (5-15 atm) syntheses at 180-200°C. They produced synthesis gas by reacting coke with steam in a water gas reaction and adjusting the proportions of carbon monoxide and hydrogen, and used a cobalt catalyst (100 Co-5 ThO_2-8 MgO-200 kieselguhr) that Ruhrchemie chemist Otto Roelen (1897-1993) developed from 1933 to 1938. Roelen's catalyst became the standard F-T catalyst because of its greater activity and lower reaction temperature, but its preparation was expensive, costing RM 3.92 per kg of cobalt. For this reason Ruhrchemie recovered the cobalt and thorium from the spent (used) catalyst by treatment with nitric acid and hydrogen gas at a cost of RM 2.97 per kg of cobalt, and re-used them in preparing fresh catalysts[13]. This gave a total catalyst cost of RM 6.89 per kg of cobalt or nearly 30 percent of the total F-T production cost. By 1937-38 the combined annual capacity of the four F-T plants increased to 300,000 metric tons (2.17 million barrels) and with the completion of five additional plants, total capacity rose to 740,000 metric tons (5.4 million barrels) at the outbreak of World War II in September 1939. Production at the nine F-T plants peaked at 576,000 metric tons (4.1 million barrels) in 1944 [14].

Figure 4. Otto Roelen

The older F-T plants operated at 1 atm whereas three of the five newer plants were medium pressure 5-15 atm syntheses. Converter design differed depending on the reaction pressure, but all the plants had inefficient externally cooled converters that dissipated the high heat of reaction (600 kilocalories per m³ of synthesis gas consumed) and controlled the reaction temperature by arranging the cobalt catalyst pellets in a fixed bed within the converter and circulating pressurized water through the converter. Synthesis gas entered at the converter's top at the rate of 650-700 m³ per hour per converter and flowed down through the catalyst bed, hydrocarbon products passed out the bottom. The medium pressure synthesis gave a slightly higher yield and extended the catalyst's life from 4-7 months to 6-9 months.

For the 1 atm synthesis the converter (tube and plate design) was a rectangular sheet-steel box 5 m long, 2.5 m high, 1.5 m wide, containing about 600 horizontal water cooling tubes interlaced at right angles with 555 vertical steel plates or sheets. The complicated grid-like arrangement over which the synthesis gas flowed from top to bottom eliminated any localized heat buildup in the converter. Each steel plate was 1.6 mm thick, a space of 7.4 mm separated adjacent plates. The cooling tubes were 40 mm in diameter, 40 mm apart, and led to a boiler (steam drum) for recovery of the heat released in the synthesis. One boiler recovered the heat released from two converters. An empty converter weighed 50 metric tons. The catalyst pellets, which filled the space between the tubes and plates and occupied a volume of 12 m³, weighed 3 metric tons of which 900 kg were cobalt.

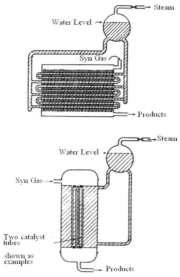

Figure 5. Tube and plate 1 atm converter (upper), concentric double tube medium pressure converter (lower).

The medium pressure converter (concentric double tube) had a simpler design. It consisted of a 50-metric ton vertical cylindrical steel shell 6.9 m high, 2.7 m internal diameter, 31 mm thick walls, and contained 2,100 vertical cooling tubes. Each cooling tube was 4.5 m long and double in construction, consisting of an outer tube of 44-48 mm diameter fitted with a concentric inner tube of 22-24 mm diameter. A top and bottom weld (T-connections) between the converter's horizontal face and an outer tube connected an inner tube with a boiler that allowed cooling water to circulate from the boiler to the main space in the shell around the outer tubes and through the inner tube. One boiler recovered the heat released from four converters. The catalyst pellets filled the annular space between the concentric tubes and occupied a volume of 10 m^3.

In the 1 atm synthesis, water sprays in packed towers directly cooled the hot hydrocarbon vapors and gases (primary products or primary oils) leaving the bottom of the converter. The vapors condensed to give light oil (C_5-C_{12}, boiling point range 25-165°C), middle oil (C_{10}-C_{14}, boiling point range 165-230°C), heavy oil (C_{20}-C_{70}, boiling point range 230-320°C), and hard and soft wax (C_{20}-C_{30}, boiling point range 320-460°C and above).

The cooled gases (propane, butane) passed to an absorber for their removal and recovery with activated charcoal and subsequent liquefaction. In the medium pressure synthesis about 35 percent of the primary products left the converter as hydrocarbon liquids. Passage through a tubular-type steel alloy condenser liquefied the hydrocarbon vapors. The remaining hydrocarbon gases, after expansion to atmospheric pressure, underwent recovery and removal with activated charcoal in an absorber.

The biggest converter used in German F-T plants had a production capacity of only 2.5 metric tons per day (18 barrels per day) so that a small, 70 metric ton per day (500 barrels per day) plant had 25 or more converters, requiring considerable amounts of material and manpower for its construction and operation. All the plants operated their converters in stages. The 1 atm plants had two stages, operating two-thirds of the converters in the first stage and one-third in the second. Some of the plants placed the condensers and absorbers between the stages, others placed only condensers. All the plants had absorbers after the second stage converters and condensers. During the last two years of the war the medium pressure plants switched from two stages to three stages, successively operating one-half, one-third, and one-sixth of their converters. They had condensers between each stage and absorbers after the final stage converters and condensers.

Average plant yield for the 1 atm synthesis was 130-165 g of liquid hydrocarbons per m^3 of synthesis gas, or about 80 percent of the theoretical maximum yield. Annual production per converter was 500-720 metric tons. For the middle pressure synthesis the corresponding yields were 145-160 g per m^3 and 600-750 metric tons. The medium pressure synthesis also extended the catalyst's life from four-seven months to six-nine months.

Product refining, especially by fractional distillation, was the same for both syntheses. Low-grade gasoline which made up the light oil fraction, had a 45-53 octane number, which after blending with 20 percent benzol and adding 0.02-0.04 percent lead tetraethyl, increased to 70-78 and provided the German army with motor gasoline. High-grade diesel oil with a 78 cetane number (middle oil fraction) and some of the heavy oil fraction, after blending with 50 percent petroleum oil, served as aviation fuel for the German air force. Further treatment of most of the heavy oil at IG Farben's Leuna plant after its opening in 1927 gave the inexpensive synthetic detergent *Mersol*; cracking and polymerizing the remaining heavy oil and some of the soft wax gave good quality lubricating oil. Oxidizing the rest of the soft wax produced fatty acids for conversion to soap and small quantities of edible fat. The German wax industry used most of the hard wax for electrical insulation, the manufacture of polishes, and as a paper filler [15].

The most efficient F-T plants recovered only 30 percent of the total heat energy input as primary products and another 25 percent as steam and residual gas. The net heat energy required for the production of one metric ton of primary products was equivalent to 4.5 metric tons of coal (1 lb coal = 12,600 BTU) [16].

3. Germany's energy plan

The growth of the German synthetic fuel industry remains inseparably linked to events taking place there in the 1930s and 1940s. A special relation existed between the industry and Nazi government, and without it Germany's emerging synthetic fuel industry might have collapsed. The small German oil industry had remained reasonably free from government interference and had benefited from a tariff increase in April 1930 that raised the duty on imported oil from RM 77.40 per metric ton (8.5¢ per US gallon) to RM 129 per metric ton (14.3¢ per US gallon).

Government policy began to change with the German banking crisis that followed the failure of the Kredit Anstalt on 3 May 1931 and the Darmstaedter National Bank on 15 July 1931 and led the Weimar government to impose a number of controls and regulations which the Nazi government expanded and intensified beginning in 1933. The Weimar government established Supervisory Boards to allocate raw materials and placed these boards under control of the Reichswirtschaftsministerium (Ministry of Economics) which in 1939 renamed them Reichsstellen (Reich Offices). The Reichsstelle für Mineröl (Office of Mineral Oil) regulated the oil industry, additional regulations came from the Reichsstelle für Rohstoffamt (Office of Raw Materials) and its subdivision Wirtschaftsgruppe Kraftstoffindustrie (Economic Group for Liquid Fuels). All liquid fuel producers reported their production and import figures and any new Plant and refinery construction to the oil regulatory boards. In addition to regulatory boards, the government established four industry associations that had responsibility for the production and allocation of the fuels under their control.

Table 1. Government-Industry Assocations

Association for Crude Oil Production and Refining (REV)
Association for Hydrogenation, Synthesis and Low Temperature Carbonization (ARSYN)
Assocation of German Benzol Producers (ARBO)
Association for Allocation of German Bituminous Coal Tar Products (AVS)

The government-industry relation also resulted in risk-free partnership agreements between the government and any industry, such as coal and chemical, involved in synthetic fuel production. The earliest of these was the Fuel Agreement (Benzinvertrag) that IG Farben, the only company then producing synthetic fuel, and the Reichswirtschaftsministerium signed on 14 December 1933. It required IG Farben to produce at least 300,000-350,000 metric tons (2,490,000 barrels) of synthetic gasoline per year by the end of 1935 and to maintain this production rate until 1944. The agreement set the production cost, which included depreciation, five percent interest on IG Farben's investment, and a small profit, at 18.5 pfennig per liter (29¢ per US gallon). The government not only guaranteed the production cost but agreed to pay IG Farben the difference between that cost and any lower market price, and to buy the gasoline if no other market emerged. Alternatively, IG Farben had to pay the government the difference between the production cost of 18.5 pfennig per liter, which was at that time more than three times the world market price, and any higher price obtained on the market. Because of increasing petroleum costs, as well as improvements in the hydrogenation process, IG Farben paid RM 85 million to the government by 1944 [17].

Eight months after signing the fuel agreement with IG Farben the government took two additional steps to assist the synthetic fuel industry. The first was the establishment on 24 August 1934 of Wirtschaftliche Forschungsgesellschaft (WIFO, Economic Research Company), a completely government-owned company capitalized at RM 20,000 and charged with the construction and operation natural and synthetic of liquid fuel storage depots. German fuel producers sent WIFO their lubricating oil, and their aviation grade products for blending and leading, which WIFO stored and eventually distributed mainly to the air force and minimally to the army.

In 1938 WIFO had a storage capacity of 630,000 metric tons (820,000 m^3) of motor and aviation gasoline and 84,000 metric tons (110,000 m^3) of lubricating oil. It actually stored 500,000 metric tons of aviation gasoline, most of it in bombproof underground locations within Germany. The government in 1938 planned to increase Germany's total storage to 6,000,000 metric tons of liquid fuel and lubricating oil by 1943 and projected the following contributions: WIFO 2,900,000 metric tons; German industrialists 1,250,000 metric tons; and the navy the remaining 1,800,000 metric tons. The navy had underground storage tanks and a smaller number of surface tanks in the North Sea and Baltic Sea areas and in the German interior.

The government's lofty projection fell short. Germany's total storage reached 2,400,000 metric tons of liquid fuel on 21 June 1941. WIFO's contribution, 500,000 metric tons of aviation gasoline, was a significant amount that represented about one-third the total 1940 US production of aviation gasoline (40,000 barrels per day). It almost equaled Germany's refining capacity of 420,000 metric tons per month or 5,000,000 metric tons per year, about half of it refined in the Hamburg and Hannover areas.

Two months after establishing WIFO the German government took the second step when it forced the establishment of Braunkohlen Benzin AG (Brabag) to promote and carry out commercial-scale synthesis of synthetic liquid fuel and lubricating oil from coal and tar. Brabag was an association of IG Farben and nine central German brown coal producers (Compulsory Union of German Lignite Producers) that accounted for 90 percent of Germany's brown coal. At the time of its formation on 26 October 1934, it had a capitalization of RM 100 million financed entirely with a fifteen year loan that the German brown coal producers guaranteed. Gesellschaft für Mineralölbau GmbH, a division of Brabag established two years later in November 1936 by the ten brown coal producers, carried out the design and engineering of the Brabag plants, using technical information that the government required IG Farben, Ruhrchemie, and other synthetic fuel producers to provide as a result of entering into licensing agreements with the government. Brabag and Mineralölbau built and operated three coal hydrogenation and one F-T plant during the 1930s and 1940s.

Additional government commitment to the synthetic fuel industry, and indicative of the supportive government-industry relation, emerged at a Nazi party rally in Nürnberg on 9 September 1936. At that time Adolf Hitler (1889-1945) announced his Four Year Plan to make the German military ready for war in four years and the economy independent and strong enough to maintain a major war effort. Hitler put Hermann Göring (1893-1946) in charge of the plan, gave him the title Commissioner General for the Four Year Plan, and had Göring officially approve the plan in May 1937. With Hitler's war strategy requiring large supplies of petroleum, a petroleum-independent Germany became the Four Year Plan's major thrust. Of the 289 projects scheduled for the period 23 October 1936 to 20 May 1937 at a cost of RM 1,369 million, 42 percent costing RM 570 million were synthetic fuel projects. In fact in 1936 Hitler urged the petroleum industry, including synthetic fuel produced by both coal and tar hydrogenation and F-T synthesis, to become independent of foreign production in eighteen months and called for synthetic fuel production to increase from 630,000 metric tons in 1936 to 3,425,000 metric tons in 1940.

As an incentive toward the synthesis of petroleum from coal the German government in December 1936 raised the tariff on imported petroleum from a 1931 levy of RM 219.30 per metric ton (24.4¢ per US gallon) to RM 270.90 per metric ton (30.1¢ per US gallon). By this time only four coal hydrogenation and two F-T and plants were operating with a combined production far less than required for petroleum independence. The high tariff enabled the synthetic fuel plants to show a profit even though they were highly inefficient and had production costs much greater than the cost of natural petroleum.

In the beginning, private capital coming from bank loans and from the synthetic fuel companies' own funds, stock, and bond issues provided practically all of the financing for the plants. But by 1939, as the cost of the program increased significantly and private capital dried up, the German government provided more and more of the funding. A report of 21 March 1939 showed that of the RM 132 million spent on synthetic fuel in 1939, the government provided RM 70 million to Minerölbau for the purchase of plant equipment. Additional government support came in the form of guaranteed purchases of synthetic fuel at prices high enough to allow for short term amortization of plant costs. Total synthetic fuel production from the seven coal hydrogenation and seven F-T plants operating in September 1939 was 1,280,00 metric tons increasing to almost 1,900,000 metric tons in May 1940. It exceeded Germany's refining of crude oil from natural sources (1,256,000 metric tons) and imports mainly from Romania (1,085,000 metric tons) in 1939.

The Reichsamt für Wirtschaftsausbau (Office of Economic Development) constantly revised the Four Year Plan. Its general concerns were the raw material and manpower requirements and the never ending iron and steel shortages, and in particular for the synthetic fuel industry the anticipated shortages of aviation gasoline and fuel oil. The first revision at Karinhall, Göring's vast country estate in the Schorfheide (Berlin- Postdam) on 12 July 1938 gave priority to hydrogenation plants for the production of aviation gasoline and to bituminous coal distillation for the production of fuel oil. The Welheim hydrogenation plant, which began production of aviation gasoline and fuel oil for the navy in 1937-38, already had received priority; and the Brüx hydrogenation plant, which produced diesel oil, later benefited from the revised plan.

The 1938 revision also dealt with the number of workers required for the construction, operation, and maintenance of the synthetic fuel plants. It called for 30,000 construction workers in 1938, 57,600 workers on 1 July 1939, and increasing from a projected 70,000 workers on 1 October 1939 to 135,000 workers during the last quarter of 1941. The actual construction force numbered about 35,000 at the outbreak of war, about 70,000 by mid-1941, and peaked at 85,000 in spring 1943.

Steel production also failed to meet the Four Year Plan's requirements. A second Karinhall revision of 1 January 1939 called for the production of 4.5 million metric tons of steel by the end of 1943. With this amount of steel Germany expected to expand existing plants and construct new plants to increase synthetic fuel production from 3.7 million metric tons in 1938 to 11 million metric tons per year by 1944. According to the US Strategic Bombing Survey's postwar report the required 4.5 million metric tons of steel equaled the amount necessary to build a fleet 3.5 times the size of the British navy that existed on 1 January 1940 [18].

4. Commercial developments of the F-T synthesis

The first of the commercial-size F-T plants to produce synthetic fuel was the Steinkohlen-Bergwerk Rheinpreussen plant located in Mörs-Meerbeck (Homberg, Ruhr) near the Rheinpreussen coal mine. Gütenhoffnungshütte, controlled by the Haniel Group, completed the plant in late 1936. Most of the synthetic fuel plants had scientists or engineers with doctorates in either chemistry or chemical engineering as managers or directors as was the case at Rheinpreussen where plant manager Struever, H. Kobel, and W. Dannefelser, directed a work force of 750. Liquid fuel synthesis took place at 1 atm, 190-195°C, and in two stages with 60 of the plant's 90 tube and plate converters operating in stage one and the other 30 operating in stage two. Rheinpreussen designed its own coal coking ovens for the production of coke and coke (coal) oven gas, a hydrogen-carbon monoxide-methane mixture used for cracking (reacting) with steam at 1,200°C in a Koppers gasifier to increase the hydrogen content of the gas mixture. Combining this mixture with twice as much water gas, produced by reacting coke with steam, gave the synthesis gas of proper proportions, 2 H_2 : 1 CO. Rheinpreussen's annual capacity was 25,000-30,000 metric tons (later increased to 70,000 metric tons) of gasoline and diesel oil (primary oils) and paraffin wax. An alcohol plant produced another 3,000 metric tons of propyl and butyl alcohol [19].

The mining company Gewerkschaft Viktor AG (Klocknerwerke AG), a subsidiary of Wintershall AG, constructed the second commercial-size F-T plant at a cost of RM 30 million in Castrop-Rauxel (Ruhr) also in late 1936. The plant site adjoined Gewerkschaft Viktor's coal mine and was the location of a synthetic ammonia plant. Gewerkschaft Viktor designed its own coal coking ovens and gasifier that was similar to a Koppers gasifier. It produced synthesis gas by cracking coke oven gas with steam and mixing the cracked gas with water gas obtained from coke. The plant's 63 tube and plate converters operated in two stages at 1 atm and according to plant manager Braune had an annual capacity of 30,000-40,000 metric tons of gasoline and diesel oil [20].

Ruhrchemie's Ruhrbenzin AG plant in Oberhausen-Holten was the third commercial-size F-T plant constructed in the 1930s. Ruhrbenzin, established in September 1935 with a capitalization of RM 4.5-6 million and increased to RM 15 million in 1940, planned in 1936 to complete construction of a plant annually producing 30,000 metric tons (increased to 62,000 metric tons in 1942) of gasoline, diesel oil, and lubricating oil. Production did not begin until 1937. The plant differed from the Rheinpreussen and Viktor plants in having two independent synthesis systems: a two-stage 1 atm synthesis with 48 tube and plate converters and a three-stage 10-15 atm synthesis with 72 concentric double tube converters. Water gas, prepared from coke, one-third of which after an iron-catalyzed reaction with steam at 500°C, gave a mixture containing 61 percent hydrogen and 5 percent carbon monoxide. Adding the mixture to the remaining two-thirds water gas provided synthesis gas for conversion to gasoline, diesel oil, and lubricating oil. Of the nine F-T plants that eventually came into operation the Ruhrbenzin plant was the most inefficient. It lost RM 2.6 million in 1939 which Ruhrchemie's president and managing director Friedrich Martin, chief designer Willke, and plant superintendent Navelling attributed to the constant experimentation with the plant's reaction conditions and procedures [21]. Oberhausen-Holten became the research and development center for the catalytic studies of Roelen, Leonard Alberts, Walter Feisst, and others. Its catalyst plant supplied the six F-T plants in the Ruhr area with the standard cobalt catalyst, producing about 3,000 metric tons per year. Brabag's plant in Ruhland-Schwarzheide and beginning in 1938 the Wintershall plant in Lützkendorf also produced the standard cobalt catalyst [22].

Brabag, which also operated three coal hydrogenation plants, completed construction of Brabag II, the fourth F-T plant in Ruhland-Schwarzheide in 1937. Brabag II was a two-stage 1 atm plant and had an annual capacity of 25,000-30,000 metric tons of gasoline and diesel oil. Later expansion, which increased the number of tube and plate converters to 262 and maximum annual production to 162,000 metric tons (200,000 metric tons capacity), made it Germany's largest F-T plant. Brown coal briquettes, gasified in Didier-Bubiag retorts, each with a capacity of 638,000 m^3 (22 million cubic feet) per day, and Koppers gasifiers, each with a capacity of 26,100 m^3 (900,000 cubic feet) per hour, provided 20 percent and 80 percent of the synthesis gas. Purification of the synthesis gas by passing it though towers containing pellets of iron oxide and sodium carbonate to remove sulfur and other impurities was relatively simple because of the brown coal's low sulfur content. Erwin Sauter, A. Wagner, W. Sapper, and catalyst specialist Karl Meyer directed the plant's operation [23].

In addition to the ongoing catalytic research, both Ruhrchemie at its research center in Oberhausen-Holten and Fischer at the KWI investigated the F-T medium pressure synthesis hoping to improve F-T efficiency and economics. The studies showed that medium pressure gave a slightly higher yield of gasoline and diesel oil per m^3 of synthesis gas, extended the catalyst's life from 4-7 months to 6-9 months without any reactivation, and yielded a higher proportion, about 45 percent versus 18 percent, of heavier hydrocarbons such as soft and hard wax for the production of lubricating oil and chemicals. The middle pressure synthesis also had a higher operating cost. Consequently, only two of the five F-T plants constructed in 1938 and 1939 before World War II began were medium pressure syntheses. A third plant was a combination atmospheric-medium pressure synthesis [24].

The first of the newer F-T plants was the Wintershall subsidiary, Mitteldeutsche Treibstoff plant, constructed in Lützkendorf in late 1938 in the Geiseltal brown coal mining district of central Germany. Mitteldeutsche had 132 tube and plate converters that operated in two stages at 1 atm, but a maximum of 77 converters operated at one time. The plant performed poorly except for its last two years of operation in 1943-44 when annual production reached 30,000 metric tons of gasoline and diesel oil or about 40 percent of its maximum. A synthesis gas problem caused its poor performance. Mitteldeutsche used the first commercial-size Schmalfeldt generator that plant director H. Schmalfeldt had designed for the production of synthesis gas from the direct gasification of powdered brown coal. The coal had a very high sulfur content, and until plant engineers installed activated charcoal absorbers in the purification system to remove the sulfur and eliminate the catalyst's poisoning (a standard procedure in F-T plants), the catalyst lasted only two months instead of the usual 4-7 months [25].

Friedrich Krupp AG in Essen joined the expanding group of synthetic fuel producers in 1937 when it established Krupp Treibstoffwerk GmbH in Wanne-Eickel (Essen) with a capitalization of RM 20 million and a RM 10 million loan. Erich Combles general manager and assistant general manager H. Fischer directed the 900 workers who operated the only combination atmospheric-medium pressure plant. Krupp-Lurgi gasifiers of 40 metric tons per day capacity converted coke, obtained mainly from high-temperature coal carbonization, to water gas, one-third of which underwent catalytic conversion to synthesis gas. Synthesis gas first passed through one set of 72 tube and plate converters at 1 atm for conversion to gasoline and diesel oil. Residual synthesis gas, after flowing through standard tubular condensers, activated charcoal absorbers, and compressed to 10-15 atm, traveled through a second set of 24 medium pressure converters to complete the conversion. Of the 24 medium pressure converters, 16 were of a new design called *tauschenrohren*, in which single tubes of 72 mm internal diameter and fitted with fins of sheet steel, replaced the standard concentric double tube converter. The new converter design increased catalyst capacity by 5 percent but left carbon deposits in the converter.

Maximum production of gasoline and diesel oil at the plant reached 54,000 metric tons in 1943, maximum annual capacity was 130,000 metric tons [26].

Chemische Werke Essener Steinkohle AG in Essen, established in early 1937 as a partnership of Essener Steinkohlen Bergwerke AG and Harpener Bergbau AG in Dortmund with a capitalization of RM 12 million and a RM 10 million loan, constructed the second largest and the most efficient of the 1 atm plants. Plant manager Gabriel and assistant manager E. Tengelmann directed the 600 plant workers. Gasifying coke in water gas generators and cracking the resulting coke oven gas produced synthesis gas for conversion to gasoline and diesel oil in 124 tube and plate converters operating in two stages. The high efficiency of the Essener plant, according to postwar Allied investigations, appeared to depend on the purity of its synthesis gas, the equal distribution of the catalyst between the two stages, and the frequency of reactivating the catalyst by treating it with nitric acid and hydrogen gas. Gabriel and Tengelmann believed, however, that the constant composition of the synthesis gas and the plant's freedom from interruptions and breakdowns, which most likely resulted because of all the above factors, were the major reasons for the plant's successful operation from the time of its start up in 1939. Essener Steinkohle's maximum annual production was 86,500 metric tons of gasoline and diesel oil [27].

The last of F-T plants were the medium pressure operations of Hoesch-Benzin GmbH in Dortmund (Ruhr) and Schaffgotsch Benzin GmbH in Deschowitz-Beuthen, Odertal (Upper Silesia), both of which began operation in 1939. Hoesch-Benzin, a subsidiary of Bergwerksgesellschaft Trier GmbH (owned by Hoesch-Köhn-Neussen AG), had a capitalization of RM 3 million and a work force of 800 under the direction of plant manager H. Weitenhiller and plant superintendent Werres. The Hoesch plant converted coke to water gas and then cracked the water gas with additional steam to produce synthesis gas. Its 65 concentric double tube converters converted synthesis gas to gasoline and diesel oil in two stages and added a third stage during the war. Operating efficiency, measured by production per converter per month, was the highest of all the plants, its production reaching a maximum of 51,000 metric tons per year [28].

Plant manager A. Pott, formerly director-general of Ruhrgas AG, supervised the Schaffgotsch Benzin plant operation. Pintsch generators produced synthesis gas from hard coke and coke oven gas, and until mid-1943 synthesis gas conversion to gasoline and diesel oil occurred in two stages. The addition of a third stage at that time resulted in a plant similar to the Hoesch-Benzin plant. Schaffgotsch had 68 converters, 50 of them wide-tube 22-23 mm diameter converters that contained single catalyst tubes rather than the concentric double tubes used in the other medium pressure F-T plants. Its engineers claimed that their modified design increased catalyst capacity by 10 percent and that their converters functioned particularly well in the second and third stages despite having to drill interior carbon deposits in order to remove the catalyst for reactivation. Schaffgotsch achieved a maximum annual production of 39,200 metric tons of gasoline and diesel oil. Its annual capacity was 80,000 metric tons [29].

F-T plant construction ended with the outbreak of the war, resulting in standardization of plant apparatus and operation, although from 1943 to 1945 research continued on designing better converters and finding cheaper iron catalysts to replace scarce wartime supplies of cobalt compounds. Ruhrchemie, which conducted 2,000 investigations, Rheinpreussen, KWI, IG Farben, Lurgi, and Brabag developed six iron catalysts, and all gave satisfactory results in comparative tests carried out in September 1943 at Brabag's Schwarzheide plant. The Reichsamt für Wirtschaftsausbau (Office of Economic Development), which arranged for the tests, never decided on the best iron catalyst, concluding only that all six were inferior to cobalt catalysts. None of the commercial-size plants used iron catalysts.

The three new converter designs developed during the war operated at 20 atm, used iron catalysts, and were internally cooled compared to the inefficient externally cooled fixed bed converters in the existing F-T plants. Their design summaries appear below.

Table 2. Converter Designs 1942-45

Heat Temperature Medium	Process
Gas	IG Farben's fixed bed hot gas recycle process
Oil	IG Farben, Ruhrchemie and Rheinpreussen fluid bed oil slurry process
	IG Farben fixed bed, oil circulation process

The first of these new converters removed the heat the synthesis released in the gas stream by recirculating residual gas through a wide shallow bed containing a powdered iron-one percent borax catalyst. IG Farben developed this converter design and successfully tested it on a small scale with 5 liters of catalyst for 10 months. Large-scale tests were unsuccessful because of the catalyst's overheating. The converter had high energy requirements and cost more to operate than the older design converters. For these reasons IG Farben abandoned its development before the end of the war. The gasoline produced had a 68-70 octane number that additional refining increased to 75-78 and 84.

Both of the remaining converter designs used oil as the heat transfer medium. The fluidized bed or oil slurry process forced water gas through a ceramic plate at the bottom of a cylindrical converter that contained a catalyst of iron with carbonate or borate suspended in a high boiling heavy synthetic oil. The tests were small scale and aimed at the production of C_{20}-C_{70} olefins in the gas oil boiling point range (232-426°C) for use in chemical syntheses.

The other oil-cooled converter had the iron catalyst (iron oxide and other metallic oxides) arranged in a fixed bed and removed the heat of reaction by circulating oil through the catalyst bed. IG Farben tested this converter extensively in a pilot plant of 8-10 metric tons capacity, synthesizing a gasoline with a 62-65 octane number [30].

Emphasis also shifted at this time from the production of fuels and lubricants to the production of olefins (unsaturated hydrocarbons), waxes, alcohols, and other organic compounds. The oxo synthesis from the German *oxiering*, meaning ketonization, was the most important result of this research. In the oxo synthesis, straight-chain olefins such as C_2H_4 and C_3H_6, reacted with carbon monoxide and hydrogen at 110-150°C, 150 atm, and with a cobalt catalyst to form an aldehyde that had one more carbon atom in the chain. Hydrogenating the aldehyde under the same conditions gave the corresponding alcohol. Roelen of Ruhrchemie patented the process in 1938, and in 1940 Ruhrchemie and IG Farben cooperated in its development. Their objective was the production of long-chain alcohols (C_{12} to C_{18}) for conversion to detergents (sulfate esters), but the process had general applicability to all olefin-like compounds. Ruhrchemie, IG Farben, and Henkel et Cie, organized a new company called Oxo-Gesellschaft and in 1944 completed construction of an oxo plant in Sterkrade-Holten that had an annual capacity of 12,000 metric tons of alcohols and a production cost of 78.23 pfennig per kg. Allied bombing in August - October 1944 permanently prevented the plant from beginning production.

Information on F-T plant construction and operating costs has come from two main sources: captured German synthetic fuel documents and their summaries and interrogation of

German synthetic fuel scientists, such as Martin, Ruhrchemie's managing director; Heinrich Bütefisch, an IG Farben director and government economic advisor on wartime petroleum production; and F-T plant managers and operating personnel. The collective information indicates that capital and production costs were high. A F-T plant cost approximately RM 30 million. Production cost, including catalyst, water gas manufacture, synthesis of primary products and all other costs was 23.5-26 pfennig per kg (RM 240-330 per metric ton, $13.2-18.4 per barrel, 31-44¢ per gallon) for both the 1 atm and medium pressure operation.

Hoesch-Benzin's medium pressure plant with an annual production of 40,000 metric tons had a capital cost of RM 26 million (RM 650 per metric ton per year) and a production cost in 1942 of 25.81 pfennig per kg of products. The Essener Steinkohle 1 atm plant with 80,000 metric tons annual production had a capital cost of RM 32 million (RM 400 per metric ton per year) and a production cost of 23.71 pfennig per kg of synthetic products. Ruhrchemie's combined atmospheric-medium pressure plant with an annual production of 42,000 metric tons had a capital cost of RM 15 million i 1940 (RM 380 per metric ton per year) and a production cost in 1939-40 of 23.57 pfennig per kg of products. The nine F-T plants provided 12-15 percent of Germany's total synthetic fuel production during their nine years of operation [31].

Table 3. Fischer-Tropsch Plants. (Source: Compiled from information Report on the Petroleum and Synthetic Oil Industry of Germany (London 1947) and High-Pressure Hydrogen at Ludwigshafen-Heidelberg, FIAT, Final Report No. 1317 (Dayton,Oh, 1951))

Plant	Location	Raw material (Coal)	Production in 1944 (metric tons)	Products	Pressure (atm)	Started Operation
Ruhrbenzin AG	Oberhausen-Holten Sterkrade-Holten), Ruhr	Bituminous	62,200	Gasoline motor fuel, Lubricating oil	Atmospheric (1 atm) and medium (5-15 atm)	Construction started by Nov. 1935, in operation 1937
Steinkohlen-Bergwerk Rheinpreussen	Moers-Meerbeck (Homberg), Neiderrhein	Bituminous	19,700	Gasoline, diesel oil, Hard and soft paraffin wax, Oils for fatty acids	Atmospheric	Construction started by Nov. 1935, in operation late 1936
Gewerkschaft Viktor, Klocknerwerke-Wintershall AG	Castrop-Rauxel, Ruhr	Bituminous	40,380	Primary oils (gasoline and diesel oil)	Atmospheric	Construction started by Nov. 1935, in operation second half of 1936
Braunkkohle-Benzin AG (Brabag)	Ruhland-Schwarzheide (north of Dresden)	Lignite	158,500	Primary oils	Atmospheric	Construction started by Nov. 1935, in operation in 1937
Mitteldeutsche Treibstoff und Ol Werke (subsidiary of Wintershall AG)	Lutzkendorf-Mucheln (Leipzig area)	Lignite	29,320	Primary oils	Atmospheric	1938
Krupp Treibstoffwerk	Wanne-Eickel, Ruhr		39,802	Primary oils	Atmospheric and medium	Late 1938
Chemische Werke Essener Steinkohle AG	Kamen-Dortmund (Bergkamen), Ruhr	Bituminous	86,580	Primary oils	Atmospheric	1939
Hoesch-Benzin GmbH	Dortmund, Ruhr	Bituminous	51,000	Primary oils	Medium	March 1939
Schaffgotsch Benzin GmbH	Deschowitz-Beuthen, Odertal (upper Silesia)	lignite	39,200	Primary oils	Medium	Plant complted in 1939, in operation in 1941

5. Summary of commercial development in Germany 1927-45

Despite substantial government support and Hitler's 1936 call for petroleum independence, the synthesis of petroleum from coal and tar never completely solved Germany's liquid fuel problem. Bureaucratic confusion, material shortages, and later Allied bombing limited its effectiveness. Production, nevertheless, increased dramatically under the Four Year Plan and its renewal in October 1940. In 1933, only three small synthetic fuel plants were operating, Ludwigshafen, Leuna, and Ruhrchemie Oberhausen-Holten, the last a F-T plant, that produced mainly diesel oil and petrochemicals. At that time, Germany's petroleum consumption was about one-half of Great Britain's, one-fourth of Russia's, and one-twentieth that of the United States. Yet, even at such low consumption, domestic resources were inadequate. Total consumption of liquid fuels, including 274,000 metric tons of lubricating oil, in 1932 was 2,755,000 metric tons, 73 percent of which (2,020,000 metric tons) Germany imported mainly from the United States.

For gasoline consumption the situation was the same, Germany consumed 1,460,000 metric tons, two-thirds of which (930,000 metric tons) it imported. By September 1939 when World War II began seven coal hydrogenation (plus Ludwigshafen) and eight F-T plants were in operation and were beginning to contribute increasingly to Germany's domestic liquid fuel supply. When plant construction ceased in 1942 twelve coal hydrogenation and nine F-T plants converted coal and coal tar into gasoline, diesel oil, and other petroleum products.

From the first coal hydrogenation plant that began operation at Leuna on 1 April 1927, the twelve coal hydrogenation plants in early 1944 reached a peak production of over 3,170,000 metric tons (23 million barrels) of synthetic fuel. Two million metric tons (14 million barrels) after adding lead tetraethyl were high quality motor and aviation gasoline approaching 100 octane number. In World War II these plants provided 95 percent of the German air force's aviation gasoline and 50 percent of Germany's total liquid fuel requirements [32]. Extensive Allied bombing of the Leuna, Böhlen, Zeitz, Lützkendorf, and Brüx plants in May 1944 significantly reduced production from 236,000 metric tons to 107,000 metric tons in June. With the repeated attacks on the Ruhr plants at Welheim, Scholven, and Gelsenberg, production fell dramatically to 17,000 metric tons in August, then stopped completely in March 1945.

The nine F-T plants contributed another 585,000 metric tons of primary products to the war effort, or 12-15 percent of Germany's total liquid fuel requirements. Their production fell significantly because of Allied bombing, decreasing from 43,000 metric tons in the first four months of 1944 to 27,000 metric tons in June, to 7,000 metric tons in December, and to 4,000 metric tons in March 1945 [33].

The average cost of hydrogenating coal or tar was high, 19-26 pfennig per kg (RM 190-260 per metric ton) or the equivalent of 26-34¢ per US gallon ($11.2-14.4 per barrel) of gasoline. The average cost of primary products at the F-T plants was a comparable 23.71-25.81 pfennig per kg (RM 240-330 per metric ton). These figures were more than double the price of imported gasoline, but for Germany, with only a limited supply of natural petroleum, no alternative remained during the war other than the construction of synthetic fuel plants. In this way Germany utilized its naturally abundant supplies of bituminous and brown coal [34].

6. Labor force in the synthetic fuel plants

Faced with a growing labor shortage as the war dragged on, German industrial firms, including synthetic fuel producers such as IG Farben, Brabag, Sudetenlandische Treibstoffewerke AG, and Hydrierwerke Pölitz AG, increasingly supplemented their labor force with paid coerced (forced) laborers and (or) concentration camp inmates (slave laborers) of many nationalities. French, Belgian, Polish, British, Serbian, Czech, Hungarian, and Russian laborers, Jews and non-Jews, worked in the prewar plants in Ludwigshafen in western Germany and Leuna in eastern Germany and in several of the synthetic fuel plants constructed after the war had started. Pölitz in northern Germany (July 1940), Lützkendorf in central Germany (1940), Wesseling in western Germany near Bonn (August 1941), Brüx in Bohemia (October, 1942), and the Blechhammer plant (1942) and Heydebreck saturation plant in Upper Silesia (April 1944) used forced laborers and (or) concentration camp inmates. IG Farben's labor force contained about 9 percent forced laborers and concentration camp inmates by 1941, the number increased to 16 percent in 1942, and to 30 percent of all workers in its synthetic fuel plants near the war's end. In addition to the forced laborers and concentration camp inmates, some free foreign workers came from Germany's allies, mainly Italy and Romania.

Table 4. German Coal Hydrogenation Plants 1927-45. (Source: Compiled from information in High-Pressure Hydrogenation at Ludwigshafen-Heidelberg, FIAT,Final Report #1317 (Dayton, OH, 1951), p. 112)

Plant Location	Process	Pressure (atm) Liquid/vapor Phase	Final Products	Plant Capacity and Production, metric Tons per year Liquid products Including LP gas, 1944
Ludwigshafen/Oppau				
Leuna	Liquid & vapor phase	250/250	Gasoline, diesel Oil, LP gas	620,000 (640,000)
Böhlen	Liquid & vapor phase	300/300	Gasoline, diesel Oil, LP gas	220,000 (275,000)
Magdeburg	Liquid & vapor phase	300/300	Gasoline, diesel Oil, LP gas	220,000 (275,000)
Scholven	Liquid & vapor phase	300/300	Gasoline, LP gas	220,000 (240,000)
Welheim	Liquid & vapor phase	700/700	Gasoline, fuel oil	130,000 (145,000)
Gelsenberg	Liquid & vapor phase	700/300	Gasoline, LP gas	400,000 (430,000)
Zeitz	TTH process & Vapor phase	300	Diesel oil, wax, Gasoline, lubricating Oil, LP gas	250,00 (250,000)
Pölitz	Liquid & vapor Phase	700/300	Gasoline, fuel oil (diesel Oil), LP gas	700,000 (750,000)
Lützkendorf	Liquid & vapor Phase	700/700	Gasoline, diesel oil, Fuel oil	50,000 (12,000)
Wesseling	Liquid & vapor Phase	700/300	Gasoline, diesel Oil, LP gas	200,000 (230,000)
Brüx	Liquid & vapor Phase	300/300	Gasoline, diesel Oil, LP gas	400,000 (360,000)
Blechhammer	Liquid & vapor phase	700/300	Gasoline intended, Fuel oil	60,000 (65,000)

All skilled and unskilled foreign workers in a specific industry (automotive, coal, steel) earned the same wage as an unskilled German worker in that industry, about 64.1 pfennig per hour or RM 38 for a 60 hour week, but received RM 18-25 after deductions for taxes, room and board. A skilled German worker received 81 pfennig per hour or about RM 49 for a 60 hour week [35]. At the coal hydrogenation and F-T plants the average wage for all workers involved in synthetic fuel production was RM 1.30 per hour, a considerably higher amount. This was the wage plant officials told postwar Allied investigating teams they used to calculate synthetic fuel production costs [36].

The never-completed IG Farben synthetic fuel plant at Auschwitz (Auschwitz III) or Oswiecim, in south Poland west of Cracow, was a different story. Free German and Polish workers as well as forced eastern European workers contributed to its construction, but the largest group of workers was the approximately 300,000, concentration camp inmates that included Germans, Greeks, Dutch, Czechs, Hungarians, Poles, and Russians, most of whom were Jews. IG Farben paid all its unskilled workers 30 pfennig per hour (RM 0.30 per hour) or RM 3 for a 10 hour work day and all skilled workers 40 pfennig per hour (RM 0.40 per hour) or RM 4 per day after deductions for taxes, room and board. This was the same wage as the 1944 average industrial wage for unskilled and skilled foreign workers for a six-day week (RM 18-25) with one difference. Free and forced foreign workers received these wages, but the total wage each concentration camp inmate earned went instead to the SS (Schutz-Staffel) for taxes and expenses (room and board and clothing). In effect, IG Farben was paying the SS for the labor it provided [37].

Table 5. Summary of German Oil Availability from Various Sources at the beginning of 1944. Source: Compiled from Information in High-Pressure Hydrognation at Ludwigshafen-Heudelberg, FIAT, Final Report #1317 (Dayton, OH, 1951), p. 112.

Type of Oil Production	Annual Rate of Production in Metric tons by						Total
	Hydrogenation	F-T Synthesis Plants	Refining of German and Austrian petroleum	Brown coal and Biruminous coal tar distillation	Benzole	Imports from Rumania and Hungary	
Aviation fuel	1,900,000	---	---	---	50,000	100,000	2,050,000
Motor spirit	350,000	270,000	160,000	35,000	330,000	600,000	1,745,000
Diesel oil	680,000	135,000	670,000	110,000	---	480,000	2,075,000
Fuel oil	240,000	---	120,000	750,000	---	---	1,110,000
Lubricating oil	40,000	20,000	780,000	---	---	---	840,000
Misc.	40,000	160,000	40,000	50,000	---	---	290,000

Auschwitz, however, never produced a drop of synthetic fuel. Construction started in 1941, and it remained largely unfinished at the time Soviet troops overran it on 27 January 1945. Its scheduled production of 24,000 metric tons per year, making it the smallest of the coal hydrogenation plants, was only a fraction of the Leuna plant which produced at its rated capacity of 620,000 metric tons of synthetic liquids per year. Auschwitz cost 25,000-30,000 lives and RM 900 million for all operations including the never-completed synthetic rubber plant, but it was a miserable failure [38].

To determine an accurate production cost in the operational synthetic fuel plants, or in any of the wartime plants that used forced laborers and concentration camp inmates, remains very complicated, mainly because of difficult-to-measure factors and incomplete data. First of all, even though concentration camp inmates received no wages, the cost of their guarding, housing, even their near-starvation feeding involved some expense. The production cost calculation also requires knowing the number of years each plant was in operation; what percent of workers in each plant was forced or concentration camp; how long each worker worked at the plant; and worker efficiency which, according to Fritz Sauckel, Reich Commissioner for Labor, ranked Polish forced workers one-half as efficient, and concentration camp inmates one-third as efficient as German workers. Total German synthetic fuel production in fact fell to its lowest in the last months of World War II when the number of coerced and concentration camp inmates reached a maximum.

Other factors to consider are plant operation time versus shutdown time because of bombings and equipment malfunctions and what reduction in production cost resulted completely from technical improvements in each plant [39]. Some of this information, such as the composition of the workforce in a few of the plants, is available. The Leuna plant as of 1 October 1944 employed 34.9 percent foreign workers; in Ludwigshafen, which was a research facility and only a small producer of synthetic fuel, the foreign work force numbered 36.6 percent. Auschwitz, by the same date, had 55.1 percent foreign workers, 26.6 percent concentration camp inmates both foreign and German, and 18.3 percent free German workers [40].

Postwar court settlements, such as the 1957 Braunschweig court case settlement between a Jewish concentration camp inmate and the Bussing Company of Braunschweig, have provided additional information on wartime labor. Bussing manufactured trucks for the German army and during the war it had used foreign inmates from the Neuengamme concentration camp. Because the inmates received no compensation for their wartime work the court set the wage at RM 1.00 per hour (before deductions) for a 10 hour work day arguing that it was the scale established according to wartime wage controls [41].

In another postwar settlement IG Farben and the Conference on Jewish Material Claims Against Germany, a consolidation of twenty-three major Jewish organizations, and reached an agreement to provide compensation for IG Farben's use of unpaid concentration camp inmates. By 1958, IG Farben had arranged to pay DM 27 million to the Jewish Material Claims Conference [42]. Its settlement followed an earlier 1952 agreement between the Federal Republic of Germany and the Material Claims Conference in which the German government paid DM 450 million ($105 million) to the Material Claims Conference and also sent DM 3 billion ($700 million) worth of goods such as petroleum and steel to Israel over a ten-year period. The German government estimated that its payments would have to continue beyond the year 2000 and its total payments would reach DM 100 billion ($40 billion) [43].

7. Conclusion

Germany had the first technologically successful synthetic fuel industry producing eighteen million metric tons (130 million barrels) from coal and tar hydrogenation and another three million metric tons from the F-T synthesis in the period 1939-1945. After the war ended German industry did not continue synthetic fuel production because the Potsdam (Babelsberg) Conference of 16 July 1945 prohibited it [44]. The Allies maintained that Germany's Nazi government had created the industry for strategic reasons under its policy of autarchy and that in postwar Germany there were, economically, better uses for its coal than synthetic fuel production. Four years later on 14 April 1949, the Frankfurt Agreement ordered dismantling of the four coal hydrogenation plants in the western zones, all of which were in the British zone [45].

Shortly after the formal establishment of the West German government in September 1949, a new agreement, the Petersberg (Bonn) Agreement of 22 November 1949, quickly halted the dismantling process in an effort to provide employment for several thousand workers.[46] The West German government completely removed the ban on coal hydrogenation in 1951, although by this time Ruhröl GmbH (Mathias Stinnes) had deactivated the Welheim plant, and the plants in Scholven, Gelsenberg, and Wesseling, after design modifications, were hydrogenating and refining crude oil rather than hydrogenating coal.

The Soviets (Russians) dismantled the Magdeberg plant located in their zone and the three plants in Poland at Pölitz, Blechhammer, and Auschwitz. They used parts from the Magdeberg and Auschwitz plants to reconstruct a plant in Siberia that had an annual production capacity of one million metric tons of aviation fuel and a second plant in Kemerow-Westbirien that also produced aviation fuel from coal. The Pölitz and Blechhammer plants provided scrap iron. Three other plants in their zone, at Leuna, Böhlen, Zeitz, and the Sudetenland plant at Brüx (Möst), which the Soviets gave to Czechoslovakia, continued with coal and tar hydrogenation, and after modification, refined petroleum into the early 1960s. Some dismantling and conversion to synthetic ammonia production for fertilizers occurred at the Leuna plant which by 1947 the Soviets had renamed the Leuna Chemical Works of the Soviet Company for Mineral Fertilizers. The last of the coal hydrogenation plants in the Soviet Zone at Lützkendorf did not resume production after the war.

Three of the F-T plants continued operation after the war. Schwarzheide in the Soviet Zone, which had a labor force of 3,600, produced gasoline for Soviet civilian and military consumption. Gewerkschaft Victor in Castrop-Rauxel and Krupp Treibstoffwerk in Wanne-Eickel in the British zone, as of February 1946 were producing oils and waxes from fatty acids and using them to make soaps and margarine [47]. The six other plants remained inoperative. Today none of the 21 synthetic fuel plants produces synthetic fuels.

The German synthetic fuel industry succeeded technologically because in the 1920s Pier at IG Farben developed suitable sulfur-resistant catalysts for the hydrogenation of coal and tar and divided the process into separate liquid and vapor phase hydrogenations, improving both economics and yield. A short time later Fischer and his co-workers at the KWI prepared the cobalt catalysts and established the reaction conditions that made the F-T synthesis a success. But neither coal-to-oil conversion process could produce a synthetic liquid fuel at a cost competitive with natural petroleum. Coal hydrogenation and the F-T synthesis persevered and survived because they provided the only path Germany could follow in its search for petroleum independence. Despite the unforeseen and unfortunate social and political environment in which the German synthetic fuel industry arose, Germany remains the only nation that attempted and developed a synthetic fuel industry [48].

- **References**

Error! Main Document Only. Most of the information on the Fisher-Tropsch and coal hydrogenation plants has come from the Allied investigative teams that went to Germany during World War II's closing months. These teams, such as United States Technical Oil Mission (TOM) and the British Intelligence Objectives Subcommittee (BIOS), examined the thousands of technical reports Allied troops captured at the synthetic fuel plants, interviewed many of the German synthetic fuel scientists, and sent their information to the Combined Intelligence Office Subcommittee (CIOS) in London for translating and abstracting. CIOS prepared 141 microfilm reels, and after moving its operation to the United States produced another 164 reels. CIOS, BIOS, TOM, and Field Intelligence Agency Technical (FIAT) also printed and released more then 1,400 reports on the German synthetic fuel plants, many of which are on TOM microfilm reels.

In addition to the 1,400 investigative reports several exhaustive summaries of the reports are available. The most important of these are the Ministry of Fuel and Power, *Report on the Petroleum and Synthetic Oil Industry of Germany* (London, 1947) and the Joint Intelligence Objectives Agency, *High-pressure Hydrogenation at Ludwigshafen-Heidelberg*, FIAT, *Final Report 1317* (9 vols., Dayton, Ohio: Central Air Document Office, March 1951). The Ministry's *Report* deals with the Fischer-Tropsch synthesis and the coal hydrogenation process whereas the Joint Intelligence's *High-pressure* discusses only coal hydrogenation. A third comprehensive source is Henry H. Storch, Norma Golumbic, and Robert B. Anderson, *The Fischer-Tropsch and Related Syntheses* (New York, 1951). It also relies heavily on the captured German World War II synthetic fuel documents. These are the best and most comprehensive sources, and I have relied on them extensively.

During the early 1970s after the Arab oil embargo and crisis of 1973-74 Richard Wainerdi and Kurt Irgolic established the German Document Retrieval Project at Texas A&M University. They set as its objective the collecting, translating, and organizing of the thousands of German World War II documents and reports that the Allied intelligence teams brought to the United States and now were scattered around the country in various government repositories, archives, and even with members of the TOM. The German Document Retrieval Project, of which I was a member, accomplished its objective, and as a result Texas A&M's archives contain what is very likely the most comprehensive collection of information on Germany's World War II synthetic fuel industry. I have used this collection in this and other papers I have written on the history of synthetic fuel. This paper's citations on the plant descriptions are from the Ministry's and the Joint Intelligence's summaries, with other sources included when required for greater detail or clarification. Many of the documents are now on line at Syntroleum's Fischer-Tropsch Archive, *www.fischertropsch.org*.

[1]. Anthony N. Stranges, "From Birmingham to Billingham: High-Pressure Coal Hydrogenation in Great Britain," *Technology and Culture*, 20 (1985): 726-757.

[2]. Anthony N. Stranges, "Canada's Mines Branch and its Synthetic Fuel Program for Energy Independence," *Technology and Culture*, 32 (1991), 521-554; Stranges, "Synthetic Fuel Production in Japan: A Case Study in Technological Failure," *Annals of Science* 50 (1993): 229-265.

[3]. Anthony N. Stranges, "The US Bureau of Mines' Synthetic Fuel Programme," *Annals of Science*, 54 (1997): 29-68.

24

[4]. Anthony N. Stranges, "Synthetic Petroleum form Coal Hydrogenation: Its History and Prsent State of Development in the United States," *Journal of Chemical Education*, 60 (1983): 617-625.

[5]. Anthony N. Stranges, "Germany's Synthetic Fuel Industry 1927-45" in *The German Chemical Industry in the Twentieth Century*, edited by John E. Lesch, Dordrecht/Boston/London: Kluwer Academic Publishers, 2000.

[6]. A third process, the distillation or carbonization of coal at either a high temperature (HTC) of 700-1000°C or a low temperature (LTC) of 500-700°C produces petroleum. The process is not a synthesis but a decomposition and gives small yields of only gallons per metric ton of coal rather than barrels. It is a derived process and was never a major contributor to Germany's liquid fuel requirements. Its simplicity, distilling the petroleum from coal, not its yield has resulted in its use.

[7]. Anthony N. Stranges, "Friedrich Bergius and the Rise of the German Synthetic Fuel Industry," *Isis*, 75 (1984): 43-67.

[8]. BASF, German Patent 293,787 (8 March 1913); BASF, British Patent 20,488 (10 September 1913); BASF, French Patent 468,427 (13 February 1914); BASF (Alwin Mittasch and Christian Schneider), US patent, 1,201,850 (17 October 1916).

[9]. Franz Fischer and Hans Tropsch, "Über die Reduktion des kohlenoxyds zu Methan am Eisenkontakt under Druck," *Brennstoff-Chemie,* 4 (1923): 193-197; Fisher and Tropsch, "Über die Herstellung synthetischer Ölgemische (Synthol) durch Aufbau aus Kohlenoxyd und Wasserstoff," ibid., 4 (1923): 276-285; Fischer and Tropsch, "Methanol und Synthol aus Kohlenoxyds als Motorbetreibstoff," ibid., 6 (1925), 233-234; Fischer, "Liquid Fuels from Water Gas," *Industrial and Engineering Chemistry,* 179 (1925): 574-576; Fischer and Tropsch, German Patent 484,337 (22 July 1925); Fischer and Tropsch "Die Erodölsynthese bei gewöhnlichem Druck aus den Vergangsprodukten der Kohlen," *Brennstoff-Chemie,* 7 (1926): 97-104; Fischer and Tropsch, German Patent 524,468, (2 November 1926); Fischer and Tropsch, "Über Reduktion und Hydrierung des Kohlenoxyds," *Brennstoff-Chemie,* 7 (1926): 299-300; Franz Fischer, "Über die synthese der Petroleum Kohlenwasserstoffe," *Brennstoff-Chemie,* 8 (1927): 1-5 and *Berichte,* 60 (1927), 1330-1334; Fischer and Tropsch, "Über das Auftreten von Synthol bei der Durchführung der Erdölsynthese unter druck und über die Synthese hochmolekular Paraffin Kohlenwasserstoffe aus Wassergas," *Brennstoff Chemie,* 8 (1927): 165-167; BASF, British Patents 227,147, 228,959, 229,714, 229,715 (all 28 August 1923); Fischer, "Zwölf Jahre Kohlenforschung," *Zeitschrift angewandte* Chemie, 40 (1927): 161-65; "Zur Geschichte der Methanolsynthese," *Zeitschrift angewandte Chemie,* 40 (1927): 166; BASF (Alwin Mittasch, Matthias Pier, and Karl Winkler), German Patent 415,686 (application 24 July 1923, awarded 27 July 1925); BASF, US Patent 1,558,559 (27 October 1925); BASF (Mittasch and Pier), US Patent 1,569,775 (12 January 1926); BASF, French Patents 571,285 (29 September 1923); 571,354, 571,355, and 571,356 (all 1 October 1923); 575,913 (17 January 1924); 580,914 (30 April 1924); 580,949 (1 May 1924); 581,816 (19 May 1924); 585,169 (2 September 1924); Fischer and Tropsch, "Über die direketer synthese von Erdöl-Kohlenwasserstoffen bei gewöhnlichem Druck," (Erste Mittelung), *Berichte,* 59 (1926): 830-831; ibid. 832-836; Fischer and Tropsch, "Uber Einige Eigenschaften der aus Kohlenoxyd bei gewohnlichem Druck Hergestellten Synthetischen Erdöl-Kohlenwasserstoffe," *Berichte,* 59 (1926): 923-925. Henry H. Storch, Norma Golumbic, and Robert B. Anderson, *The Fischer-Tropsch and related syntheses* (New York, 1951): 115.

[10]. Franz Fischer and Hans Tropsch, German Patent 484,337 (22 July 1925); Fischer and Tropsch publications (ref. 9); Fischer, "The synthesis of petroleum," International Conference on Bituminous Coal, *Proceedings* (Pittsburgh, 1926): 234-246; Storch, Golumbic, and Anderson, *Fischer-Tropsch* (ref. 9: 116-117.

[11]. Storch, Golumbic, and Anderson, *Fischer-Tropsch* (ref. 9), 135; Franz Fischer, Helmet Pichler, and Rolf Reder, "Überblick über die Möglichkeiten der Beshaffung geeigneter Köhlenoxyd-Wasserstoff-Gemische für die Benzinsynthese auf Grund des heutigen Standes von Wissenschaft und Technik," *Brennstoff Chemie,* 13 (1932): 421-428; Franz Fischer, Otto Roelen, and Walter Feisst, "Über die nunmehr

erreichten technischen Stand der Benzinsynthese," ibid., 13 (1932): 461-468; Franz Fischer and Herbert Koch, Über den Chemismus der Benzinsynthese und über die motorischen und senstigen Eigenschaften der dabei auftretenden Produkte (Gasol, Benzin, Dieselöl, Hartparaffin), ibid., 13 (1932): 428-434; Herbert Koch and Otto Horn, "Vergleichende Untersuchung über das motorische Verhalten eines synthetischen Benzins nach Franz Fischer (Kogasin I) und eines Erdöl-Benzins, ibid., 13, (1932): 164-167.

[12]. Storch, Golumbic, and Anderson *Fischer-Tropsch* (ref. 9): 337; "Synthetic petrol by the Fischer process," *Gas World,* 105 (1936): 362-363; H.H. Storch, "Synthesis of Hydrocarbons from Water Gas," H. H. Lowry, ed., *Chemistry of Coal Utilization* (3 vols., New York, 1945 and 1963), *2*: 1797-1845, on 1800. The motor fuel's estimated average cost was 22 pfennig per kg. According to Arno Fieldner and American engineers the wartime cost of either the F-T or the hydrogenation process was 20-30¢ per US gallon. See Arno C. Fieldner, "Frontiers of fuel technology," *Chemical and Engineering News,* 26, *23* (7 June 1948): 1700-1701. For the conversion of metric tons to barrels use 1 metric ton = 7.2 barrels.

[13]. Ministry of Fuel and Power, *Report on the Petroleum and Synthetic Oil Industry of Germany* (London, 1947), 82-90; "Germany's Home Production of Motor Fuels," *Gas World,* 104 (9 May 1936): 421; Franz Fischer, "The Conversion of Coal into Liquid Fuels," *Chemical Age,* 35 (24 October 1936): 353-355; "The Fischer Process," *Chemical Age,* 35 (31 October 1936): 367.

[14]. Storch, Golumbic, and Anderson, *Fischer-Tropsch* (ref. 9): 336-338; *Report* (ref. 13); G. Wilke, "Die Erzeugung und Reinigung von Synthesegas für die Benzinsynthese," *Chemische Fabrik,* 11 (1938): 563-568; "Substitute Motor Fuels in Europe," *Petroleum Press Service,* 5 (1938): 301-304; Fieldner, "Frontiers" (ref. 12). Average exchange rate from 1934 to 1941 was RM 1 = 40¢, *Banking and Monetary Statistics, 1914-1941*, Federal Reserve System (Washington DC, 1947), 671.

[15]. *Report* (ref. 13): 82-90.

[16]. Ibid., 91 (Table LV). The German Health Office officially approved the synthetic fat as fit for human consumption, but the Nazi government suppressed the findings of certain (unnamed) university scientists which threw considerable doubt on the fat's safety. Synthetic fat always contains esters of branched-chain fatty acids some of which are toxic (see p. 94). C. C. Hall, "Oils and Waxes from Coal," *Chemical Age,* 55 (9 November 1946): 569-570.

[17]. *Trials of War Criminals before the Nürnberg Military Tribunals* (Washington DC, 1953): see vol. 7, the IG Farben Case (case 6), testimony of defendant (Carl) Krauch. The Fuel Agreement is *Nürnberg Industrialists Document* NI-881, reel 9, 14 December 1933. See also Wolfgang Birkenfeld, "Leuna, 1933," *Tradition,* 8 (1963): 107-108; Wolfgang Birkenfeld, *Der synthetische Treibstoff 1933-1945* (Göttingen, 1964): 23-34; Peter Hayes, *Industry and Ideology* (Cambridge, 1987): 115-120.

[18]. The United States Strategic Bombing Survey, *The German Oil Industry Ministerial Report Team* 78, Oil Division, 1st ed. 5 September 1945, 2nd ed. January 1947, 19-20.

[19]. *Report* (ref. 13): 89; T. E. Warren, *Inspection of Hydrogenation and Fischer-Tropsch Plants in Western Germany during September 1945*, FIAT, *Final Report no. 80, item no. 30* (London: British Intelligence Objectives Subcommittee 1945): 1-28 on 16-18; *Germany, Liquid Fuels, v-Synthetic Oil Plants-Fischer-Tropsch Synthesis*, (typed) Report 75687, Reference numbers 5.01-5.09; "Coal Hydrogenation-Germany, Hydrocarbons Synthesis," *Petroleum Press service,* 6 (1939): 529-532.

[20]. *Report* (ref. 13): 89; Warren, *Inspection* (ref. 19): 23-24; *Germany, Liquid Fuels* (ref. 19).

[21]. *Report* (ref. 13): 89; Warren, *Inspection* (ref. 19): 14-15; *Germany, Liquid Fuels* (ref. 19).

[22].*Report* (ref. 13): 85.

26

[23]. *Report* (ref. 13): 90; "Coal-Hydrogenation-Germany" (ref. 19); "Coal Hydrogenation-Germany, Hydrocarbons Hynthesis," *Petroleum Press Service,* 7 (1940): 31-32; *Germany, Liquid Fuels* (ref. 19).

[24]. *Report* (ref. 13): 87 (Table L); Hall, "Oils" (ref. 17).

[25]. *Report* (ref. 13): 90; *Germany, Liquid Fuels* (ref. 19).

[26]. *Report* (ref. 13): 89; Warren, *Inspection* (ref. 19): 21-22; *Germany, Liquid Fuels* (ref. 19); "Motorkraftstoff von Kohl," Teer und Bitumen, 35 (1937): 231-232.

[27]. *Report* (ref. 13): 89-90 (Table LIV); Warren, *Inspection* (ref. 19): 16; *Germany, Liquid Fuels* (ref. 19); "Coal Hydrogenation, Germany" (ref. 19); "Motorkraftstoff," (ref. 26).

[28]. *Report* (ref. 13): 90; Warren, *Inspection* (ref. 19); "Motorkraftstoff" (ref. 26).

[29]. *Report* (ref. 13): 90-91; *Germany, Liquid Fuels* (ref. 19).

[30]. *Report* (ref. 13): 96-100.

[31]. Ibid., 92-95, 101-102.

[32]. *Report* (ref. 13): 1-2, 88; Strategic Bombing Survey (ref. 18).

[33]. Strategic Bombing Survey (ref. 18).

[34]. *Report* (ref. 13): 1-2, 50, 61, 67, 68, 93. The prewar exchange rate is from *Statistical Abstract of the United States* 1939 (Washington, 1939): 208. RM 1 = 25¢.

[35]. Edward Homze, *Foreign Labor in Nazi Germany* (Princeton, 1967): 171.

[36]. *Report* (ref. 13): 61 (Table XXXV).

[37]. Homze, *Foreign Labor* (ref. 35): 171; Joseph Borkin, *The Crime and Punishment of IG Farben* (New York, 1978): 117. RM 10 = DM1 = 40¢ US, currency conversion law of 1948.

[38]. *Report* (ref. 13): 48 (Table XXVI).

[39]. Jurgen Kuczynski, *Germany: Economic and Labor Conditions under Facism* (New York), 1968): 213.

[40]. Homze, *Foreign labor* (ref. 35): 238-239.

[41]. Ibid.

[42]. Benjamin B. Ferencz, *Less than Slaves* (Cambridge, MA, 1979): 52.

[43]. Ibid., xvi-xvii.

[44]. "Principles to govern the treatment of Germany in the Initial Control Period," no. 848, The Conference of Berlin (The Potsdam Conference), 1945, *Foreign Relations of the United States, Diplomatic Papers* (17 vols., Washington, DC, 1949-1964-1968): **2** (1960), 750-753, on 752.

[45]. "Multilateral: German industries," *United States Treaties and other International Agreements,* part I, 1951 (35 vols., Washington DC, 1950-84), *2* (1952): 962-972, on 963.

[46]. "Multilateral: incorporation of Germany into European Community of Nations," *United States Treaties and other International Agreements*, part II, 1952 (35 vols., Washington DC, 1958-84), *3* (1954): 2714-2722, on 2716.

[47]. Birkenfeld, *Der synthetische Treibstoff* (ref. 17): 213-215; E. E. Donath, "Hydrogenation of Coal and Tar," H. H. Lowry, ed., *Chemistry of Coal Utilization* (3 vols., New York, 1945 and 1963), supplementary volume, 1041-1080, on 1042-1044; "German Synthetic Petrol and the Moscow Conference," *Petroleum Times,* 51 (1947): 430, 446; "Present Position and Future Role of Möst (Brüx) Synthetic Oil Plant," *Petroleum Times,* 50 (1946): 852; Strategic Bombing Survey (ref. 18).

[48].   SASOL in South Africa developed its F-T coal-to-oil process after World War II. Its Sasolburg plant began liquid fuel production in 1955, a second plant at Segunda (SASOL Two) opened in 1980, and a third plant SASOL Three in 1982. By 2001 SASOL was providing 29 percent of South Africa's motor fuel requirements, both gasoline and diesel, as well as industrial chemicals.

Fischer-Tropsch Synthesis, Catalysts and Catalysis
B.H. Davis and M.L. Occelli (Editors)
© 2007 Elsevier B.V. All rights reserved.

Synthetic Lubricants: Advances in Japan up to 1945 Based on Fischer-Tropsch Derived Liquids

Edwin N. Givens[1], Stephen C. LeViness[2] and **Burtron H. Davis**[1]

[1]*Center for Applied Energy, Research, University of Kentucky, 2540 Research Park Dr., Lexington, KY, 40511*
[2]*Syntroleum Corporation, 880 W. Tenkiller Rd., Tulsa, OK, 74015*

A review of Japanese technology regarding lubricants from Fischer-Tropsch derived liquids that was developed before and during World War II is presented. Extensive studies were performed on cracking FT liquids to make charge stock for an $AlCl_3$ polymerization plant to make an aircraft lubricating oil. The physical properties and oxidation stability of these oils will be compared with U. S. oils available at that time.

1. Introduction

The U. S. Naval Technical Mission to Japan was established immediately after surrender of Japan in 1945 to survey technological developments of interest to the Navy and Marine Corps in Japan and areas occupied by Japan during WWII. This involved seizure of material, interrogation of personnel, examination of facilities and preparation of reports to appraise the status of that technology.[1] One objective among many was to evaluate Japanese fuel and lubricant technology which was of interest to the U. S. Navy. Included was an evaluation of Fischer-Tropsch operations and supporting research, although Japanese technology did not hold promise of commercial advances as did those of Germany.

A unique feature of the Japanese Navy was that they built and operated one of the largest fuel and lubricant research institutions in the world and, at the same time, they built and operated two of the largest refineries in Japan. Based upon the survey, it was concluded that Japanese technical ability had been underestimated. Much of the work in these naval laboratories was devoted toward aviation gasoline and lubricants, especially toward the end of the war when frantic efforts were devoted toward boosting the rapidly dwindling supply of aviation gasoline.

The principal sources of lubricants for the Japanese Navy during World War II was as follows:

1940-1942 - Imported finished lubricants much of which was derived from California crudes

1943 -1945 - Omaga and Rhodessa crudes

No mention was made that any lubricants produced from Fischer-Tropsch liquids were utilized. However, during the period 1943 to 1945, diesel fuel used by both the Navy and Army contained FT liquids. The FT diesel oil, most of which was obtained from the Miike Synthetic Oil Plant at Omuta, was utilized by the Navy by blending with 90% of Tarakan oil, to meet diesel fuel specifications. The Army used about 50% of the FT diesel oil production as fuel for diesel engines in tanks.

1.1. Lubricants from FT Liquids

The Miike Synthetic Oil Company, located in Omuta, had the most productive FT plant in Japan. It employed the low-pressure process using cobalt-thorium catalyst for production of oil from Miike coal. This plant was ordered by the army in 1941 to concentrate on the construction of a lubricating oil polymerization plant which was still under construction at the end of the war. The process to produce the lubricating oil was licensed from Ruhrchemie and most equipment was constructed by the Koppers Company. The process was based on plans purchased from UOP in 1939 which called for a Dubbs-Kogasin cracking unit and a gas polymerization unit. The cracking unit was to lightly crack paraffin oil to make charge stock for an $AlCl_3$ lubricating oil plant. This cracked distillate, having an end point of $250°C$, was then to be charged to a polymerizer with 3-5% anhydrous aluminum chloride and maintained with agitation at $60-90°C$ for 8-12 hours. The product was to be settled, dechlorinated, purified with active clay (3 hours at $150°C$, 5% by weight of clay), filtered, and topped. A light lubricating oil and an aircraft lubricating oil were to be obtained from the product. Chemical and physical properties of product produced in small scale pilot plant tests are shown in **Table 1**.

Table 1. Lubricating Oil Properties from OMUTA Process		
	Aircraft Lubricating Oil	Light Lubricating oil
Specific Gravity, 15/4°C	0.85-0.87	0.82-0.83
Flash Point, °C	240	145
Pour Point, °C[1]	20	25
Neutralization Number	0.05	0.06
Viscosity (S.U.S.)		
at 100°F	1550	104
at 210°F	125	34
Viscosity Index	107	54
Carbon Residue, %	0.08	
Ash, %	0.003	
1. The pore points as reproduced from the U.S. Naval Technical Mission to Japan report X-38(N)-8, page 13 are obviously incorrectly stated because the negative signs are missing.		

1.2. Lubricant Research

Work on Fischer-Tropsch liquid as a raw material for the synthesis of lubricants was undertaken since it was believed that its highly paraffinic nature should yield a stable and high viscosity-index lubricating oil. The properties of FT derived aero-engine oil, petroleum based oils from U. S. sources and other oils synthesized by the Japanese are shown in **Table 2**. Oxygen absorption tests were performed on aero-engine oils to determine the effect of antioxidants on oil stability. The absorption of oxygen was measured by Warburg's apparatus using the British Air Ministry Test. As shown in **Figure 1**, oxygen absorption by the synthetic oil prepared from Fischer-Tropsch oil was higher than for Texaco #120 or the aero-engine oil being used at the time. The viscosity ratios of these oils correlated with the amount of oxygen absorbed in the oil. The viscosity ratio of the oil prepared from FT liquid as shown in **Figure 1** was 3.4.

Table 2. General Properties of Some Aero-Engine Oils

	Phillips #120	Texaco #120	Polymerized Oil from Cracked Distillate of sweated Wax	Polymerized Oil from Cracked Distillate of Crude Wax	Blend of Natural and Synthetic Oils	Polymerized oil of Fischer Oil
Specific Gravity (d^{15}_4)	0.8965	0.8938	0.8534	0.8782	0.8820	0.8653
Flash Point (°C)	240	230	-	224	232	-
Viscosity (S.U.S., 100°F)	1794.2	1649	1187.3	1440.2	1573.2	790.6
Viscosity (S.U.S., 210°F)	121.1	117	137.6	123.0	125.4	79.2
Viscosity Index	93.4	95.4	127.9	110.0	106.9	100.8
Conradson's Carbon (%)	0.70	1.06	0.10	0.47	0.46	0.09
Pour Point (°C)	-13	-11	-25	-34	-27	-22
Stability, Viscosity Ratio	1.33	1.17	2.13	1.78	1.64	3.43[1]
Stability, Conrad. C. (%)	1.80	1.60	0.7	1.58	1.50	1.0
Elemental Analysis C%	80.00	85.83	80.14	86.01	86.22	84.20
H%	12.80	13.38	13.88	13.39	13.65	13.56
S%	0.15	0.05	0.02	-	-	0.02
Mean Molecular Weight	614	694	1095	865	740	684
Empirical Formula	$C_{44}H_{88}$	$C_{50}H_{92}$	$C_{78}H_{151}$	$C_{62}H_{115}$	$C_{53}H_{100}$	$C_{48}H_{92}$

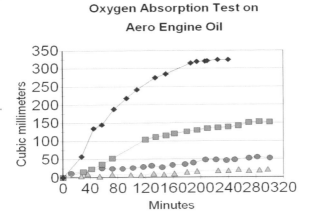

Oxygen Absorption Test on Aero Engine Oil

Figure 1.   Oxygen uptake by aero-engine oil for: ◆, synthetic oil from FT; ■, synthetic oil from cracked crude; ●, aero-engine oil in use; ▲, Texaco #30.

Oxidation inhibitors were tested with FT derived oils. Among those reported, triphenyl phosphite, tricresyl phosphite, tin oleate, chromium oleate, individually and in combination, are shown in **Table 3**. (In addition, copper soaps were also tested with FT derived liquids as mentioned in the text, as well as discussed below.)   The best antioxidant was found to be a combination of 0.5% triphenylphosphite and 0.5% chromium oleate. They found that the oxidation inhibitors suitable for natural oils were not, as a rule, suitable for the synthetic oils. For example, for aircraft engine oil produced from petroleum, tricresyl phosphite was the best anti-oxidant, as shown in **Table 4**. For synthetic aircraft engine oil prepared from paraffin wax, copper soaps were the most effective. Although copper stearate was not included in those presented in **Table 3**, its effect on retarding oxygen absorption at 150°C on synthetic polymerized oil prepared from Fischer oil is shown in **Figure 2**. Unfortunately, no data are presented showing the effect of the mixture of triphenyl phosphite and chromium oleate on oxygen absorption in synthetic aero-engine oil prepared from FT oil even though the text indicated this was the most effective additive.   Likewise, in the report, no supporting engine data for the synthetic oil so compounded is provided.

Table 3. The Effect of Antioxidants on Synthetic Aero-Engine Oil from Fischer Oil				
			Oxidation Test	
No.	Addition Compound	Conc.(%)	Viscosity Ratio	Conradson Carbon (%)
1	None	0	3.43	1.1
2	Triphenyl phosphite	0.5	2.60	1.1
3	Tricresyl phosphite	0.5	2.63	1.4
4	Triorthocresyl phosphite	0.5	2.64	1.3
5	Tin oleate	0.5	3.00	1.5
6	Chromium oleate	0.5	2.79	1.4
7	Triphenyl phosphite-Tin oleate	0.5/0.5	2.71	1.6
8	Triphenyl phosphite-Chromium oleate	0.5/0.5	1.87	1.2
9	Tin oleate-Chromium oleate	0.5/0.5	2.62	2.2

Table 4. The Effect of Antioxidants on Oxidation Stability of Aero-Engine Oil			
Addition Compounds	Amount (%)	Viscosity Ratio	Conradson Carbon (%)
None	0	1.56	1.90
Tricresylphosphite	0.5	1.17	1.23
Tricresylphosphate	1.0	1.23	0.84
Dibenzyldisulfide	0.5	1.49	1.34
Copper oleate	0.1	1.70	-
Copper stearate	0.1	1.58	1.96
Stearonitrile	0.5	1.26	0.96
Stearophenone	1.0	1.25	0.96
Trilaurylphenylphosphate	1.0	1.57	1.74
Tricresylphosphite/Tricresylphosphate	0.5/0.5	1.22	-
Tricresylphosphite/Tricresylphosphate	0.3/0.7	1.25	-
Tricresylphosphite/Tricresylphosphate	0.2/0.2	1.24	-
Tricresylphosphite/Tricresylphosphate	0.1/0.2	1.33	-

2. Conclusions

The Japanese produced FT product at the Miike Synthetic Oil Company located in Omuta and were in the process of constructing an adjoining a lubricating oil facility that was not completed before the end of the war. Laboratory evaluation of lubricants prepared from FT oils were performed in the laboratory. Although the U. S. Naval Technical Mission Reports provide data derived from these evaluations, the data are sparse and obviously incomplete likely due to the fact that the technical files of the Japanese Research Institute were destroyed in August 1945. The available data show that for the FT derived lubricating oils the best antioxidant was a combination of triphenylphosphite and chromium oleate.

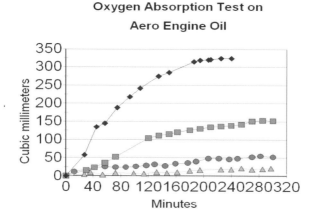

Figure 2.   Oxygen uptake by aero-engine oil for: ◆, synthetic oil from FT + Cu stearate (150mg); ■, synthetic oil from FT (50mg); ●, synthetic oil from FT (20mg); ▲, synthetic oil from FT + Cu stearate (20mg).

3. Acknowledgments

Financial support for this work has been provided by the Syntroleum Corporation, Marathon Oil Company and ConocoPhillips.

Footnotes

1.   The review prepared by the U. S. Naval Technical Mission is undoubtedly incomplete since the technical files of this institute, which employed some 3,200 men and comprised over 70 modern buildings, were burned 1 August 1945 by order of the Director of the Depot. The Technical Mission ordered approximately 100 of the technical personnel to return to the institute to reproduce from notebooks, personal files, memory and other sources, reports in English covering all of their research activities during the war period.

Fischer-Tropsch Synthesis, Catalysts and Catalysis
B.H. Davis and M.L. Occelli (Editors)

A history of the BP Fischer-Tropsch catalyst from laboratory to full scale demonstration in Alaska and beyond

Joep J.H.M. Font Freide[a], John P. Collins[b], Barry Nay[a], and Chris Sharp[c]

[a]BP Exploration Operating Company, Chertsey Road, Sunbury on Thames, Middlesex, TW167LN
[b]BP Exploration America Inc., 501 Westlake Park Blvd., Houston, TX 77079
[c]BP Chemicals, Hedon, Saltend, East Yorkshire HU128D5

1. Introduction

This paper describes the development of BP's Fischer-Tropsch (FT) catalyst from the early days of laboratory scale preparations and micro-reactor tests to commercial scale manufacture and operation at BP's Gas to Liquids (GTL) demonstration facility in Nikiski, Alaska. A detailed description of the catalyst development activities, preparation methods, and experimental facilities is provided by Font Freide and coworkers [1]. The initial research was focused on catalyst development for a fixed bed reactor design. Recent activities include the commercial scale fixed bed tests in progress at Nikiski and development of a novel slurry-based reactor technology.

2. Catalyst Development

The story of the BP FT catalyst began at BP's Sunbury Research Centre in the early 1980's when the search for a non-iron FT catalyst was initiated. Cobalt was chosen as the active metal for reasons of cost and availability. Flow-sheeting studies were used to define targets for catalyst performance. These targets required a catalyst capable of giving a single pass conversion of greater than 70 % with a C_5+ productivity of greater than 150 g l_{cat}^{-1} h^{-1} and selectivity to C_5+ of greater than 80 %. For a viable fixed bed commercial process a catalyst life of 4 years was thought to be the minimum required.

Laboratory evaluations of the FT catalysts were initially conducted in fixed bed micro-reactors. These have an internal diameter of approximately 9 mm and are fitted with 3.2 mm thermo-wells. Catalysts were tested as 250 – 500 micron particles usually diluted with an inert material of the same particle size. Activations were normally carried out in-situ by reduction with hydrogen prior to beginning testing.

The micro-reactor studies progressed in tandem with testing in a small pilot plant that was constructed to allow evaluation of formed catalysts that could be used in a commercial reactor, instead of the pressed powders tested in the micro-reactors. This larger scale testing was carried out in a 1" internal diameter reactor (120 ml cat.) fitted with a 5 zone heating system to ensure more isothermal operation and typically allowed 1-3 mm particles to be evaluated. The reactor tube internals were configured in such a way as to minimise the radial temperature profile and ensure that the reactor operated in a thermally stable regime.

Many of the cobalt catalysts made during the initial screening studies showed good activities for CO conversion and high selectivity's to liquid transportation fuels (> C_5 hydrocarbons) but were found to be intolerant of CO_2. It was a considerable problem since the natural gas is likely to contain CO_2 and CO_2 is also a by-product of syngas production. In addition the cobalt catalyst itself will make some CO_2 via water gas shift. It was known that some methanol catalysts, not only tolerate CO_2 but also require it in the syngas feed. For this reason it was decided to investigate a cobalt-based FT catalyst employing a similar catalyst formulation.

The first of these new cobalt catalysts were made in 1986 by co-precipitation techniques using aqueous solutions with ammonium bicarbonate as the precipitant in a similar way to the methods used for methanol synthesis catalysts. The new catalysts were immediately found to be very active and selective catalysts for the conversion of syngas into hydrocarbons. A particularly attractive feature was their low methane make and tolerance of CO_2. The CO_2 tolerance was ascribed to the interplay between the support and the cobalt phase both in the oxidized and reduced forms. The general belief is that the support stabilizes the cobalt phase such that the catalyst can be operated at the higher temperatures, required to maintain activity despite competitive adsorption by CO_2, without any loss in stability. Other investigators e.g. Shell have used similar strategies [2].

Early catalyst life studies indicated a steady deactivation of the catalyst, regardless of preparation method. Improvements in preparation, formulation and activation eventually lead to an increase in catalyst life. Methane selectivity of > 10 % with larger catalyst particles was ascribed to bed geometry and diffusional problems [3]. Many different particle sizes and catalyst shapes were investigated including pellets, spheres and extrudates. Similar procedures and reactor internals to those used in the 1" reactor were developed to allow testing of large particles in the small micro-reactors. Results correlated extremely well with those achieved in the 1" tube and thus allowed all future testing of full size catalyst pellets to be undertaken in fixed bed micro-reactors, with confidence.

It became clear early in the development program that the target of a 4 year catalyst life was only likely to be met if the catalyst could be regenerated in-situ. Investigations showed that the catalyst was best regenerated using conventional wax removal, oxidation and then re-reduction techniques. However, the success of regenerations was highly variable with many failing for no apparent reason. An in-depth correlation study revealed that small changes in the chemical composition of the catalyst had a drastic effect on the ability of the catalyst to regenerate fully. It was found that parts per million levels of certain contaminants in the Fischer-Tropsch catalyst could be massively detrimental to regeneration effectiveness. The source of these contaminants was traced to the base component chemicals and also the water used in various stages of the preparation. Elimination of the impurities improved the effectiveness of the catalyst regeneration and it was possible to completely regenerate a catalyst after 8000 hours on stream

3. In-House Pilot Manufacturing Facility.

In 1989 a pilot manufacturing facility was built for the scale up of a variety of catalysts. The purpose of this facility was to allow in-house production of catalysts at a scale of up to 100kg via co-precipitation, impregnation, extrusion, pelletization or granulation.

Concern over the high projected cost of the co-precipitated FT catalyst led to a search for cheaper alternatives. Impregnation of cobalt *via* cobalt nitrate salts onto bulk support material gave catalysts of similar performance to co-precipitated catalysts in the laboratory, particularly when a support with high surface area was used. Therefore, manufacturing efforts focused on development of an aqueous impregnation route for large scale catalyst production.

Initially, the aqueous impregnation routes were beset with problems due to cobalt hydrolysis reactions that were amplified by the presence of the support material. The rate and extent of these reactions was highly dependent on, for instance, cobalt source, metal concentration, temperature and time. However, after much work, the chemistry was understood well enough to allow aqueous solutions to be used for impregnation. The ultimate performance of catalysts in the FT reaction (activity, selectivity and stability) could be directly related to the structure of the catalyst precursors. Detailed recipes were required to ensure that the correct precursors were made during the impregnation and subsequent calcination steps.

Extruded catalysts were made by means of a Winkworth extruder. This could produce a range of shaped (tri-lobe, quadri-lobe, star shapes etc) and cylindrical extrudates at 10 kg h^{-1} scale with 1 – 4 mm diameter. A wide variety of lubricants and binding additives were investigated to aid the extrusion process and improve the crush strength of the resulting extrudates. Finding additives that improved the physical properties of the extrudate but did not interfere with the FT chemistry proved to be a challenge but eventually suitable additives were found.

Drying and subsequent calcination of the catalyst pre-cursors were found to be critical steps in producing a catalyst with desirable characteristics for FT (activity, selectivity and life). Drying of large batches of catalyst was carried out in an APV Mitchell oven which could be charged with up to 100 kg of catalyst for fan-assisted air drying at temperatures below 200°C. The use of multiple trays allowed thin bed depths to be achieved resulting in more constant drying characteristics.

A variety of different equipment was used for calcinations. An OMI 8CA-135 belt furnace with an 8 inch wide conveyor belt and heated length of 3.5 m was temperature controlled in seven zones (plus pre-heat and cooling zones). Belt speeds ranged from 2.5 to 25 cm/min and it could reach a temperature of 1100°C with air or nitrogen atmospheres. Three sampling points for oxygen concentration allowed continuous oxygen monitoring and throughput would generally be in the range 10 –100 kg depending on residence time. An EFCO furnace comprised two chambers with an 800°C upper limit that could hold purged boxes of 100 liter capacity. The boxes used up-flow through a fine mesh onto which the catalyst (ca 25 kg) was charged before loading into the oven. Various calcinations gases could be used and the apparatus included on-line analysis and logging of inlet and outlet streams for CO, CO_2, NO_x, NH_3, CH_4, O_2, humidity content and total hydrocarbon content, plus flow pressure and temperature recordings.

4. Pilot Plant Operations.

In 1992 a purpose built FT pilot plant at BP Chemicals Saltend site near Hull in North East England was modified and started up for the first time with a prototype of the new FT catalyst. The plant employed a 6 meter salt-cooled tube of commercial diameter designed to simulate a single tube in a commercial multi-tubular reactor. Initial tests employing formed powder granular catalysts in this unit indicated similar performance to that observed at Sunbury in micro-reactor tests. This powder granule formulation was translated into a shaped extrudate catalyst offering acceptable pressure drop characteristics. The new extrudate catalyst exhibited the high activity/selectivity and cycle time > 7000 hours normally achieved using powder catalyst, and was regarded as a suitable candidate for a commercial fixed-bed process.

5. Catalyst Specifications.

In order to progress from prototype catalysts to a version capable of commercial scale production, not only does the catalyst preparation route need to be specified in considerable detail but also, the specification of the finished catalyst needs to be defined. A set of physical and chemical characteristics needed to be determined which fully defined the catalyst and then, for each of these characteristics an acceptable value, or more usually a range of values, was set. Physical parameters used to specify the finished catalyst were as follows:

- BET surface area as measured by nitrogen porosimetry
- Mean pore diameter
- Skeletal density
- Mercury intrusion volume
- Mercury pore area
- Mercury particle density
- Mercury skeletal density
- Extrudate diameter, length
- Attrition resistance and bulk crush strength

The definition of the chemical specification was more complex. For instance not only did the cobalt loading and the levels of numerous elements, some present in only in trace amounts, need to be specified but also measures of the state of the cobalt oxide were also required. Techniques used to "fingerprint" the active phases included X-ray diffraction (XRD) to determine crystallite phase sizes and temperature programmed reduction (TPR). The TPR experiments were run with hydrogen and conditions were chosen to produce a characteristic trace with sharp peaks. The temperatures corresponding to these peaks were indicative of the "reducibility" of the cobalt and proved to be strongly correlated with catalyst performance under process conditions. Electron microscope techniques were used to determine the distribution of the active phase throughout the extrudates to ensure it was uniform. The Catalyst Specification proved to be essential for moving into the next phase of development.

6. Commercial Manufacture.

Based on the success of in-house catalyst manufacturing efforts and the encouraging performance in FT evaluations, it was decided in 1991 to investigate production of large quantities of catalyst at the tonnage scale. This was beyond the capabilities of the equipment available in-house so it was decided to approach catalyst manufacturers with a view to selecting one to move forward with. Proposals were sought from six leading manufacturers and discussions were held with three companies.

One manufacturer was selected, who was willing and capable of working to BP's timescale. Key aspects of the manufacturing proved to be sourcing the correct raw materials, validating and guaranteeing their quality, as well as the quality of the water used. It took a year of painstaking analysis to confirm which trace impurities introduced in the commercial manufacturing route were detrimental to catalyst performance and to eliminate these from the production method. Two batches were made at the two ton scale and tested in the Hull pilot plant (1993), as well as the Sunbury micro-reactors in 1997. The basic recipe of the catalyst remained unaltered during this period.

7. Nikiski Demonstration Plant.

The sanction for the 300 bpd GTL demonstration plant in Nikiski, Alaska was given in July 2000. In addition to demonstration of the fixed bed FT process and catalyst described in this paper this facility will also demonstrate the Compact Reformer syngas generation technology jointly developed by BP and Davy Process Technology. A commercial hydrocracking process is the final step leading to the synthetic crude product.

A total of 45 tonnes of the FT catalyst was manufactured for the demonstration plant including extra catalyst as back up. Prior to manufacture of the full 45 tonnes a 2 ton test batch was made and validated at the Davy Process Technology R&D Centre in Stockton, England during 2000. Once validation had been signed off, production of the full 45 tonne order commenced. The full order was delivered to Nikiski during the latter half of 2001. Validation tests carried out on representative samples taken during the manufacture confirmed that the production had gone to plan. This validation was done in Stockton as well as in BP's new facilities at Sunbury during 2001 and 2002.

Loading of the catalyst into the multi-tube reactor (several thousand narrow tubes) was a challenge. Hence a specialist catalyst loading company was contracted in to carry out a loading trial. The trial was successful with minimal fines production during both transportation and the actual loading itself. This success was mirrored when the catalyst was loaded at Nikiski during October of 2001. In less than a week the loading was completed and verified through pressure drop testing.

The Nikiski plant underwent extensive commissioning and prestart-up tests during 2002 and early 2003 and the first product was made on 21st July, 2003. It was a truly historic day in BP's quest to deliver an economic Fischer-Tropsch process to the market. The Nikiski test program has continued through 2003 and 2004 with the aim of demonstrating the following features of the FT process:

- Steady operation for an extended period
- A managed catalyst activity decline with a 4 year target life
- Compatible operation with the Compact Reformer
- Minimal loss in performance after multiple start ups and shutdowns

Progress has been steady since production of the first liquid product. Extended operation has been achieved quite easily. Catalyst performance has been relatively stable through the multiple start ups and shutdowns and during extended operation. Encouraging results have also been achieved with regard to the carbon number distribution of the FT wax with alpha values ranging from

0.92 to 0.95. While continued operation is needed to achieve the test program goals, there have already been many lessons learned that will help pave the way for commercial plants.

8. Novel Reactor Technology Development.

An attractive feature of the fixed bed operating mode employed at Nikiski is its operational simplicity since there are no moving parts and no need to separate catalyst from the product. A disadvantage is the lower catalyst productivity that is obtained compared to slurry reactors due to heat and mass transfer limitations. In collaboration with Davy Process Technology, BP began development work on a novel slurry-based Fischer-Tropsch reactor technology in 2000. Key aspects of the reactor operation involve the mixing of syngas and liquid catalyst suspension in a high shear zone followed by syngas conversion to liquid hydrocarbons in a post mixing zone [4]. The intensive mixing of syngas and the catalyst suspension allows smaller catalyst particle sizes to be employed compared with conventional slurry and fixed bed technologies reducing the formation of unwanted by products such as methane. Mass transfer is also improved due to the production of micron-sized gas bubbles that enhance the contact between the gaseous reactants, liquid medium, and solid catalyst particles.

As in any slurry based process, complex hydrodynamic behavior complicates the scale up to a commercial size unit. Therefore, computational modeling supported by cold flow x-ray mapping tests on a pilot reactor prototype has been used to assist the reactor development efforts. Operation of a 2 bpd Fischer Tropsch pilot plant began in 2003 and has continued through 2004 at the BP Chemicals Saltend site.

References

(1) Font Freide, J.H.M.; Gamlin, T.D.; Graham, C; Hensman, R.J.; Nay, B.; Sharp, C. *Topics in Catalysis*, **2003,** 26(1-4), 3.
(2) EP 142 887, 9th November, 1984, Process for the preparation of hydrocarbons.
(3) Iglesia, E.; Reyes, S.C.; Soled, S.L., *Advances in Catalysis*, **1993**, 39, 199.
(4) US 6635682 B2, 21st October, 2003, Process for converting synthesis gas into higher hydrocarbons.

Fischer-Tropsch Synthesis, Catalysts and Catalysis
B.H. Davis and M.L. Occelli (Editors)
© 2007 Published by Elsevier B.V.

Loading of Cobalt on Carbon Nanofibers

Melene M. Keyser and Frans F. Prinsloo

Sasol Technology R&D, KlasieHavenga Road 1, Sasolburg, 1947

Abstract

This paper reports the loading of oxidized carbon nanofibers with cobalt employing a deposition precipitation method with urea as precipitating agent. The catalysts were characterized by using thermogravimetric analysis (TGA), chemisorption, Fourier-Transform Infrared (FTIR), X-ray photoelectron spectroscopy (XPS), zeta potential, X-ray diffraction (XRD), transmission electron microscopy (TEM), scanning electron microscopy (SEM), and temperature programmed (TPR) analyses. An oxidation treatment with nitric acid introduced ketone, carboxylic, phenolic, nitro and lactone groups on the carbon nanofiber surface. An ion exchange reaction between the carboxylic and phenolic groups and cobalt species has been illustrated.

The interaction of the precipitated cobalt oxides with the carbon nanofiber support is influenced to a large extent by heat treatment in an inert atmosphere. Heat treatment of the carbon nanofibers at 573 K resulted in an increase in the interaction between the cobalt particles and the support. Under these conditions a small amount of cobalt carbide and cobalt metal was detected by the XRD and XPS analyses. Heat treatment at 873 K resulted in a further increase in the interaction between the metal and the support leading to increasing amounts of cobalt carbide and cobalt metal.

A narrow metal particle distribution with a homogeneous coverage of the support with the metal particles was observed from the TEM images.

Key Words : catalytically grown carbon; catalyst support; thermal analysis; infrared spectroscopy; X-ray photoelectron spectroscopy; X-ray diffraction; transmission electron microscopy.

1. Introduction

It is well-known that the catalyst support facilitates the preparation of a well-dispersed, high surface area catalytic phase, stabilises the active phase against loss of surface area and significantly influences the morphology, adsorption, and activity/selectivity properties of the active phase [1,2,3,4]. Despite numerous advantages, carbon as a support for heterogeneous catalysts has received little attention as compared to oxidic supports [5].

We are interested in the properties of carbon nanotubes/fibers as support for transition metal catalysts in heterogeneous catalysis. Carbon nanotubes and nanofibers have similar advantages to activated carbon and carbon black if utilized as support in heterogeneous catalysis [6,7,8] but they outperform activated carbon with regard to reproducibility, filterability, stability under oxidising and reducing conditions and abrasiveness [6]. It was established that carbon nanofibers interweave during growth, resulting in the formation of mechanically strong tangled agglomerates [6]. The agglomerates facilitate an open pore volume, a pore size distribution, a predominant mesoporous structure, high filterability and high mechanical strength, thus rendering carbon nanofibers suitable as support for metal particles in heterogeneous catalysis [6]. In addition, the mesoporous structure of carbon nanofibers significantly reduces diffusion transport limitations, which usually occur when activated carbon is utilized as a support.

One issue that needs to be addressed is the manner in which precipitation of the metal oxide onto the carbon support is brought about. The majority of carbon supported catalysts in the literature were prepared according to the pore-volume impregnation method [1,9,10,11]. Previously we have prepared carbon nanotube supported iron catalysts according to this method and found that the metal oxide precipitated mainly in the mesoporous structure of the agglomerates [12]. A non-homogeneous coverage of the nanotube surface with the metal oxides was obtained. Building onto this, we have decided to study a homogeneous deposition precipitation method with urea as precipitation reagent [13,14].

The surface oxygen functional groups are of great interest for the preparation of carbon supported metal catalysts. Acidic groups are introduced by dry or wet oxidation of the carbon surface. They increase the hydrophilicity of the carbon surface rendering it more accessible to the aqueous solution of the metal precursor during impregnation. Accordingly various authors have reported an increase in metal dispersion with an increase in concentration of the surface oxygen functional groups [15,16,17,18]. Ebbesen et al. [17] and Ang et al. [18] utilized XPS and EDX and demonstrated that, if carboxylic groups are evenly spread on the surface of the carbon material, it is possible to obtain a very uniform coverage with the metal ion.

The temperatures at which the oxygen surface functional groups decompose are also important. For platinum/C catalysts it was observed that if the metal reduction/heat treatment temperature exceeds the decomposition temperature of the majority of the surface oxygen functional groups the metal dispersion also increased [15,16].

In this paper, oxidation of carbon nanofibers by nitric acid in combination with a deposition precipitation method was investigated as a modus operandi for the preparation of highly dispersed carbon nanofiber supported, cobalt oxides. A variety of thermal and spectroscopic techniques such as thermogravimetric analysis (TGA), chemisorption, X-ray photoelectron spectroscopy (XPS), transmission electron microscopy (TEM), scanning electron microscopy (SEM), infra-red (FTIR) and temperature programmed (TPR) were utilized to study the possible influence of the surface oxygen functional groups on the metal dispersion.

2. Experimental

2.1 Reagents

Cobalt(II)acetate tetrahydrate was obtained from Fluka and used without further purification. In all other cases reagent grade chemicals were used. Acetylene (99.999 %), argon (99.999 %), and hydrogen/argon mixtures (10 % H_2) were obtained from Fedgas and used without further purification.

2.2 Synthesis of carbon nanofibers

The synthesis of the carbon nanofibers was performed at atmospheric pressure via the decomposition of acetylene over a stainless steel substrate (SUS310).

The stainless steel reactor (volume of the reactor = 340 cm^3) was heated from room temperature to 873 K in 2 hours under a flowing stream of argon (1100 ml min^{-1} STP). Maintaining this temperature the carbon nanofibers were allowed to grow over a period of 30 minutes in a mixture of argon (1100 ml min^{-1} STP) and acetylene (733 ml min^{-1} STP). The temperature was then lowered to room temperature under a flowing stream of argon (1100 ml min^{-1} STP). The product was stirred overnight at ambient conditions in a mixture of hydrofluoric and hydrochloric acid (HF:HCl = 0.45) to dissolve metals and metal oxides. Afterwards the nanofibers were thoroughly washed with de-ionized water and dried overnight. The product yield was ~ 25 grams per trial. In order to render the surface of the carbon support more hydrophilic, the carbon nanofibers were refluxed in concentrated nitric acid (10 mol dm^{-3}) at 353 K for 4 hours.

2.3 Catalyst preparation

10 grams of carbon nanofibers were dispersed with 300 cm^3 of de-ionized water for 30 minutes in an ultrasonic bath. This was then slurried with a solution of 250 cm^3 deionized water and 12% m/m cobalt. It was transferred to a 1000 cm^3 3-neck indented flask fitted with a top stirrer and pH electrode and an additional 150 cm^3 of deionized water was added to the slurry. The temperature was brought to 363 K after which urea was added with vigorous stirring for 6 hours. The slurry was filtered, washed lightly and allowed to dry in a vacuum oven at 373 K overnight. The catalysts were heated (20 K min^{-1}) in a nitrogen stream for 1 hour at a desired temperature.

The final samples were analyzed by inductively coupled plasma emission spectrometry (Vista AZ CCD simultaneous ICP-AES) to determine their elemental composition. The metal contents of the catalysts are given in Table 1. Catalysts with high and low metal loading have been prepared. To study the influence of temperature on the metal dispersion and metal particle diameter, the loaded catalysts were calcined at 573 K and 873 K respectively.

2.4 SEM and TEM analyses

The SEM analyses was performed on a JEOL JSM 6000 F scanning electron microscope. Sample preparation for the SEM and TEM images were performed following the procedure described by Goldstein et al. [19]. Accordingly, 20 – 50 μg of sample was placed together with 1 cm^3 of 100 % ethanol in a red top vacutainer. The mixture was then dispersed for 15 – 30 seconds in a high frequency ultrasonic bath (Transsonic 460 Elma). A drop (~5 μl) of the resulting suspension was then placed onto a polished graphite stub and the ethanol was evaporated at 333 K for 10 – 15 minutes.

For the TEM analyses a drop of the suspension was placed on a carbon coated formvar support on a copper grid. Transmission electron microscopy measurements were made with a Philips EM301 electron microscope operating at 100 kV.

Table 1   Heat treatment temperature and metal loading of the carbon nanofiber supported cobalt catalysts

Catalyst	Heat treatment temperature (K)	wt % cobalt
DPCo10	-	9.0
DPCo10300	573; heating rate = 25 K min^{-1}	9.0
DPCo10600	873; heating rate = 25 K min^{-1}	9.0
DPCo4	-	3.8
DPCo4300	573; heating rate = 25 K min^{-1}	3.5
DPCo4600	873; heating rate = 25 K min^{-1}	3.7

2.5 FTIR analyses

FTIR spectra were collected on a Nicolet Magna-IR 550 Spectrometer Series II. The samples were prepared by mixing pre-dried samples of the catalyst and potassium bromide using a dilution ratio of KBr/catalyst ≈ 500. The IR parameters were set at a resolution of 2 cm^{-1} and the spectra were obtained by averaging 25 scans. The FTIR spectra were recorded between 500 cm^{-1} and 3000 cm^{-1}.

2.6 Zeta potential measurements

Zeta potential measurements were performed on a Malvern Zetaseizer 2C instrument. The sample was dispersed in 5×10^{-3} mol dm^{-3} NaClO$_4$ and agitated until the pH reading on the pH meter was stable. This reading was recorded and the pH was adjusted using HClO$_4$ or NaOH and left for 15 minutes to equilibrate at pH intervals of approximately 1. Once the pH was obtained the sample was allowed to settle for 2 minutes and the supernatant was used for the zeta potential measurements.

2.7 TPR study

TPR analysis was performed on a locally manufactured instrument. A sufficient amount of sample (~ 0.02 g) was placed on a quartz wool plug paced in a ¼ inch U-shaped quartz tube. The sample tube was connected to the analyses ports and carrier gas (argon) was allowed to flow over the sample (46 cm^3 min^{-1} STP). The detector base line was allowed to stabilize. At this stage a 10 % hydrogen/argon mixture was introduced over the sample (H$_2$ flow rate = 4.6 cm^3 min^{-1} STP) while simultaneously closing the argon gas flow. The furnace was ramped from room temperature to $\sim$1073 K at 10 K min^{-1}.

2.8 XPS study

XPS analyses was performed on a Quantum 2000 Scanning SK Microprobe instrument. For the XPS analyses the samples were mounted on a sample plate, introduced into the XPS chamber and evacuated to $<2 \times 10^{-10}$ Pa. Elemental surveys and narrow scans were conducted with Al K$\square$ X-rays on the uncalcined samples after which the temperature was increased to 573 K and 873 K (heating rate = 20 K min^{-1}). Semi-quantitative data were calculated from the survey scans. The peak area ratios reported in this paper are fractions, which were calculated from the areas under the XPS peaks. The narrow scan data were used to determine probable compounds for the Co and C peaks. This was accomplished from a Gauss-Lorenzian peak fit of the appropriate photoelectron peaks. Wide spectra were recorded to obtain a semi-quantitative analyses of all elements present on the surface (except H and He) and high-resolution narrow spectra were recorded to identify the oxidation states and/or compounds.

2.9 BET and pore size distribution measurement

Nitrogen adsorption measurements were performed at 77 K using a static volumetric apparatus (Micromeritics ASAP 2010 adsorption analyzer). The samples were degassed at 473 K for 24 hours and 10^{-5} Pa before analyses.

2.10 Chemisorption measurements

Chemisorption measurements were performed on an ASAP 2010C, Micromeritics instrument. Before chemisorption measurements, the sample was heated under flowing H$_2$ (flow rate = 46.6 cm^3 min^{-1} STP) at a heating rate of 1

K min^{-1} until 343 K before the final reduction temperature and then at a heating rate of 0.5 K min^{-1} to the final reduction temperature (623 K). At this temperature it was reduced in flowing hydrogen (flow rate = 46.6 cm^3 min^{-1} STP) for 16 hours. After reduction the catalyst was degassed for two hours at 10^{-7} Pa at the reduction temperature in order to eliminate the chemisorbed hydrogen. The temperature was then lowered to 293 K below the reduction temperature while maintaining the vacuum and degassed for a further 10 minutes. Thereafter the temperature was lowered to the desired analyses temperature (493 K) where the chemisorption measurements were performed. To evaluate the chemisorbed monolayer uptake v_M, the isotherm was fitted to a Langmuir isotherm.

Estimated metal particle sizes based on H_2 uptake were calculated assuming spherical geometry and an adsorption stoichiometry of $H/Co_s = 2$. The procedure for the calculation of the metal diameter is described elsewhere [20].

2.11 TGA analyses

TGA analyses was conducted on a Perkin Elmer TGA. After placing 10 – 20 mg sample in a platinum crucible it was heated at 20 K min^{-1} up to 1173 K under a constant flow of nitrogen (flowrate = 183 ml min^{-1} STP). The weight loss as a function of temperature was monitored.

2.12 Catalytic testing

Fischer-Tropsch synthesis was performed in a Berty micro reactor. The carbon monoxide, hydrogen and argon and hydrogen flows were controlled with Brooks flow-controllers. The reactor, gas outlet and hot trap lines were heated by means of electrical heaters; each of which was controlled by separate temperature controllers. Inserting a thermocouple into the catalyst bed controlled the catalyst bed temperature.

The bottom of the reactor was equipped with a stainless steel sleeve. A small amount of fibrefrax was placed on top of the stainless steel sieve. The reactor was filled with 5 grams of catalyst, sealed of and checked for pressure leaks. Fischer-Tropsch synthesis was performed at a space velocity (H_2+CO) of 185 cm^3 min^{-1} g^{-1} (STP). Before the Fischer-Tropsch synthesis the catalyst was reduced in situ under a constant flow of hydrogen (space velocity = 185 cm^3 min^{-1} g^{-1} STP) at 563 K. During the Fischer-Tropsch synthesis 10% argon (103 cm^3 min^{-1} STP) was passed through the reactor as internal standard.

The ampoule technique was used to withdraw samples of the feed and tail gases at 24-hour intervals.

2.13 Analysis of the Fischer-Tropsch products

A Dow Mac chromatograph equipped with two thermal conductivity detectors was employed for analysis of the CO, CO_2, H_2, CH_4, and Ar concentrations in the feed and tail gas samples. For the analysis of H_2 on the first thermal conductivity detector helium was used as carrier gas. On the second thermal conductivity detector the CO, CO_2, CH_4, and Ar concentrations were determined with helium as carrier. The GC was connected to a printer/integrator. The products were separated with molecular sieve (5 Å) and Porapak Q columns.

For the analysis of the organic products a Perkin Elmer gas chromatograph equipped with a flame ionisation detector (FIDGC) was employed. Helium was used as carrier gas. Beforehand the ampoule was preheated to ~200°C before it was broken, in order to ensure that the hydrocarbons are evaporated before analyses. To accomplish this, as well as enabling the automatic injection of the ampoule content into the capillary column, use was made of a pneumatic ampoule breaker. Once the ampoule was placed in the pneumatic ampoule breaker it was flushed with helium, whilst the ampoule heats up to 200°C. A 6-port valve was then switched allowing for the flow of the hydrogen carrier gas into the ampoule breaker. As soon as the pre-column pressure was stabilised and the GC reached approximately –50°C the ampoule was broken pneumatically. The separation of the hydrocarbons was achieved with a Petrocol DH 150 (150 m long x 0.25 mm ID coated with a 1 μm methyl silicon stationary phase film) capillary column. The temperature program started off at –50°C, held for 10 minutes, after which it was increased at 10°C per minute to –20°C, and then immediately raised at 2°C per minute to 280°C.

The TCD detectors were calibrated with gas mixtures, which contained known concentrations of Ar, N_2, CO, CH_4 and CO_2.

3. Results and discussion

3.1 Characterization of the carbon nanofibers

3.1.1 TEM and SEM images.

A SEM image (Fig. 1a) of the carbon nanofibers show that the decomposition of acetylene over the stainless steel substrate results in a non-homogeneous mixture of relatively straight and coiled carbon nanofibers which have interwoven during growth and formed agglomerates (lower left hand side of Fig. 1a). The nanosized graphite platelets are stacked in various orientations with respect to the fiber axis. The bulk of the sample consists of the coiled carbon nanofibers. It was not possible to establish beyond doubt whether the coiled carbon nanofibers are hollow inside strongly suggesting that the carbon layers to be stacked perpendicular to the carbon nanofiber axis [21]. The mixture also contains a very small percentage of carbon nanotubes with a fishbone structure (Fig. 1b) (graphitic planes are oriented at an angle varying between 40° and 10° to the fiber axis) and some short, thick fibers (see for example middle left hand side of Fig. 1a). The average outer diameter of the carbon nanofibers determined from the SEM images turned out to be 100 ± 30 nm and the average length > 50 μm.

3.1.2 Nitrogen adsorption.

The nitrogen adsorption isotherm of the carbon nanofibers (Fig. 2) is of Type II in the BDDT classification, which is normally obtained for carbons, which are predominantly mesoporous. The BET surface area (calculated with the data between $0.05 < p/p_0 < 0.2$) equals 91 m^2 g^{-1} and the BJH mesopore surface area equals 79 m^2 g^{-1}. The general view is that a low surface area of the support would limit the metal loading if the purpose is to finely disperse the metal catalyst [22]. As was discussed in the introduction the highly mesoporous structure of the carbon nanofibers significantly reduces diffusion limitations.

a.

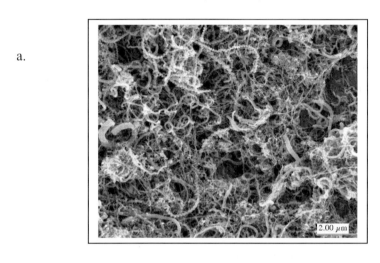

b.

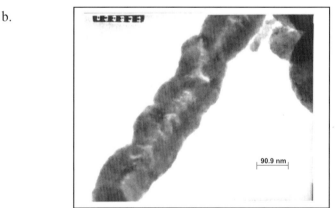

Fig. 1.   SEM and TEM images of carbon nanofibers out of acetylene at 873 K over a stainless steel substrate. a. SEM image; b. TEM image, x 220 000.

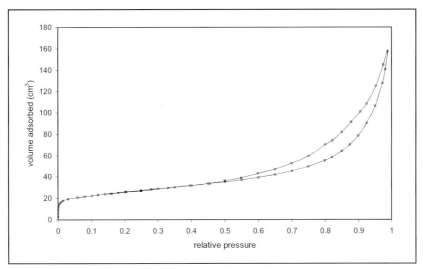

Fig. 2. Adsorption-isotherm of N_2 by carbon nanofibers synthesized via the catalytic decomposition of acetylene over a stainless steel substrate at 873 K.

3.1.3 TGA analyses .

The TGA analyses provides an indication of the temperatures where the surface oxygen functional groups decompose. The TGA and corresponding differential curve of the oxidized carbon nanofibers are given in Fig. 3. The peak at 360 K is due to the desorption of physisorbed water [23]. With the TGA instrumentation it is not possible to establish beyond doubt whether the shoulder in the region of 520 K is due to the desorption of more strongly bound water and/or CO_2 forming group(s). An extensive literature TPD study [23] revealed that both strongly bound H_2O and CO_2 forming groups desorb in this temperature range and it was interpreted in terms of the dehydration of phenolic hydroxyl groups (H_2O desorption) and carboxylic groups (CO_2 desorption). There are also peaks and shoulders at 620 K, 770 K, 925 K and 1020 K and 1120 K. TPD studies [23,24,25] on activated carbons oxidized with HNO_3 showed that, in general, CO_2 desorbs between 373 K < T < 973 K and as much as four peaks have been observed in this region [23] whereas CO desorbs between 500 K < T < 1273 K and in this region six peaks have been observed [23]. Thus, in the present case, it is safe to conclude that the peaks at 620 K and 770 K are due to CO and CO_2 forming groups, while the peak and shoulder at 1020 K and 1120 K are only due to CO forming groups. Although the shoulder at 925 K could either be due to CO_2 or CO forming groups a shoulder at this temperature has previously been observed in the CO TPD profile of an HNO_3 oxidized activated carbon [23]. Thus it appears necessary to calcine at least at 770 K in order to decompose the majority of the CO_2 forming groups.

3.1.4   FTIR and XPS analyses

In order to study the chemical nature of the surface oxygen functional surface groups FTIR and XPS were employed. An FTIR spectrum obtained for the oxidized carbon nanofibers is depicted in Fig. 4. The absorption at 1384 cm^{-1} is due to nitrate adsorbed on the carbon nanofibers and can be discarded. The peak at ~ 1720 cm^{-1} can be assigned to the C=O stretching vibration of carboxyl or carbonyl groups, while the shoulder at ~ 1740 cm^{-1} is associated with carboxylic anhydride or lactone groups [26,27,28]. The carbonyl group is highly polar and stretching of this bond would result in a relatively large change in dipole moment. There are also phenolic groups (1233 cm^{-1}) [29] present on the surface of the carbon nanofibers. The band at 1600 cm^{-1} has been observed by many authors [30] and has been assigned to aromatic stretching coupled to highly conjugated carbonyl groups (C=O) [26]. The bands below 950 cm^{-1} are characteristic of out-of-plane deformation vibrations of C-H groups in aromatic structures [26]. The presence of the asymmetric (1538 cm^{-1}) and symmetric (1338 cm^{-1}) NO$_2$ stretch vibrations indicates that the HNO$_3$ treatment introduced NO$_2$ groups on the surface of the carbon nanofibers.

If the oxidized carbon nanofibers are calcined at 573 K, the shoulder at ~ 1740 cm^{-1} and the band at ~ 1720 cm^{-1} becomes less pronounced (Fig. 4). The phenolic and asymmetric and symmetric NO$_2$ stretching vibration peaks have disappeared completely. Thus ketone, carboxylic, phenolic and nitro groups decompose if the carbon nanofibers are calcined at 573 K. The peak that remains at ~ 1720 cm^{-1} is most likely due to lactones because these groups decompose between 600 and 950 K [31]. If the temperature is increased further to 873 K, the shoulder and peak at ~ 1740 cm^{-1} and ~ 1720 cm^{-1} have disappeared completely. It follows that the majority of the surface oxygen functional groups decompose at 573 K.

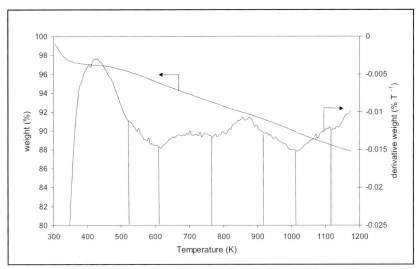

Fig. 3.   TGA curve and differential plot of the oxidized (uncalcined) carbon nanofibers.

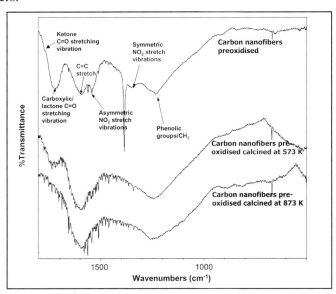

Fig. 4.   FTIR spectra of the oxidized and oxidized/calcined carbon nanofibers.

XPS is primarily a surface technique, as the escape depth of the photoelectrons ranges from 2 to 5 nm, and it would therefore primarily yield information on the composition of the most external surface of the carbon nanofibers. For this reason should it be possible to obtain the surface oxygen content from XPS [28]. In this regard, it was suggested that 80-90 % peak area of the O_{1s} peak originates from the contribution of surface atoms, or atoms of the 3.5 - 4 nm thick outer shell and the C_{1s} peak from C atoms located at the surface or in the 4 - 5 nm outer shell [32].

From the wide spectra, the ratio of the atomic percentage of the O_{1s} and C_{1s} peaks can be calculated (Table 2). The O_{1s}/C_{1s} ratio decreases with increasing heat treatment temperature and reinforces the FTIR results.

Table 2   XPS study of the heat treatment of the oxidized carbon nanofibers

Catalyst	O_{1s}/C_{1s}
Oxidized carbon nanofibers – no heat treatment	0.48
Carbon nanofibers heat treated at 300°C	0.25
Carbon nanofibers heat treated at 600°C	0.10

The XPS data in the C_{1s} region for the oxidized and calcined carbon nanofibers as well as for a high purity graphite sample is given in Fig. 5. As often done in the literature it is possible to deconvolute the spectra into a set of Gauss-Lorenzian peaks [26]. We found that the Gauss-Lorenzian C_{1s} peak fit for the oxidized carbon nanofibers (C=O (carboxyl and carbonyl), C-OH (alcohol and phenol), aromatic, aliphatic groups, the shake-up satellite (290.6 eV) and plasmon (291.6 eV) were considered) did not follow the expected decrease in the peak area ratio of the compounds with a C-OH functionality with increasing heat treatment temperature as observed by the FTIR analysis. Although the deconvoluted peak area ratios of the compounds with carboxylic and carbonyl functionalities decreased with increasing temperature treatment of the data in this manner is dubious because the asymmetric line profile of the graphitic C_{1s} signal is disregarded. If the asymmetric line profile is however removed from the main peaks (Fig. 5) small peaks and shoulders at 285.5, 286.0 (alcohol, ether, phenol), 288.3 (carbonyl, quinone), 288.8 (carboxyl groups) and 289.8 eV (carbonate, CO_2) can be isolated. In correspondence with the FTIR data these peaks become smaller with increasing temperature.

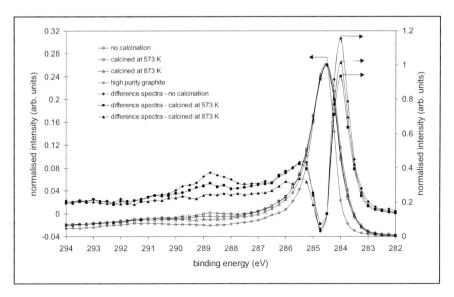

Fig. 5. XPS data in the C_{1s} region for the calcined and uncalcined carbon nanofibers and high purity graphite. The spectra after removal of the asymmetric main contribution are shown on the secondary y-axes.

3.1.5 Zeta potential measurements

The dependence of the zeta potentials on pH is shown in Fig. 6. At very low pH, the ionic strength of the slurry becomes very high resulting in a rapid rise in temperature of the carbon nanofiber suspension during the zeta potential measurements. This, in turn, results in erratic motion of the particles, which hinders accurate zeta potential measurements [33]. For this reason it is only possible to estimate the IEP's, which occurred at low pH. All the zeta potentials were negative and the estimated isoelectronic point for the unoxidized carbon nanofibers occurred at pH $\sim$ 1.8 while that for the pre-oxidized carbon nanofibers is estimated at pH $\sim$ −3; indicating the introduction of a significant amount of acidic groups during the pre-oxidation step.

58

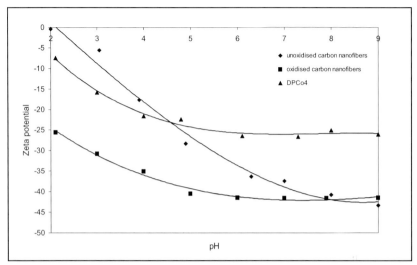

Fig. 6. Variation of Zeta potential of unoxidized and oxidized carbon nanofibers and oxidized carbon nanofibers loaded with cobalt (no heat treatment).

3.1.6 Interaction of precipitated cobalt oxide(s) with the surface of the carbon nanofibers

As already discussed, anchoring of cations (e.g. $Co(H_2O)_6^{2+}$) is facilitated via a negatively charged functional group like for example a surface carboxyl or carbonyl group. Very often ion exchange with the carboxylic groups according to equation 1 is postulated [5,17,18].

$$C\text{-}COOH + M^+X^- \rightarrow C\text{-}COO^-M^+ + HX \qquad\qquad 1.$$

In this respect, it is important to note that, depending on the pH, six soluble complexes, i.e., $Co(H_2O)_6^{2+}$, $Co(H_2O)_5OH^+$, $Co(H_2O)_4(OH)_2$, $Co(H_2O)_3(OH)_3$, $Co(H_2O)_2(OH)_4$ and polynuclear Co_2OH^{3+}, $Co_4(OH)_4^{4+}$ could exist in aqueous solution [34]. An electrostatic interaction between the negatively charged surface oxygen functional groups and these positively charged monomeric species cannot be ruled out.

Besides the existence of these different soluble complexes in aqueous solution, the solid phase $Co(OH)_2(cr)$ also exists [34]. Literature solubility products, pK_{sp}, for $Co(OH)_2(cr)$ have been reviewed elsewhere [34]. This data showed that $Co(OH)_2(cr)$ is scarcely soluble in aqueous solution and that mainly $Co(OH)_2(cr)$ precipitates if the pH is raised to values above pH = 9. Thermodynamic calculations showed that there is virtually no effect on the solubility of $Co(OH)_2(cr)$ if the temperature is increased to 363 K.

After the homogeneous deposition precipitation of the cobalt species the pH was pH approximately 6.7. In this regard it is important to note that the first pK_a

for the base hydrolysis of $Co(H_2O)_6^{2+}$ in aqueous solutions is 9.7 ($I = 1$ mol dm^{-3}; T = 298 K) [34] and $Co(OH)_2(cr)$ could not have been precipitated at pH ~ 6.7. It is more likely that the precipitate is the basic cobalt acctate salt, *viz.*, $Co(OAc)_2(1-x)(OH)_{2x}$ [35,36,37,38,39]. It has been established that this type of compound decomposes to CoO if heat treated in an argon atmosphere at 553 K – 653 K [38,39].

The onset pH for the formation of the $Co(H_2O)_4(OH)_2$ is ~8.5 [34] and, consequently, is it possible that, under the present experimental conditions, significant amounts of $Co(H_2O)_6^{2+}$ and $Co(H_2O)_5(OH)^+$ will be present in solution during the preparation of the catalysts. Furthermore, if the carbon nanofibers were not oxidized before the catalyst preparation, very low amounts of cobalt was loaded on the carbon nanofibers support. Thus in correspondence with the zcta potential results electrostatic interaction between the negatively charged carbon nanofiber surface and the positively charged cobalt species will be needed for the preparation of the catalysts.

An FTIR spectrum of the carbon nanofibers after catalyst preparation (dried at 393 K) (Fig. 7) yields more specific information on the sites that interact with the ion exchanged and or precipitated cobalt species. After catalyst preparation, modification of the infrared absorption carboxylic and phenolic bands is evident. The aromatic stretching band at 1600 cm^{-1} also shifts to lower wave numbers. These observations illustrate an interaction of the metal with the surface carboxylic and phenolic groups. Interaction between the precipitated cobalt with the aromatic ring substituents or conjugated double bonds on the carbon surface is highly likely and have also been proposed by others [5,40]. The increase in the IEP from pH ~ -3 to pH ~ 1 after the cobalt species have interacted with the carbon nanofibers (Fig. 6) further reinforces an electrostatic attraction between negatively charged surface oxygen functional groups and the cobalt species during the catalyst preparation procedure.

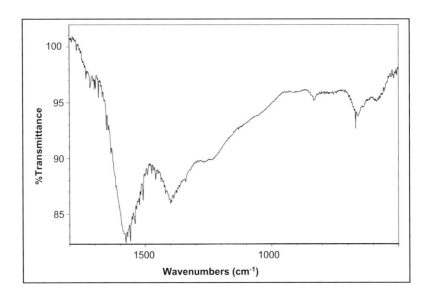

Fig. 7.   Infrared spectra of the DPCo4 catalyst (dried at 373 K).

3.1.7   Degree of dispersion and surface coverage

It has been shown that hydrogen chemisorption on cobalt is activated and that the extent of activation depends upon the support and metal loading [1,3,41]. This is due to an increase in the activation barrier for dissociative adsorption if the temperature is lowered and/or the degree of metal-support interaction is increased. This is in turn controlled by the preparation method and conditions. It is expected that, contrary to supports such as alumina and silica, the presence of zero-valent cobalt in the catalyst will be favorable due to the inertness of the carbon support [5]. It has been observed that the adsorption temperature for maximum hydrogen uptake is 423 K if cobalt (10 wt %) is supported on activated carbon [1]. For these reasons, the H_2 chemisorption experiments in the present case were performed above this temperature, i.e., 493 K.

The metal particle diameters calculated from the H_2 chemisorption data are given in Table 3 together with percentage dispersion, metal surface area and total amount of chemisorbed gas.

Immediately evident from the data in Table 3 is the small metal particle diameters and high metal dispersion.   Chemisorption data for a literature catalyst prepared with activated carbon as a support is also given in Table 3 [3]. It follows that the metal dispersion is lower and the metal diameter is larger for the carbon nanofiber supported catalysts; most probably due to the micropores of the activated carbon.

The metal dispersion does not change much as a function of heat treatment temperature.   As was already shown by the FTIR results, heat treatment at 573 K results in the decomposition of the majority of the surface oxygen functional groups and after heat treatment at 873 K no surface oxygen functional groups could be detected.   According to Rodriguez-Reinoso [5] the decomposition of the surface oxygen functional groups increases the strength of the π sites because the electron withdrawing effect exerted by these groups has been neutralized. In line with this reasoning a worse dispersion is expected when the support's surface is free of surface oxygen functional groups due to repulsive forces between the electropositive cobalt species and the more basic carbon surface. Thus processes other than merely the decomposition of the surface oxygen functional groups and the resulting influence on the strength of the π sites must be considered in order to explain the observed trends. As will be discussed later on heat treatment at 573 K and 873 K results into the formation of cobalt carbides and cobalt metal indicating that the carbon nanofibers have reducing activity and, at sufficiently high temperatures, the cobalt oxides transfer their oxygen atoms to the support, which is in turn oxidized [11].

Table 3. Hydrogen chemisorption on carbon nanofibers supported cobalt catalysts[a]

Sample	d (nm)	V_m (cm^{3}g^{-1})	Metal SA (m^{2}g^{-1})	H/Co	d_{TEM} (nm)
DPCo10	9	1.87	7.09	0.11	9.1±1.9
DPCo10300	8	1.90	7.18	0.11	9.0±1.2
DPCo10600	8	1.91	7.26	0.11	9.8±2.0
DPCo4	8	0.88	3.2	0.12	7.3±1.9
DPCo4300	7	0.92	3.4	0.14	7.2±1.7
DPCo4600	6	1.16	4.3	0.17	8.6±1.3
Co/activated carbon (3 wt % Co)	2	2.91	1.9		

[a]Experimental conditions: $T_{reduction}$ = 623 K; $T_{analysis}$ = 493 K; linear gas flow rate was kept constant at 46.6 cm^3 min^{-1} (STP).
d = metal diameter; V_m = monolayer adsorbed gas; SA = surface area.

It is important to note that there is no large influence of the metal loading on the metal dispersion. The limits of conventional impregnation, incipient wetness impregnation and ion exchange methods for the preparation of supported metal catalysts often relate to poor reproducibilities, a broad particle-size distribution and low to medium loading of the active phase. In contrast, the homogeneous deposition precipitation method with urea as pecipitation reagent is particularly appropriate to avoid the particle size effect. This is due to the slow and homogeneous introduction of the precipitating agent (hydroxyl anion) via the rate-determining urea hydrolysis at 363 K. By maintaining the concentration of the cobalt and hydroxyl species between that of the solubility and super solubility curve is it possible to perform precipitation of small metal particles exclusively on the surface of the carbon nanofiber support.

In this regard it is noteworthy that according to the FTIR and XPS results the majority of the surface oxygen functional groups have decomposed in the case of the DPCo4 and DPCo10 catalysts at the chosen reduction temperature (623 K). Any difference in dispersion between the catalysts with no prior heat treatment and those heat treated at 573 K is a kinetic effect i.e., the latter catalysts have been treated for longer times at temperatures (heat treatment time at 573 K plus heat treatment time at 623 K) where the less temperature stable surface oxygen functional groups decompose with an accompanying increase in the degree of graphitization. Thus, it is expected that an influence of the surface oxygen functional groups, if any (see discussion further on), on the metal dispersion will be larger for the catalysts heat treated at 573 K than for the catalysts with no prior heat treatment.

TEM and SEM images of the carbon nanofiber supported cobalt catalysts are given in Fig's 8 and 9 respectively. The cobalt particles generally appear as dark dots on the surface of the carbon nanofiber support in the TEM images. In the backscatter SEM images they appear as bright spots due to the more effective backscattering (than by carbon atoms) of the electrons by the cobalt atoms. In general the SEM and TEM images revealed a homogeneous coverage of the carbon nanofiber's surface with the cobalt particles. Noteworthy, however is the preferential interaction of the cobalt particles with the fibers with the relatively straight and helicoidal structures (Fig. 8a). The individual metal particle diameters were measured with the aid of Adobe PhotoShop 6.0 software. The resulting average particle diameters are also given in Table 3. This table learns a small influence of the heat treatment temperatures on the metal dispersion.

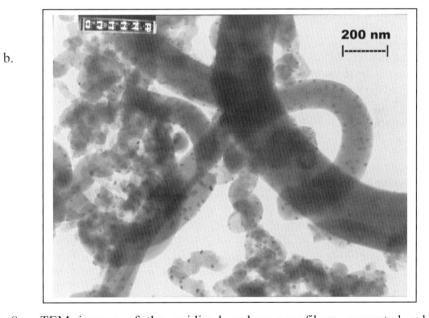

Fig. 8.   TEM images of the oxidized carbon nanofibers supported cobalt catalysts. a. calcined at 573 K (x130 000). b. Calcined at 873 K (x43 000).

64

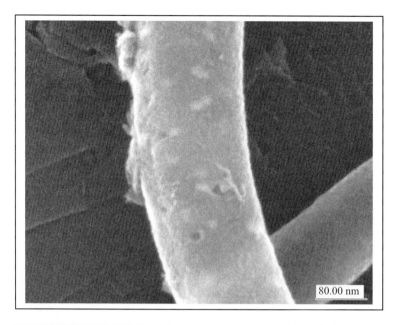

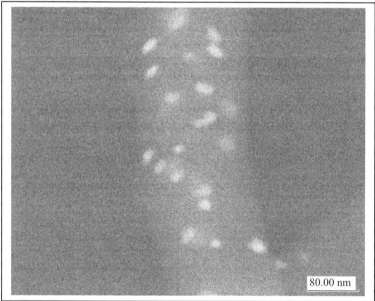

Fig. 9.   SEM image with corresponding backscatter image of the DPCo4300 catalyst.

3.1.8   Surface complexes

Difference curves from the TGA data were determined by subtracting the differential curve for the oxidized carbon nanofibers from that for the carbon nanofibers supported cobalt catalysts. The TGA difference curve for the carbon nanofiber supported catalysts (Fig. 10) suggests the formation of CoO at ~ 570 K [38]. The latter is stable in air at ambient temperatures and above 1173 K. As already mentioned transfer of oxygen atoms by the cobalt oxide species to the support and the consequent oxidation of the support most probably explains the peak at ~ 880 K .

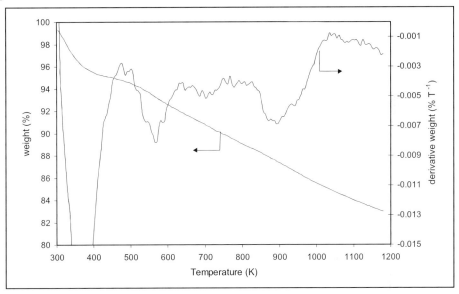

Fig. 10. TGA curve and differential plot of the DPCo4 catalyst (N_2 atmosphere).

Representative TPR profiles for the catalysts are given in Fig. 11. The large peak that can be observed above 780 K is due to the catalyzed hydrogasification of the carbon nanofibers at these temperatures [11,11,12]. For the DPCo4 and DPCo4300 catalysts, the second peak at ~ 650 K is in correspondence with the one-stage reduction of CoO to Co. The shoulder at ~ 740 K could be due to the decomposition of the more temperature stable surface oxygen functional groups and/or cobalt species that are more difficult to reduce. A TPR profile of the oxidized carbon nanofibers also showed a peak in this region strongly suggesting the former to be the case.

For the DPCo4600 and DPCO10600 catalysts, it is expected that the heat treatment at 873 K will convert the CoO to Co_3O_4. One is therefore tempted to interpret the two peaks at ~575 K and ~493 K due to the reduction of Co_3O_4 (cannot be due to the decomposition of surface oxygen functional groups due to the heat treatment at 873 K). However, no Co_3O_4 was detected by XPS or XRD (see discussion below) contradicting this possibility. Because the separation between these two peaks are relatively high it can rather be suggested that they are the consequence of the reduction of cobalt oxides, which have different interaction strengths with the support. The particles of higher size will interact to a lesser extent with the support and would be reduced first. The bulk of the catalyst was reduced at ~ 800 K (peak overlaps with the methane peak). This peak is at a much higher temperature than the peak maximum at ~ 650 K (observed for the other two catalysts) suggesting that heat treatment at 873 K results into a strong interaction between the carbon nanofiber support and the cobalt with the consequent formation of cobalt species (most probably Co_2C [42]) that are more difficult to reduce than CoO.

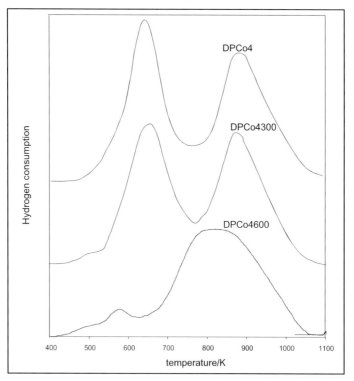

Fig. 11. TPR profiles of the carbon nanofibers supported cobalt oxide catalysts.

XRD analyses (Fig. 12) of the DPCO4300 and DPCO4600 catalysts confirms the formation of cobalt carbides and cobalt metal. The peaks at $\sim30.4^{\circ}$ and $\sim50.6^{\circ}$ are due to crystalline graphitic carbon and the peaks at $\sim59.3^{\circ}$ and $\sim89^{\circ}$ can be assigned to cobalt carbide (one of the cobalt carbide peaks overlap with the graphitic carbon peak at $\sim50.6^{\circ}$). In the case of the DPCo4600 catalyst, cobalt was also detected (peaks at $\sim52^{\circ}$, $\sim60.08^{\circ}$ and $\sim91^{\circ}$) which could be due to cobalt particles wrapped in the graphitic carbon support [43]. Although not detected by XRD, the presence of amorphous or microcrystalline CoO phase in both catalysts cannot be totally excluded.

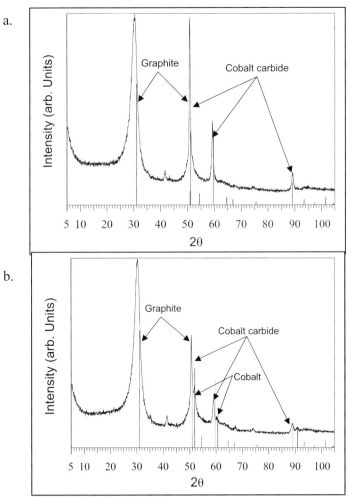

Fig. 12. XRD patterns of the carbon nanofibers supported cobalt oxide catalysts. a. DPCo4300. b. DPCo4600.

Table 4. Comparison of Co_{2p} binding energies for the carbon nanofiber supported cobalt

Catalyst	Binding Energy (eV)	Probable compound	Peak area ratio
DPCo4	782.0	$Co(OAc)_2(1-x)(OH)_{2x}$	1.0
DPCo4300	781.4	CoO	1.0
DPCo4600	No Co_{2p} peaks		
DPCo10	782.0	$Co(OAc)_2(1-x)(OH)_{2x}$	1.0
DPCo10300	781.4	CoO	0.83
	778.9	Co metal	0.17
DPCo10600	781.3	CoO	0.64
	778.9	Co metal	0.36

XPS was employed to further investigate the nature of the precipitated cobalt species. The binding energies and peak area ratios for cobalt metal and different cobalt oxides are summarized in Table 4. The data confirms the existence of a divalent cobalt oxide species on the carbon nanofiber surface for all the catalysts (the Co_{2p} peak was absent for the DPCo4600 catalyst). Unfortunately the XPS technique does not distinguish between CoO and $Co(OH)_2$ because, in general, CoO has a binding energy between 780.3 and 780.4 eV while $Co(OH)_2$ has a binding energy between 779.8 and 780.3 eV but, based on the TGA results it is expected that the compound at 781.4 eV is CoO. The shoulder at ~ 786.2 eV is a shake-up peak. No Co_3O_4 could be detected for the catalysts but a large amount of cobalt metal was detected for the DPCo10600 catalyst.

From the wide spectra, the total carbon, oxygen and cobalt content of the catalysts can be calculated (Table 5). Although the total metal content did not change significantly if the catalysts are calcined at 573 K and 873 K respectively the surface cobalt concentration decreased upon heat treatment at 873 K to such an extent that it was hardly possible to detect any cobalt species on the surface of the support from the XPS spectra. This indicates that the outer atomic layers changed upon heat treatment at 873 K. This could partly be due to the decomposition of the majority of the surface oxygen functional groups at this temperature but as already mentioned other effects should also be considered .

Until now the influence of the surface oxygen functional groups on the metal dispersion have not been addressed. It is well-known that thermal desorption of oxygen containing groups increases the number of active sites on a carbon surface [44,45]. Thus, it is possible that the dominant role of the surface functional groups is to provide active sites and thus enhancing the reactivity of the gasification reaction between the carbon material and the oxygen atom [46,47,48,49] from the CoO.

3.1.9 Catalytic testing

The catalytic results are given in Table 6 (the conversion for the catalyst heat treated at 873 K was about six times lower than that for DPCo10 and DPCo10300 and is therfore omitted from the table). The main product formed is methane. The data in Table 6 shows that, at similar conversion, the olefin selectivity increases and the activity remains constant with increasing heat treatment temperature. This demonstrates the important influence of the surface-oxygen functionalities on the product selectivity in the Fischer-Tropsch synthesis.

By way of comparison data for the activated carbon supported catalytic system is also included in Table 6. Table 6 learns a much better performance for the activated carbon supported catalyst if the overall purpose is to obtain high olefin selectivities. Considering the highly microporous structure of the support this result is unexpected.

At this stage it is difficult to explain the influence of the carbon surface chemistry on the product selectivity. The effect of the polarity of the carbon nanofiber support (at increasing treatment temperatures, the carbon nanofiber surface changes gradually from polar to non-polar due to the removal of the oxygen-containing surface groups) and the effect of the support on the Fermi-level of the metal particles (shifts to a higher binding energy with decreasing electron richness of the support oxygen atoms) are some factors to be considered in order to explain the observed trend.

Table 5. XPS study of the heat treatment of the carbon nanofibers supported cobalt catalysts

Catalyst	Element	Peak area ratio
DPCo4	C	0.86
	O	0.12
	Co	0.024
DPCo4300	C	0.845
	O	0.127
	Co	0.028
DPCo4600	C	0.925
	O	0.089
	Co	-
DPCo10	C	0.61
	O	0.29
	Co	0.10
DPCo10300	C	0.72
	O	0.18
	Co	0.066
	N	0.029
DPCo10600	C	0.87
	O	0.084
	Co	0.019
	N	0.023

Table 6. Catalytic data for carbon nanofiber/activated carbon supported cobalt catalysts (72 hours TOS).

Catalyst	10^5Activity/μmol g^{-1} catalyst	(CO + CO$_2$) conversion/%	S_{CH4}/%	$C_{2=}/C_2$	$C_{3=}/C_3$	α_{olefin}	$\alpha_{hydrocarbon}$	CO$_2$ yield/%
ActC	2.48	58.7	79	0.83	0.34	0.33	0.26	1.09
DPCo10	3.25	44.3	89	0.27	0.66	0.23	0.22	11.5
DPCo300	3.47	46.9	95	0.58	1.28	0.29	0.23	18.0

4. Conclusions

Carbon nanofiber supported cobalt catalysts were prepared according to a deposition precipitation method utilizing urea as precipitating reagent. XPS,

FTIR and TGA analyses revealed that oxidation of the carbon nanofibers with nitric acid introduced ketone, carboxylic, phenolic, nitro and lactone groups on the carbon nanofiber surface. During the preparation of the catalysts, an ion exchange reaction between the carboxylic and phenolic groups is favored. The small supported cobalt particles have high dispersions and small metal particle diameters.

The interaction of the precipitated cobalt oxides with the carbon nanofiber support is influenced to a large extent by the heat treatment temperature. Heat treatment of the carbon nanofibers at 573 K resulted in an increase in the interaction between the cobalt particles and the support. Under these conditions a small amount of cobalt carbide and cobalt metal was detected by the XRD and XPS analyses. Heat treatment at 873 K resulted in a further increase in the interaction between the metal and the support leading to increasing amounts of cobalt carbide and cobalt metal. The increase in the interaction between the carbon support and the cobalt species can be explained in terms of the reducing activity of the carbon nanofibers support and, at sufficiently high temperatures, the cobalt oxides transfer their oxygen atoms to the support, which is in turn oxidized. The decomposition of the surface oxygen functional groups results in an increase in the number of active sites on a carbon surface and it is claimed that there dominant role is to provide active sites for the gasification reaction between the carbon material and the oxygen atom from CoO.

The TEM analyses revealed a narrow metal particle distribution with a homogeneous coverage of the support with the metal particles.

The carbon surface chemistry has an influence on the product selectivity in the Fischer-Tropsch synthesis.

Acknowledgments

The authors acknowledge Sasol Technology R&D for financial support, the CSIR for performing the XPS analysis, the Potchefstroom University for Christian Higher Education for performing the FTIR measurements and the University of Pretoria for performing the TEM and SEM measurements.

References
[1]   J.M. Zowtiak, C.H. Bartholomew, J. Catal. 83 (1983) 107.
[2]   E. Iglesia, S. Soled, R.A. Fiato, J. Catal. 137 (1992) 212.
[3]   Reuel R, Bartholomew C, J. Catal.85 (1984) 78.
[4]   S . Bessell, Appl. Catal. A. 96 (1993) 253.
[5]   F. Rodríquez-Reinoso, Carbon. 36 (1998) 159.

[6]   J.W. Geus, A.J. Van Dillen, M.S. Hoogenraad, Mater. Res. Soc. Symp. Proc. 368 (1995) 87.

[7]   M.S. Hoogenraad, R.A.G. Van Leeuwarden, G.J.B. Van Breda Vriesman, A.J. Van Dillen, J.W. Geus, Stud. Surf. Sci. Catal. 91 (1995) 263.

[8]   K.P. De Jong, J.W. Geus, Catal. Rev.-Sci. Eng.42 (2000) 481.

[9]   A. Guerrero-Ruiz, A. Sepulveda-Escribano, I. Rodriques-Ramos, Applied Catal. 120 (1984) 1681.

[10]   A.A. Chen, M. Kaminsky, G.L. Geoffrey, M.A. Vannice, J. Phys. Chem. 90 (1986) 4810.

[11]   L.M. Gandia, M. Montes, J. Catal. 145 (1994) 276.

[12]   F.F. Prinsloo, E. van Steen. Catal. Today 71 (2002) 327.

[13]   K.P. De Jong, Current Opinion in Solid State & Materials Science. 4 (1999) 55.

[14]   E.B.M. Doesburg, K.P. De Jong, J.H.C. Van Hooff, Stud. Surf. Sci. Catal.123 (1999) 433.

[15]   C. Prado-Burguete, A. Linares-Solano, F. Rodríquez-Reinoso, C. Salinas-Martínez de Lecea, J. Catal.115 (1989) 98.

[16]   C. Prado-Burguete, A. Linares-Solano, F. Rodríquez-Reinoso, C. Salinas-Martínez de Lecea, J. Catal.128 (1991) 397.

[17]   T.W. Ebbesen, H. Hiura, M.E. Bisher, M.J. Treacy, J.L. Shreeve-Keyer, R.C. Haushalter, Adv. Mat. 8 (1996) 155.

[18]   J.M. Ang, T.S.A. Hor, G.Q. Xu, C.H. Tung, S.P. Zhao, J.L.S. Wang, Carbon. 38 (2000) 363.

[19]   J.I. Goldstein, D.E. Newbury, P. Echlin, D.C. Joy, A.D. Romig, C.E. Lyman, C. Fiori, E. Lifshin, Scanning electron microscopy and X-Ray microanalyses : A text for biologists, materials scientists and geologists, 2nd ed., Plenum Press, New York, 1992, p. 820.

[20]   P.A. Webb, C. Orr, Analytical methods in fine particle technology. Micromeritics Instrument Corporation : Norcoss, 1997, p. 301.

[21]   Y. Soneda, M. Makino, Carbon. 38 (2000) 475.

[22]   A. Linares-Solano, F. Rodriguez-Reinoso, C. Salinas-Martinez de Lecea, O.P. Mahajan, P.L. Walker, Carbon 20 (1982) 177.

[23]   S. Haydar, C. Moreno-Castilla, M.A. Ferro-Garcia, F. Carrasco-Marín, J. Rivera-Utrilla, A. Perrard, J.P. Joly, Carbon. 38 (2000) 1297.

[24]   C.W. Kruse, A.A. Lizzio, J.A. DeBarr, C.A. Feizoulof, Energy & Fuels. 11 (1997) 260.

[25]   I. Mochida, K. Kuroda, S. Miyamoto, C. Sotowa, Y. Korai, S. Kawano, K. Sakanishi, A. Yasutake, M. Yoshikawa, Energy & Fuels. 11 (1997) 272.

[26]   C. Morena-Castilla, M.V. López-Ramón, F.J. Maldonado-Hódar, J. Rivera-Utrilla, Carbon. 36 (1998) 145.

[27]   G. Socrates, Infrared characteristic group frequencies, John Wiley & Sons, New York, 1980.

[28]   C. Morena-Castilla, M.V. López-Ramón, F. Carrasco-Marín, Carbon. 38 (2000) 1995.

[29]   C.E. Alciatari, M.E. Escobar, R. Vallejo, Fuel. 75 (2000) 491.

[30]   J. Surygala, R. Wandas, E. Sliwka, Fuel. 72 (2000) 409 .

[31]   Y. Otake, R.G. Jenkins, Carbon. 31 (1993) 109.

[32]   E. Papirer, R. Lacroix, J.B. Donnet, G. Nanse, P. Fioux, Carbon. 32 (1994) 1341.

[33]   K.B. Bota, G.M. Abotsi, L.L. Sims, Energy and Fuels. 8 (1994) 937.

[34]   N.V. Plyasunova, Y. Zhang, M. Muhammed, Hydrometall. 48 (1998) 153.

[35]   J.L. Doremieux, Bull. Soc. Chim. Fr. 12 (1967) 4586.

[36]   L. Poul, N. Jouini, F. Fievet, Chem. Mater. 12 (2000) 3123.

[37]   A. Goerge, J. Meese-Marktscheffel, D. Naumann, A. Olbrich, F. Schrumpf, Ger. Offen. 1996, p. 19.

[38]   T.A. Soldatova, G.L. Tudorovskaya, N.V. Novozhilova, L.S. Vashchilo, Z.M. Kuzina, Zh Neorg. Khim. 21 (1976) 163.

[39]   M.A. Mohamed, S.A. Halawy, M.M. Ebrahim, J. Therm. Anal. 41 (1994) 387.

[40]   F.J. Derbyshire, V.H.J. De Beer, G.M.K. Abotsi, A.W. Scaroni, J.M. Solar, D.J. Skrovanek, Appl. Catal. 27 (1986) 117.

[41]   R. Reuel, C. Bartholomew, J. Catal. 85 (1984) 78.

[42]   V.N. L'nyanoi, S.A. Chepur, Izv. Akad. Nauk. SSSR Met. 5 (1981) 161.

[43]   Y. Saito, T. Yoshikawa, M. Okuda, N. Fujimoto, S. Yamamuro, K. Wakoh, K. Sumiyama, K. Suzuki, A. Kasuya, J. Appl. Phys. 75 (1994) 134.

[44]   M.A. Daleya, C.L. Manguna, J.A. DeBarrb, S. Rihaa, A.A. Lizziob, G.L. Donnalsb, J. Economya, Carbon. 35 (1997) 411.

[45]   I. Mochida, K. Kuroda, S. Miyamoto, C. Sotowa, Y. Korai, S. Kawano, K. Saknishi, A. Yasutake, M. Yoshikawa, Energy and Fuels. 11 (1997) 272.

[46]   S.G. Chen, R.T. Yang, Energy and Fuels 11 (1997) :421.

[47]   C.A. Mims, J.K. Pabst, Prepr. ACS Div. Fuel Chem. 25 (1980) 263.

[48]   J.A. Moulijn, F. Kapteijn, NATO ASI Ser. Ser. E. 105 (1986) 181.

[49]   P. Salatino, O. Senneca, S. Masi, Carbon. 36 (1998) 443.

Fischer-Tropsch Synthesis, Catalysts and Catalysis
B.H. Davis and M.L. Occelli (Editors)

Production of hard hydrocarbons from synthesis-gas over co-containing supported catalysts

Sholpan S. Itkulova[a], Gaukhar D. Zakumbaeva[a], Rozallya S. Arzumanova[b], and Valerii A. Ovchinnikov[b]

[a]*D.V. Sokolskii Institute of Organic Catalysis and Electrochemistry of the Ministry of Education and Science of the Republic of Kazakhstan; IOCE, 142, Kunaev str., Almaty, 005010, Kazakhstan; e-mail: ioce.lac@topmail.kz*
[b]*Novocherkassk's Plant of Synthetic Semiproducts, Novocherkassk, Rostov oblast, 346416, Russia*

Cobalt-containing alumina supported catalysts modified by a second metal have been parallel tested in the Fischer-Tropsch synthesis under pilot conditions. It has been observed that promotion of alumina supported Co-containing catalysts by certain amounts of Pt leads to an increase in the formation of high molecular-weight hard hydrocarbons, so-called ceresins (77%). Also, it has been found that ceresin formation is strongly dependent on process conditions (pressure, temperature and space velocity). By parallel testing, the catalyst composition and process conditions for the selective production of ceresins from synthesis-gas have been determined and optimised. The physico-chemical properties have been studied by X-Ray, TEM, IR-spectroscopic, TPR and other methods. The synthesised catalysts are highly effective and stable.

1. Introduction

Ceresins are a mixture of high molecular-weight hard hydrocarbons (n- and iso-alkanes). Ceresins have a fine crystalline structure. Their temperature of melting is higher than $65°C$ and molecular mass is 500-700 that corresponds to a composition of C_{37}-C_{53}. Due to the specific properties such as hydrophobicity, solidity, dielectric and other material advantages, ceresins are used in more than two hundred branches of industry [1]. There are no ceresin in oil or their amount is very low. The extraction of these industrially important products from oil is a fairly labour-intensive process requiring a complex separation and cleaning of ceresin.

The Fischer-Tropsch synthesis (FTS) [2] is commercially used for ceresin production. The C_{36+} hydrocarbons may be produced from synthesis-gas

(H_2/CO) under certain conditions [3-5]. This process is much more effective than ceresin production from oil. Ceresin obtained by FTS is a clean and ready marketable product.

Mostly iron or cobalt supported catalysts are industrially used for Fischer-Tropsch synthesis. The efficiency of FTS is strongly dependent on catalyst properties and may be increased by the modification of the catalyst. The process selectivity may be controlled by a choosing appropriate process conditions.

We have studied Co-containing catalysts modified by a second metal (Pt) and supported on alumina in the Fischer-Tropsch synthesis. The optimal additive amount and process conditions for the selective production of ceresins have been determined by parallel pilot testing.

2. Experimental

2.1. Catalyst characterisation

Catalysts were prepared by means of impregnating alumina with a total metal content of 10%. The content of the second metal (Pt) is varied from 0.06 to 5 weight %. Temperature of the catalyst reduction was 200-400°C.

The physico-chemical properties of catalysts were studied at the Laboratory of Physico-Chemical Studies of Catalysts of the Institute of Organic Catalysis and Electrochemistry, Kazakhstan.

X-ray diffraction (XRD) measurements were performed with the catalysts using the CuK_α or CoK_α radiation of a "Dron-4" diffractometer instrument.

Electron microscopic studies of fresh and used samples were provided with units: EM-125K, Ukraine and JEM-100CX, Japan with a resolution of about 0.5 nm. Specimens for TEM-analysis were prepared by conventional methods with use both of dried suspension and extraction carbon replica. The identification of phases was performed with the help of ASTM (American Society for Testing and Materials, 1986).

The reducibility of the cobalt catalysts was measured by temperature-programmed reduction using a "Setaram" device. TPR used a temperature ramp of 12°C/min from 30 to 900°C in a flow of 5% H_2 in Ar. H_2 consumption was measured by analyzing the effluent gas with a thermal conductivity detector.

IR-spectroscopy was performed with Specord-JR-75. Infrared spectra were recorded after the CO adsorption (T=298K) and after CO+H_2 reaction (P=0.1 MPa, T=200°C, experiment time is 30-120 minutes) over 10%Co-Pt/Al_2O_3 catalysts, using a conventional KBr cell combined with a flow reactor at room temperature in the wave number range of 1300-2400 cm^{-1}.

Specific surface area measurements were performed with an Accusorb unit. The BET specific surface area and the pore volume were deduced from the N_2 adsorption/desorption isotherms at 77 K. Specific area of the 10% Co-Pt/Al_2O_3 catalysts varies from 125 to 175 m^2/g. There is no significant change between the specific area of fresh and used samples.

After reaction, the thermo-programmed hydrogenation has been carried out to determine the carbon-containing species on a catalyst surface.

The degree of Co reduction has been determined by a chemical method based on measurements of hydrogen volume extracted from the reduced catalyst at its acid treatment by using a closed system with a gas circulating unit. The degree of cobalt ion reduction has been calculated as a percent ratio of number of cobalt atoms in zero-valence state to total number of cobalt atoms in catalyst.

Also, the mechanical strength of catalysts has been measured by the standards industrially used. The loss of catalyst weight was not more than 3% after their treatment.

The last two methods were carried out at the Central Laboratory of the Novocherkassk's Plant of Semiproducts (Russia).

2.2. Catalyst testing

The process was carried out in flow with a pilot multi-tube reactor at Novocherkassk's plant (Russia). Each tube was combined with a separator and heating trap to collect a mix of liquid and solid products. To produce the suitable hydrocarbon amount for their analysis and to convert to the further industrial scale, 100 g of catalyst was placed into each reactor tube. Three series of 4 samples of 10% Co-M/Al_2O_3 catalysts with different amounts of the second metal have been parallel tested. The commercial synthesis-gas with CO/H_2 = 1/2.2 was used. Temperature and space velocity were increased step by step from 150 until 200°C and from 100 to 300 hr^{-1} respectively. Pressure was 0.9-1.0 MPa. The duration of each experiment over the bimetallic catalysts was at least 10 days; maximum duration of continuous catalyst testing was 3 months. For comparison, also the monometallic 10%Co/Al_2O_3 has been tested for 100 hours.

The activity of catalysts was compared at the same process conditions and defined as percent conversion of carbon oxide. Selectivity was calculated on the basis of the ratio of CO converted into a certain product fraction to total amount of CO converted into all products.

2.3. Product analysis

The reaction products were investigated at the Central Laboratory of the Novocherkassk's Plant. The gaseous products were analysed by gaseous chromatographs (GSs) with coal and 'tsvetochrom' columns using a thermal conductivity detector (H_2, CO, CH_4, N_2, and CO_2) and with Chromaton N-AW and modified alumina columns using a flame-ionisation detector (CO_2, gaseous hydrocarbons). After separation of a liquid and solid product mixture collected, both the hydrocarbon and aqueous fractions were analysed. The physical characteristics of hydrocarbons such as fraction composition, average molecular weight, viscosity, density, temperature of melting of hard hydrocarbons, and other properties were determined by standard methods. The following parameters for ceresin fraction extracted were defined: T_{boil} and T_{melt}. The composition and content of oxygenates in aqueous fraction was analysed by GS (FID, column- Poropack).

3. Results and discussion

3.1. Catalyst testing

The second metal (Pt) modifying the Co-M/Al_2O_3 catalyst for producing hard hydrocarbon fractions from the synthesis-gas has been found earlier. In this work aimed at optimising the composition of the Co-M/Al_2O_3 catalyst, twelve samples with varying contents of second metal from 0 to 5.0 wt.% have been studied.

In Table 1, the results of parallel testing of the 10% Co-M/Al_2O_3 catalysts in FTS at a low temperature – 177°C and space velocity –100 hr^{-1} are summarised. The dependence of ceresin yield on the content of second metal has an extreme character. The optimal content of the promoter producing ceresin was found within 0.20-0.30 wt.%. Maximal yield of high quality ceresin with $T_{boil} > 450$°C and $T_{melt} = 100$°C reaches 77.6% over the catalyst with content of promoter is 0.25 wt.%.

Table 1. Effect of content of second metal on yield of ceresins in CO hydrogenation over Co-M/Al$_2$O$_3$ catalyst (H$_2$/CO=2.2; P=0.95 MPa, T=177°C, S.V.= 100 hr^{-1})

Content of second metal, wt.%	K$_{CO}$, %	Selectivity, %	
		C$_{10+}$ fraction*	Ceresin**
0.0	53.1	12.0	-
0.06	54.0	82.1	18.0
0.125	54.0	82.4	20.0
0.20	53.0	86.4	59.2
0.25	54.0	95.0	77.6
0.30	52.0	91.0	61.3
0.45	54.0	90.2	49.0
0.50	55.0	90.1	38.0
0.75	53.0	87.2	18.3
1.00	54.0	69.7	13.0
3.00	46.0	24.5	1.3
5.00	46.0	8.7	-

* Selectivity to C$_{10+}$ fraction including ceresins (number of C atoms $\geq$ 10)
** Selectivity to ceresin fraction only

It needs noting that increasing the selectivity to ceresin can be achieved only under mild process conditions. Mild pressure, temperature, and space velocity are very important in the formation of high molecular hydrocarbons. Both high temperature and a low time contact prevent the polymerisation of CH$_2$-surface species and lead to the formation of hydrocarbons with short chains. The pressure of about 0.1 MPa has been found to be optimal for hydrocarbon chain growth.

The effect of space velocity is demonstrated by the example of 10%Co-M (9.25-0.75)/Al$_2$O$_3$. With rising space velocity from 90 to 300 hr^{-1}, increase in temperature was required from 167 to 195°C to keep the constant degree of carbon monoxide conversion at 53%. At that rate, ceresin yield decreased from 23 to 10% (Table 2). The same behaviour was observed with all the catalysts studied in this work.

Table 2. The effect of space velocity on selectivity to ceresin over
9.25%Co-0.75%M/Al$_2$O$_3$ catalyst (H$_2$/CO=2.2; P=0.95 MPa, K$_{CO}$= 53%)

T, °C	S.V., hr^{-1}	Selectivity, %	
		C$_{10+}$	ceresin
167	90	90.3	23.0
177	100	87.2	18.3
185	200	86.6	14.0
195	300	78.3	10.0

* Selectivity to C$_{10+}$ fraction including ceresins (number of C atoms ≥ 10)
** Selectivity to ceresin fraction only

It is known that Fischer-Tropsch synthesis is an exothermic process representing a complex of parallel and secondary reactions occurring on different oxidised-reduced centres. A small shift in process parameters, particularly temperature, can cause a change in reaction direction. The results of study of effect of temperature on ceresin selectivity over 10% Co-M(9.5-0.5)/Al$_2$O$_3$ are presented in Table 3. At standard pressure - 1.0 MPa and space velocity - 100 hr^{-1}, increasing temperature from 167 to 195°C was accompanied with rising carbon monoxide conversion from 45 to 100%, increasing methane formation from 4.8 to 43.0%, and decreasing selectivity to ceresin from 38.0 to 0%. However, the low process temperature is not enough for catalyst activation and appropriate CO conversion for ceresin production. Thus, at 167°C, the CO conversion is 43% and selectivity to methane and ceresin are 6.2 and 23.2% respectively. At optimum 177°C, when CO conversion is 55%, selectivity to methane and ceresin is 4.8 and 38.0% respectively.

The optimal process parameters for production of ceresin from the synthesis-gas are strongly individual and depend on the catalyst nature. Thus, at the same conditions, no ceresins are formed over the monometallic 10% Co/Al$_2$O$_3$ catalyst in contrast to the bimetallic promoted 10% Co-M/Al$_2$O$_3$ catalyst (Table 1). The promoted Co-containing catalysts are very stable. They worked with the same activity during all periods of continuous testing (> 2000 hours), while the monometallic 10% Co/Al$_2$O$_3$ catalyst lost its activity after 100 hours of testing.

Table 3. The effect of temperature on CO conversion and ceresin selectivity over 9.5%Co-0.5%M/Al_2O_3 catalyst (H_2/CO=2.2; P=1.0 MPa, S.V.=100 hr^{-1})

T, °C	K_{CO},%	Selectivity to fraction, %			
		CH_4	C_{2+}	C_{10+}	Ceresin
167	43	6.2	93.8	80.2	23.2
177	55	4.8	95.2	90.1	38.0
185	78	13.8	86.2	37.6	2.4
195	100	43.0	57.0	2.9	-

* C_{2+} fraction consisting of hydrocarbons with number of C atoms $\geq$ 2 including ceresins
* C_{10+} fraction consisting of hydrocarbons with number of C atoms $\geq$ 10 including ceresins

3.2. Physico-chemical study of 10%Co-M/Al_2O_3 catalysts

The X-Ray analysis of the bimetallic catalysts showed the amorphous structure of both fresh (before reaction) and used (after reaction) samples.

By TEM studies of the bimetallic catalysts, it has been observed that basically the metal particles are uniformly distributed over the alumina surface. Both metals are dispersed in bimetallic catalysts. In 10% Co/Al_2O_3 catalyst, the particle size is 3-7 nm (Fig.1). While introducing the second metal to 10%Co/Al_2O_3, the size of metallic particles is decreased to 1-2 nm depending on the amount of promoter (Fig.2). The further dispersion of metal particles occurs after using the catalyst in a reaction medium (CO+H_2). There is no microdiffraction from metal particles in samples used in the Fischer-Tropsch synthesis.

The absorption bands in regions of carbonate-carboxylate structures, carbonyls, CH_x, OH adsorbed groups, and physically adsorbed CO_2 have been observed by IR-studies of adsorption of CO and mix CO+H_2 on a surface of the reduced 10%Co-M/Al_2O_3 catalysts. In region of adsorbed carbonyls, the bands at 1980-2180 cm^{-1} assigned to the linear CO adsorbed on high dispersed metal centres probably in different oxidised states have been detected at adsorption of (Figure 3).

Two linearly adsorbed CO species are presented at CO adsorption in spectra of the 10% Co-M(9.5-0.5)/Al_2O_3. Their positions are slightly shifted depending on temperature, when adsorption occurs. There are no significant changes between the spectra of fresh and used samples. At adsorption of the gaseous CO+H_2 mix, the additional weak band appears at 1980-2000 cm^{-1} (Figure 3, curves a,b).

In the catalyst with lower content of second metal (0.3 wt.%), the third weak band of CO is presented in IR-spectra at CO adsorption on both fresh and used samples. They are kept at adsorption of $CO+H_2$ (Figure 3, curves c,d).

The absence of significant changes between IR-spectra of fresh and used catalysts can be an evidence of a stability of bimetallic catalysts.

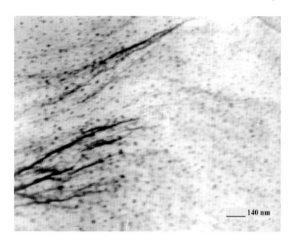

Figure 1. TEM pattern of 10% Co/Al_2O_3

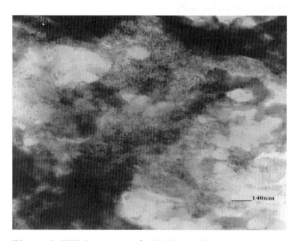

Figure 2. TEM pattern of 10%Co-M(9.5-0.5)/Al_2O_3

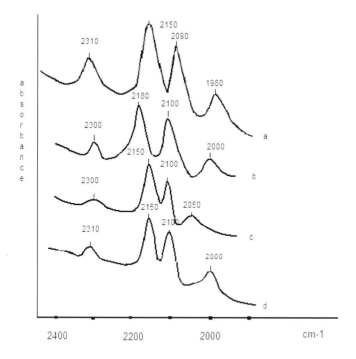

Figure 3. IR spectra of CO+H$_2$ adsorption over 10%Co-Pt/Al$_2$O$_3$ catalysts
    a)   10%Co-Pt(9.5:0.5)/Al$_2$O$_3$ (fresh)
    b)   10%Co-Pt9(9.5:0.5)/Al$_2$O$_3$ (used)
    c)   10%Co-Pt(9.7:0.3)/Al$_2$O$_3$ (fresh)
    d)   10%Co-Pt(9.7:0.3)/Al$_2$O$_3$ (used)

By TPR, it has been shown that at the addition of second metal, the temperature of Co reduction is shifted to lower value by up to 200°C depending on the amount of second metal. The comparison of monometallic and bimetallic catalysts reduced at the same temperature – 300°C demonstrates that the addition of second metal leads to increasing the degree of Co reduction from 42 to 69.0-78.1% depending on content of second metal (Table 4).

The results obtained indicate that introducing the second metal to the 10% Co-M/Al$_2$O$_3$ causes a significant change of the catalytic and physico-chemical properties. We suppose that adding the certain amount of second metal leads to the formation of bimetallic nano-particles, which are responsible for increasing the activity, selectivity to formation of high-molecular-weight hard hydrocarbons, and stability of the 10% Co-Pt/Al$_2$O$_3$ catalysts.

Table 4. The degree of Co reduction at $T_{red}=300^{\circ}C$

Content of second metal, wt. %	Degree of Co reduction, %
0.0	42.0
0.06	69.0
0.125	72.1
0.2	78.1
0.25	77.3
0.5	73.5
1.0	73.8

4. Conclusions

1. The promoter and its amount in the region of 0.2-0.3 wt.% have been to found to improve the Co-containing catalyst supported on alumina.
2. The optimal process conditions for the production of high quality ceresins from synthesis-gas over fixed-bed catalyst have been established; they are: T is 170-180°C, space velocity is about 100 hr^{-1}, CO conversion is 50-55%.
3. The developed catalyst with high activity, selectivity to ceresin formation, and stable work (more than 2000 hours) has been patented [6] and recommended for the industrial use.

5. Acknowledgement

Authors are grateful to the staffs both of the Laboratory of Physico-Chemical Studies of Catalysts of D.V. Sokolsky Institute of Organic Catalysis and Electrochemistry (IOCE, Kazakhstan) and the Central Plant Laboratory (Novocherkassk's Plant of Synthetic Semi-Products, Russia) for analysis of catalysts and products.

• References

1. P.I. Belkevich and N.G. Golovanov, Wax and Its Technical Analogues, Nauka I Technika, Minsk, 1980 (in Russian).
2. F. Fischer, H. Tropsh, Ber. Deutsch.Chem. Geselt, 69 (1926) 830.
3. G. Storch, N. Colambik, and R. Anderson, Synthesis of Hydrocarbons from Carbon Oxide and Hydrogen, Foreign Literature (translation), Moscow, 1954
4. R.A. Sheldon, Chemicals from Synthesis-Gas. Catalytic Reactions of CO and H_2, Dordrecht, 1983

5. Ya.T. Eidus Synthesis of Organic Compounds Based on Carbon Oxide, Nauka, Moscow, 1984 (in Russian)
6. G.D. Zakumbaeva, Sh. S. Itkulova, R.S. Arzumanova, V.A. Ovchinnikov, A. Selitskii. Catalyst for ceresin synthesis. Pat. Of Russia, 2054320, 7 Jul. 1992; Chem.Abstr. 125 (1996) 34015

Fischer-Tropsch Synthesis, Catalysts and Catalysis
B.H. Davis and M.L. Occelli (Editors)
© 2007 Elsevier B.V. All rights reserved.

The Function of Added Noble Metal to Co/Active Carbon Catalysts for Oxygenate Fuels Synthesis via Hydroformylation at Low Pressure

Yi Zhang, Misao Shinoda, Youhei Shiki, Noritatsu Tsubaki*

*Department of Applied Chemistry, School of Engineering,
University of Toyama, Gofuku 3190, Toyama 930-8555, Japan
TEL/FAX: (81)-76-445-6846
E-mail: tsubaki@eng.toyama-u.ac.jp*

The hydroformylation of 1-hexene was catalyzed with active carbon (A.C.)-supported cobalt catalysts under low syngas pressure. Small amount of Pt, Pd and Ru, added as promoters in Co/A.C., led to great improvement of catalytic activity for hydroformylation of 1-hexene. The promotional effect of Ru for Co/A.C. catalyst was the best in this study as the highest activity and selectivity for oxygenate formation. Meanwhile, the activity of 1-hexene hydroformylation increased with increasing Ru loading. In the Ru promoted Co/A.C. catalyst, surface Ru was significantly lower than its bulk content. This kind of unbalanced alloy formation determined the highest performance of Ru promoted Co/A.C. catalyst, via small particles but high reduction degree, keeping more CO in non-dissociative state and lowering surface hydrogen concentration.

1. Introduction

Hydroformylation, the addition of synthesis gas (CO and H_2) to alkenes, is one of the most important syngas-related reactions [1]. The hydroformylation reaction was firstly discovered on a heterogeneous Fischer-Tropsch (F-T) catalyst [2]. Oxygenates, such as alcohols and esters, can be used as clean engine and vehicle fuels due to their high efficiency and low emission. The olefins produced in the Fischer-Tropsch synthesis (FTS) can be hydroformylated to oxygenate fuels with syngas. It is of great practical importance to develop a solid catalyzed process to produce oxygenates from syngas at low pressure without any compression from FTS process. The main oxygenate from hydroformylation is aldehyde and it can easily be hydrogenated to form alcohol, the final target product. More importantly, since the high pressure in the industrial hydroformylation process using homogeneous catalyst is critical to keep high n/i ratio of the formed aldehydes, and since iso-aldehyde is acceptable if it is used as fuel after hydrogenation to branched alcohol, the reaction pressure in our experiments can be lowered significantly and high pressure is not necessary. Here 1-hexene was selected as a model compound of FTS derived olefins.

All current commercial processes are based on homogeneous catalysts, mostly using rhodium. The successful development of a heterogeneous catalyst for hydroformylation can avoid the drawbacks of homogeneous catalysis, such as, very high pressure, the catalyst separation and recovery steps. Cobalt is extensively applied in homogeneous process of this reaction due to its high activity and low cost. Supported cobalt catalysts were known to be highly active for Fischer-Tropsch synthesis (FTS). Nevertheless, supported cobalt catalysts usually show low catalytic activity and selectivity for olefin hydroformylation because of high catalytic activity for olefin hydrogenation proceeding at the same time [3]. Thus far, only limited number of papers concerning supported cobalt catalyzed hydroformylation have appeared. However, it was reported that the activity and selectivity of supported cobalt catalysts for hydroformylation can be promoted by various promoters. Qiu et al. [4] reported that the Pd promoted Co/SiO_2 catalyst was very active and selective for the hydroformylation of 1-hexene, and most of the cobalt and all noble metal, more than 90% of original cobalt metals, existed in the heterogeneous catalyst even after the reaction. The promoting effect of iridium was also found for the gas-phase hydroformylation of ethene over Co/SiO_2 catalyst [5]. Active carbon has also been shown to exhibit beneficial characteristics in carbonylation; it has been found to suppress dissociative hydrogen adsorption, to promote strong or multiple adsorption of CO and to inhibit dissociative CO adsorption [6]. These special characteristics of active carbon are also important for hydroformylation, which is governed by molecular reaction of the adsorbed CO.

In the present work, the promotional effects of noble metals, such as Pt, Pd and Ru, were studied for active carbon-supported cobalt catalyzed hydroformylation of 1-hexene.

2. Experimental

The active carbon-supported cobalt catalyst was prepared by impregnation of cobalt nitrate aqueous solution onto active carbon (Kanto Chemical Co., specific surface: 1071.7 m^2/g, pore volume: 0.43 m^3/g, pellet size: 20-40 mesh). The noble metal promoted Co/A.C. catalysts were prepared by co-impregnation of cobalt nitrate and aqueous solution of noble metal coordinated compounds with different noble metal loading. The cobalt loading of all of catalysts was 10 wt%. The noble metal promoted catalysts were named as 10Co+XM, where X was noble metal loading and M was symbol of noble metal. The details of catalyst preparation were described elsewhere [7].

Hydroformylation reaction was carried out in a magnetically-stirred autoclave with inner volume of 75 ml. The catalyst of 20-40 mesh and 1-hexene were loaded into the reactor with a stirrer. The reaction conditions were 403 K, 5.0 MPa, $CO/H_2=1$. The weight of catalyst was 0.10 g, and 1-hexene was 40 mmol. The initial reaction gas contained 70 mmol CO and 70 mmol H_2. The reaction time was 2 h. After the reaction, the reactor was cooled to 273 K and depressurized. After filtration to remove the solid catalyst, the liquid products were analyzed quantitatively by gas chromatograph (Shimadzu GC 14A) with a capillary column and a flame ionization detector (FID).

Supported cobalt crystalline size of the passivated catalysts was determined by XRD (Rigaku, RINT2000). The supported cobalt crystalline average size was calculated by   from XRD data, where L is the crystalline size, K is a constant (K = 0.9 ~ 1.1), l is the wavelength of X-ray (CuKa = 0.154 nm), and D(2q) is the width of the peak at half height.

Chemisorption experiments were carried out in a static volumetric glass high-vacuum system (Quantachrome Autosorb-1, Yuasa Ionics). Research grade gases (H_2: 99.9995%, CO: 99.99%, Takachiho Co.) were used without further purification. Before adsorption of H2 or CO, the catalysts, previously reduced by H_2 and passivated, were treated in H2 at 673 K for 1 h, followed by evacuation. H_2 adsorption isotherms were measured at 373 K, and the CO chemisorption was conducted at room temperature.

Temperature-programmed reduction (TPR) experiments were carried out in a quartz-made microreactor connected to a thermal conductivity detector (TCD) equipped with active charcoal column, using 0.2 g calcined catalysts from 373 K to 1073 K. The gas stream, 5 % H_2 diluted by nitrogen as reducing gas, was fed via a mass flow controller. After the reactor, the effluent gas was led via a 3 A molecular sieve trap to remove the produced water.

The temperature-programmed surface reaction (TPSR) experiment was carried out in a quartz-made microreactor connected to a thermal conductivity detector (TCD) equipped with active charcoal column using 0.2 g passivated catalysts. The passivated catalysts were reduced by hydrogen at 673 K for 1 h, and then the system was purged by helium stream for 1 h. After reduction of catalysts, the CO was introduced into microreactor at 15 ml/min for 10 min. After removing the physically adsorbed CO by helium stream, hydrogen of 30 ml/min passed over the catalysts, which was heated at a rate of 3 K/min. The products, consisting of CH_4, CO and CO_2, were analyzed. Temperature-programmed desorption (TPD) was conducted in the same manner as TPSR only using helium as carrier gas instead of hydrogen. The desorbed products were CO and CO_2.

3. Results and discussion

3.1. The effects of noble metals addition to Co/A.C. catalyst

Table 1 The effect of noble metals on hydroformylation of 1-hexene over Co/A.C.

| Catalyst | Conv. (%) | Selectivity (%) | | | | Yield (%) | n/iso |
		Hexane	Isomer	C7-nol	nal & nol	Oxygenate	(nal)
10Co	37.8	9.3	18.8	0.6	60.8	25.1	0.80
10Co+1Ru	97.0	5.5	10.0	4.3	79.1	81.5	0.78
10Co+1Pt	96.3	20.7	6.5	1.9	66.6	68.8	0.78
10Co+1Pd	84.0	19.5	6.0	1.2	71.1	60.8	0.94
1Ru	81.3	0.0	95.7	0.0	1.5	1.3	3.50
1Pt	6.5	49.2	40.4	0.0	0.4	0.0	2.82
1Pd	24.5	37.6	59.5	0.0	2.2	0.5	0.79

Catalyst: 20-40 mesh, 0.10 g; 1-Hexene: 40.0 mmol; Reaction temperature: 403K; Reaction pressure: 5.0 MPa; Reaction time: 2h; CO/H_2=1:1.

Table 1 shows the reaction performance of cobalt active carbon catalysts promoted by various metals. For the supported cobalt catalyst, conversion of 1-hexene gave 37.8%, and oxygenates, including aldehydes (C7-al) and alcohols (C7-ol), predominated the products, as 60.8% nal and nol selectivity. The additions of only 1wt% of Pt, Pd and Ru promoters improved the catalyst performance significantly, increasing the conversion of 1-hexene and selectivity of oxygenates, while ruthenium exhibited the best promoting effect. 97.0% 1-hexene conversion and 79.1% selectivity of oxygenate products were obtained on Co/A.C. catalyst promoted by 1wt% Ru. The 1wt% Pt promoted Co/A.C. catalyst also presented higher activity for hydroformylation of 1-hexene producing 96.3% 1-hexene conversion and 66.6% oxygenate selectivity. For the Co/A.C. catalyst promoted by Pd, the activity and selectivity were lower than those of Co/A.C. catalysts promoted by Pt or Ru in this study, even though,

Pd showed significantly promotional effect than Pt and Ru for Co/SiO$_2$ catalyst in hydroformylation of 1-hexene [4]. Cobalt free 1wt% Ru, Pt and Pd supported on active carbon were prepared and tested in hydroformylation reaction. As shown in Table 1, these kinds of catalysts showed very low activity for 1-hexene hydroformylation. For the Ru catalyst, the 1-hexene conversion was 81.3%, but the 1-hexene was only converted to isomerization products. For the Pt and Pd catalysts, both the 1-hexene conversion and oxygenates selectivity were very low. Results show in Table 1 indicate that the noble metal/active carbon catalysts themselves had no catalytic activity of 1-hexene hydroformylation to form oxygenates.

3.2. The contribution of Ru loading for cobalt active carbon catalyst.

Although supported ruthenium and supported cobalt were much less active than supported rhodium in olefin hydroformylation, the combination of these two metals has been demonstrated to exert a striking influence on the rate enhancement of oxygenate formation in F-T synthesis and olefin hydroformylation [8, 9]. In this study, the promotional effect of Ru for hydroformylation of 1-hexene was carried out with different Ru loading. Different amounts of Ru, such as 0.1 wt%, 0.5 wt% and 1.0 wt%, were added to 10 wt% Co/A.C. catalyst, respectively. Hydroformylation reaction was carried out at 403 K for 2 h. The reaction performance of various Ru-promoted Co/A.C. catalysts was shown in Table 2. The conversion of 1-hexene increased with increased Ru loading. The conversion of 0.5 wt% Ru promoted Co/A.C. catalyst was almost 100%. With the increased amount of Ru added to Co/A.C. catalyst, not only 1-hexene conversion but also oxygenate selectivity was enhanced, resulting in significantly increased oxygenate yield. As shown in Table 1, the 1 wt% Ru/A.C. showed a negligible activity for 1-hexene hydroformylation, as no oxygenate was produced and 1-hexene was mainly converted to isomers, 2-hexene etc. It was proved that active carbon supported Ru catalyst itself had no catalytic activity of 1-hexene hydroformylation to form oxygenates.

The observed rate enhancement of 1-hexene hydroformylation with Ru promoted Co/A.C. catalyst accounted for a synergy of ruthenium and cobalt, which might be explained in terms of catalysis by bimetallic particles or by ruthenium and cobalt monometallic particles in intimate contact. On the other hand, it was considered that the increased noble metal loading might increase the active site number on the surface of catalysts, contributing to the improved reaction rates. It was reported that the addition of ruthenium to the supported cobalt catalyst significantly improved the reduction degree [10]. Higher reduction degree of catalyst can provide more cobalt metal centers for the reaction and lead to higher catalytic activity. Because there was no direct relationship between the 1-hexene conversion and the amount of dissolved cobalt [11], it was considered that the remarkably increased active cobalt site was due to the addition of ruthenium.

The isomerization of 1-hexene decreased with increased ruthenium loading as shown in Table 2. It is considered that with the increased active site number on the catalyst surface, CO adsorption improved, contributing to CO insert reaction and inhibiting the isomerization of 1-hexene. The formation of oxygenate increased with increasing Ru loading, see Table 2.

Table 2 The reaction preformance of Ru promoted Co/A.C. with different Ru loading

| Catalyst | Conv. (%) | Selectivity (%) | | | | Yield (%) Oxygenate | n/iso (nal) |
		Hexane	Isomer	C7-nol	nal & nol		
10Co	37.8	9.3	18.8	0.6	60.8	25.1	0.80
10Co+0.1Ru	44.2	0.0	37.4	0.5	51.2	26.2	0.68
10Co+0.5Ru	97.8	0.0	24.5	2.6	69.6	72.2	0.85
10Co+1.0Ru	97.0	5.7	10.0	4.3	79.1	81.5	0.78

Catalyst: 20-40 mesh, 0.10 g; 1-Hexene: 40.0 mmol; Reaction temperature: 403K; Reaction pressure: 5.0 MPa; Reaction time: 2h; $CO/H_2=1:1$.

3.3. Catalysts characterization.

Table 3. The characterization of various catalysts.

| Catalyst | Chemisorption ($\mu mol/g$) | | Red. degree[c] (%) | Cobalt particle size (nm) | | Surface composition (M/Co)[e] | Disp.[f] (%) |
	H_2 Uptake[a]	CO Uptake[b]		H_2.[d]	XRD		
10Co	46.2	36.8	28.4	14.8	8.5	-	6.7
10Co+1Ru	86.3	324.3	54.8	5.4	4.2	0.05/10.0	18.6
10Co +1Pt	50.1	204.1	53.2	8.9	6.3	1.0/9.8	11.1
10Co +1Pd	20.7	52.3	56.1	22.9	18.9	1.0/9.9	4.5
10Co +0.1Ru	56.5	209.8	40.6	6.1	4.2	-	16.3
10Co +0.5Ru	77.9	304.7	45.2	4.9	4.0	-	20.4

a: H_2 chemisorption was determined at 373 K. b: CO chemisorption was determined at 313 K. c: Calculated by TPR from 323 K to 1073 K. d, f: Calculated by H_2 chemisorption. e: M/Co was in weight ratio; M was noble metal, such as Ru, Pt and Pd.

Atoms at the corners and edges of metal particles are regarded to be advantageous for CO adsorption in linear geometry, which benefit the CO insertion during the hydroformylation reaction [12]. Therefore, high dispersion or small particle size of metal on the catalyst, where the number of atoms at the corners and edges of metal particles is increased, is important for the improvement of catalytic performance. Small particle size of supported cobalt is advantageous for CO insert reaction, and the oxygenates production. Table 3 compares chemical properties of various catalysts. The supported cobalt particle size of Pt, Ru promoted Co/A.C. catalysts, which was determined by XRD, was smaller than that of unpromoted Co/A.C., similar to those from H2 chemisorption data. The Pd promoted Co/A.C. formed the largest supported cobalt particle size, resulting in the lowest cobalt dispersion, which was determined by H2 chemisorption. The reduction degree of unpromoted Co/A.C. was as low as 28.4%, while 54.8%, 53.2% and 56.1% of cobalt was reduced for cobalt active carbon catalysts promoted by 1wt% Ru, Pt and Pd, respectively. The reduction degree, H_2 and CO uptake increased with the increased Ru loading, as compared in Table 3, indicating that the active site number increased with the increased Ru loading, resulting in the increased activity of 1-hexene hydroformylation, as shown in Table 2. Based on above, the addition of noble metal to Co/A.C. catalyst significantly improved the dispersion and reduction degree of supported cobalt, contributing to the formation of more active site for hydroformylation of 1-hexene.

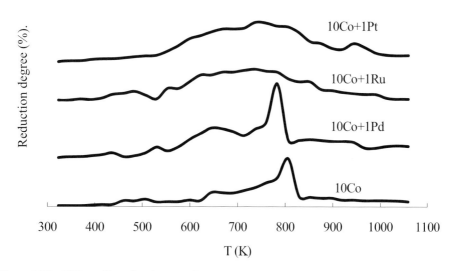

Figure 1 The TPR profiles of various catalysts.

From 323 K to 1073 K

The reduction property of Co/A.C. catalysts and 1% noble metal promoted Co/A.C. catalysts was detected by TPR. In TPR spectra of various catalysts, as shown in Fig 1, two peaks located at 647 K and 802 K existed for the Co/A.C. catalyst. The two peaks have been identified as conversion of Co^{3+} to Co^{2+} followed by the conversion of Co^{2+} to Co [13]. In the Pd promoted Co/A.C. catalyst, the first peak located at 647 K and the second peak located at 776 K lower than that of Co/A.C. catalyst. Both peaks were significantly stronger than those of Co/A.C. catalyst, indicating that the addition of Pd significantly promoted the reduction degree of supported cobalt. In the Pt and Ru promoted Co/A.C. catalyst, there was only one wide peak scattering from 532 K to 843 K, and peak intensity was weaker than that of Pd promoted Co/A.C. catalyst. Comparing the reduction degrees of various catalysts, which were calculated by TPR data from 323 K to 1073 K, the Pd promoted Co/A.C. catalyst showed the best reducibility as 56.1% reduction degree, due to its largest supported cobalt particle, as shown in Table 3, which had lighter interaction with the active carbon support. For the Pt and Ru promoted Co/A.C. catalysts, due to thire smaller supported cobalt particle, which had the strong interaction with support, the reduction degree was slightly lower than that of Pd promoted Co/A.C. but rather higher than that of unpromoted Co/A.C. catalyst, as 53.2 % and 54.8 %, respectively. It was considered that the addition of 1% Pd to Co/A.C. catalysts significantly promoted the reducibility of supported cobalt, however the highest reduction degree of this catalyst resulted in sintering of supported cobalt increasing cobalt particle size, which was disadvantageous for CO insert reaction in the hydroformylation of 1-hexene [12]. On the other hand, it was reported that Pd and Pt could not be homogeneously dispersed on the surface of active carbon, and they were thermally unstable, resulting in sintering of metal particles to form large crystalline in calcination step of catalysts preparation [14]. The surface composition of supported metal determined by EDS, as shown in Table 3. The composition of Pt and Pd promoted cobalt surface is similar to original ratio of noble metal loading to cobalt loading, indicating that the added Pt and Pd homogeneously dispersed inside the cobalt particle. It was considered that the homogeneously dispersed Pt and Pd inside the cobalt particle improved the sintering of supported cobalt during calcination and reduction steps. This forms larger cobalt particle than that of Ru promoted Co/A.C. catalyst, in which Ru is heterogeneously dispersed inside the supported cobalt particle due to the significantly lower surface composition of Ru/Co than loading ratio of Ru to Co as shown in Table 3. For 0.1 wt% Ru added catalyst or 0.5 wt% Ru added catalyst, EDS even could not detect the surface Ru. It is reported that hydrogen is much more competitive than CO for adsorption on the surface sites of Pd, and similar behavior also occurre for Pt [15]. It was considered that the hydrogen-rich circumstances on the surface of Pd and Pt contributed to hydrogenation of 1-hexene, leading to high selectivity of hexane of Pd and Pt promoted Co/A.C. catalyst, as shown in Table 1. For the Ru promoted Co/A.C. catalysts, the addition of Ru promoted the dispersion and reducibility of supported cobalt, to form more active sites on 1wt% Ru promoted Co/A.C. catalyst than on Pt or Pd promoted Co/A.C. catalyst, as compared in Table 3, which was supported by H_2 and CO chemisorption. Most of Ru is embedded inside the bulky cobalt particle, lowering the surface hydrogen concentration and suppressing hydrogenation of hexene.

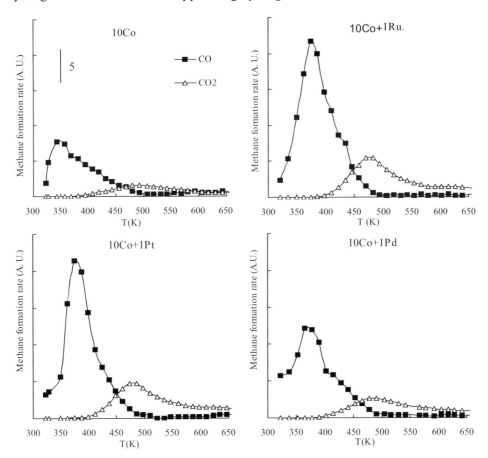

Figure 2 The TPD profiles of various catalysts.

From 323 K to 650 K

The thermal desorption pattern of the chemisorbed CO on supported metal catalysts was closely correlated to their catalytic activities for the CO hydrogenation [16]. Here TPD and TPSR were utilized to characterize various catalysts, including Co/A.C. catalyst, 1 wt% Ru, Pt and Pd promoted Co/A.C. catalyst.

The results for CO_2 and CO desorption during TPD are shown in Fig 2. The Co/A.C. and Pd promoted Co/A.C. catalyst had weak and broad CO_2 peak. The desorbed CO2 peak of these two catalysts located at 493 K and 490 K respectively. For Pt and Ru promoted Co/A.C. catalyst, the desorbed CO_2 peak located at lower temperature, 473 K and 470 K, respectively. The CO desorption peak located at 343 K, 373 K, 374 K and 365 K for Co/A.C., Ru, Pt, and Pd promoted Co/A.C. catalyst, respectively. The low temperature CO desorption peaks attribute to the CO desorbed from polycrystalline Co, especially its Co (001) surface [17]. The desorbed CO/CO_2 amount ratio was 2.24, 3.51, 3.12, and 2.29 for Co/A.C. catalyst, Ru, Pt, and Pd promoted Co/A.C. catalyst, respectively, indicating that it was the most disadvantageous for CO desorbed as CO_2 on Ru promoted Co/A.C. catalyst. That is Ru promoted Co/A.C. had low activity for cleaving CO bond, contributing to CO molecular reactions, resulting in the highest activity for hydroformylation of 1-hexene.

The TPD of CO from supported Co catalysts is a complicated process involving other reactions such as CO dissociation and CO disproportionation. It is referred that the low temperature peak of CO_2 desorption resulted from CO disproportionation [18], and the higher temperature peaks from the recombination of dissociated carbon and oxygen. For Co/A.C. and Pd promoted Co/A.C., the CO_2 desorption peak located at higher temperature than that of Ru and Pt promoted Co/A.C.; and the temperature of CO_2 desorption peak was the lowest for Ru promoted Co/A.C.. It seems that CO_2 formed on Ru promoted Co/A.C. was mainly attributed to the CO disproportionation, indicating that this catalyst was inactive to decompose adsorbed CO to form dissociated carbon and oxygen band, consequently improving the CO insert reaction in hydroformylation reaction of 1-hexene.

The reactivity of carbon species after CO adsorption at room temperature was studied by TPSR. The rate of methane formation as a function of temperature is shown in Fig 3. The peak temperature of CH_4 formation on CO TPSR was reported to closely relate to the catalytic activity of CO hydrogenation [16]. The methane formation peaks of Co/A.C. catalyst located at the lowest temperature as 462 K, indicating that this catalyst had the highest CO decomposition activity, resulting in the lowest hydroformylation activity in this study. On the other hand, the 1wt% Ru promoted Co/A.C. catalyst had the highest methane formation peak temperature which was 474 K, and its peak area was smaller than the CO desorption peak area, the same with that of 1 wt%

Pt promoted Co/A.C., which located at 466 K, unlike that of Co/A.C. and 1wt% Pd promoted Co/A.C. catalysts. The desorbed CO/CH_4 ratio in TPSR was 0.77, 0.81, 1.12 and 0.94 for Co/A.C., Pd, Ru and Pt promoted Co/A.C. catalyst, respectively. These facts suggested that CO was adsorbed in a more inactive state on the Ru promoted Co/A.C. catalyst than on the others, which was difficult to be dissociated and reacted with hydrogen to form CH_4 in TPSR. Based on above, the Ru promoted Co/A.C. had the lowest CO cleavage activity and was inactive for dissociating carbon and oxygen bond of adsorbed CO, due to its smallest particle size as compared in Table 3. This feature is very advantageous to CO molecular reactions, contributing to the CO insert reaction in hydroformylation of 1-hexene.

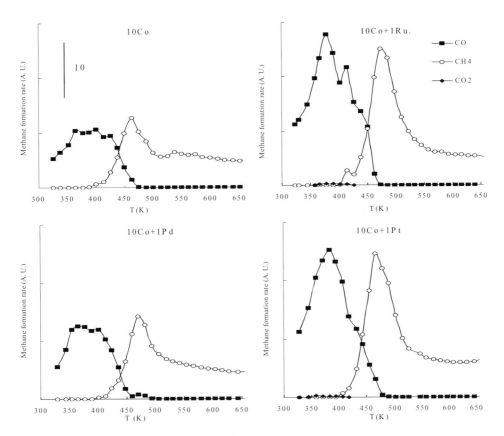

Figure 3 The TPSR profiles of various catalysts.

From 323 K to 650 K

As discussed above, the high hydrogenation activity of added Pt and Pd resulted in high selectivity of hexane, and the sintering-promotion effect of added Pt and Pd on active carbon surface formed larger cobalt particle, which was disadvantageous for hydroformylation of 1-hexene, resulting in lower oxygenate selectivity than that of Ru promoted Co/A.C. catalyst, as compared in Table 1. The Ru promoted Co/A.C. catalyst had not only high dispersion but also high reducibility of supported cobalt, contributing to form large number of active sites, which was increased with the increased Ru loading, as shown in Table 3. On the other hand, as illustrated in Fig 2 and Fig 3, the activity of CO dissociation on Ru promoted Co/A.C. catalyst was the lowest in this study, contributing to the highest activity of 1-hexene hydroformylation for this kind of catalyst, because the hydroformylation was a CO molecular insertion reaction. The increased active site number with the increased Ru loading improved the CO adsorption on Ru promoted Co/A.C. catalyst, which improved the hydroformylation reaction rate, as shown in Table 2 and Table 3. The non-dissociative CO adsorption on Ru promoted Co/A.C. catalyst contributed to CO insert reaction and inhibited the isomerization of 1-hexene. Therefore, the formation of oxygenates was promoted by increasing Ru loading.

4. Conclusions

The addition of small amount of Pt, Pd and Ru exhibited significant effects on the Co/A.C. catalyst. Especially, the Ru promoted Co/A.C. catalyst was very active and selective for the hydroformylation of 1-hexene, due to its high dispersion and reducibility of supported cobalt, as well as to the high activity of CO molecular reaction. Reverse to surface-rich, added Ru was bulk-rich in the active carbon supported cobalt particles, showing very low surface Ru density. This kind of unbalanced alloy formation, different from the uniformly-distributed Pt-Co or Pd-Co alloy, determined the highest performance of the Ru promoted Co/A.C. catalyst, via small particles but high reduction degree, keeping more CO in a non-dissociative state and lowering surface hydrogen concentration. For Ru promoted Co/A.C. catalyst, the activity and selectivity of 1-hexene hydroformylation increased with the increased Ru loading.

References

1. G.W. Parshall, Homogeneous Catalysis: The Application and Chemistry of Catalysis by Transition Metal Complexes, Wiley, New York, 1980.
2. B. Cornils, in: J. Falbe (Ed.), New Synthesis with Carbon Monoxide, Springer, New York, 1980, p. 1.
3. H. Arakawa, N. Takahashi, T. Hnaoka, K. Takeuchi, T. Matsuzaki, Y.Sugi, Chem. Lett. (1988) 1917.
4. X. Qiu, N. Tsubaki, K. Fujimoto, Catal. Comm. 2 (2001) 75.

5. K. Takeuchi, T. Hanaoka, T. Matsuzaki, Y. Sugi, H. Asaga, Y. Abe, T. Misono, Nippon Kagaku Kaihshi 7 (1993) 901.
6. K. Omata, K. Fujimoto, T. Shikada, H. Tominaga, Ind. Eng. Chem. Res. 27 (1988) 2211.
7. M. Shinoda, Y. Zhang, Y. Shiki, Y. Yoneyama, K. Hasegawa, N. Tsubaki, Catal. Comm. 4 (2003) 423.
8. E. Iglesia, S.C. Reyes, R.J. Madon, Adv. Catal. 39 (1993) 221.
9. M. Reinikanen, J. Kiviaho, M. Kroger, M.K. Niemela, S. Jaashelainen, J. Mol. Catal. A 174 (1998) 61.
10. N. Tsubaki, S. Sun, K. Fujimoto, J. Catal. 199 (2001) 236.
11. X. Qiu, N. Tsubaki, K. Fujimoto, J. Chem. Eng. Jpn. 34 (2001) 1366.
12. N. Takahashi, T. Tobise, I. Mogi, M. Sasaki, A. Mijin, T. Fujimoto, M. Ichikawa, Bull. Chem. Soc. Jpn. 65 (1992) 2565.
13. M. P. Rosynek, C. A. Polansky, Appl. Catal. 73 (1991) 97.
14. I. Furuya, T. Shirasaki, J. Jpn. Petrol. Inst. 13 (1971) 78.
15. S. J. Tauster, S. C. Fung, R. L. Garten, J. Am. Chem. Soc. 100 (1978) 170.
16. K. Fujimoto, M. Kameyama, T. Kunugi, J. Catal. 61 (1980) 7.
17. M. E. Bridge, C. M. Comrie, R. M. Lambert, Surf. Sci. 67 (1977) 393.
18. M. Watanabe, J. Catal. 110 (1988) 37.

Fischer-Tropsch Synthesis, Catalysts and Catalysis
B.H. Davis and M.L. Occelli (Editors)

EELS-STEM INVESTIGATION OF THE FORMATION OF NANO-ZONES IN IRON CATALYSTS FOR FISCHER-TROPSCH SYNTHESIS

Uschi M. Graham [a], Alan Dozier[b], Rajesh A. Khatri[a], Ram Srinivasan[a] and Burtron H. Davis[a]

[a] Center for Applied Energy Research, University of Kentucky, 2540 Research Park Drive, Lexington, KY 40511
[b] Electron Microscopy Center, University of Kentucky, Lexington, KY 40506

1. Introduction

Fischer-Tropsch synthesis (FTS) technology, which produces hydrocarbons from synthesis gas (CO reacts with H_2 to form a hydrocarbon chain extension - [CH_2]-) over iron and cobalt catalysts is very effective.[1-15] Recent studies emphasize the importance and development of a robust, active catalyst suitable for the production of transportation fuels (high-alpha catalyst).[16] FTS is a commercially viable approach to convert synthesis gas into liquid fuels and can help decrease the U.S. dependence on oil imports. Furthermore upon combustion, FTS derived diesel fuel compared with conventional diesel fuel not only yields low nitrogen oxides (NO_x), carbon monoxide (CO) and hydrocarbon (HC) emissions, but also much lower nano-particulate matter emissions which are suspected of being a leading health threat.[17-20] The economic and ecological importance of "green" FTS-derived transportation fuels is based on generation of synthesis gas by biomass conversion (renewable energy source). This route has the advantage that almost all biomass materials are suitable for gasification to produce "syngas" for the FTS process.[19, 20]

There is a continuing requirement for development of FTS-catalysts with improved properties to answer today's challenges for alternative fuel supplies.[6, 21-25] Cobalt-based FTS catalysts are claimed to have the advantage of a higher conversion rate, higher selectivity to paraffins and longer life with considerably clean synthesis gas which leads to a higher production rate and smaller reactors.

[9, 26, 27]On the other hand, Fe-based catalysts have higher selectivity to olefins and desired C_{5+} hydrocarbons, lower CH_4 selectivity, and higher tolerance for sulfur compounds.[28-34] More over, Fe-based catalysts can be operated in wider range of temperature and H_2/CO ratios including low H_2/CO ratios such as those derived from coal and biomass gasification.[19, 20] When iron is used as a catalyst for FTS, several forms of iron oxide and iron carbides (Fe_xC_y) may typically be present during the reaction. These forms include magnetite (Fe_3O_4) and various forms of iron carbides including: ε-Fe_2C, ε'$Fe_{2.2}C$, χ- $Fe_{2.5}C$, Fe_3C and Fe_7C_3 (Table 1).[35, 36] The iron catalysts are typically prepared by precipitation of an iron nitrate precursor phase and can either be supported (e.g. silica binder for excellent mixing characteristics and non-porous silica microbeads for added attrition resistance) or unsupported. Both kinds may be further enhanced by trace additions of Cu, K or other promoters.[37-39]

The overall goal of this study is to develop superior catalysts suitable for use in modern advanced slurry phase or membrane reactors. Successful investigations have produced an iron-based unsupported catalyst with high activity and extended catalyst life (improved attrition resistance). This catalyst was used by the authors to understand the phase transformations from iron oxide precursors after activation to iron carbide crystallites and to characterize the presence of characteristic amorphous carbon species that have been reported to envelop the spent Fe-FTS catalyst grains in form of surface layers.[40-42] Previous studies suggest two different phases to be considered as the "active phase" including iron oxide and a mixture of χ-and ε'-carbides (Table 1) and in some instances minor amounts of metallic iron, however the spatial distribution of the different phases is still under debate.[43, 44] It has been proposed that the complex phase transformations that accompany the activation and reaction processes of iron catalysts during FTS are governed by an intricate particle zonation process which results due to the transformation of iron oxide precursor into stoichiometric and disordered iron carbide, magnetite and both crystalline and amorphous carbon phases.[8] Previous work also demonstrated that iron carbide particles are surface-coated by a distinctive amorphous carbon layer, but the interface between iron carbide and amorphous carbon was not sufficiently addressed because many of the described zones are only nanometers wide and are rather difficult to analyze or are frequently concealed.[41, 44, 45] Speculation about the composition of the interfacial area between iron carbide core and outer amorphous carbon zone has been fueled by the industries' need to answer questions about the location and state of active sites in the FTS catalyst material. Formerly it has been suggested that the amorphous carbon layer itself may act as a catalyst.[44, 46] The amorphous carbon layer may act as a screen or shield that gradually leads to a cover-up of active sites in the carbide layer.[44]

Also, it has been revealed in recent studies that the width of the amorphous carbon layer can vary significantly with precursor materials for catalyst preparation. The three most commonly utilized iron oxide catalyst types include (a) unsupported crystalline iron oxide, (b) precursor iron oxide that is well mixed with a silica powder and, (c) iron oxide precipitated on silica-bead surfaces.[47, 48] These types of iron oxide sample catalysts were previously studied for CO-hydrogenation at elevated temperature regimes (250 °C) with the goal of obtaining greater amounts of graphitic carbon species in the vicinity of the FTS catalyst grains. This was necessary in order to study and compare the reactivity of the graphitic forms with that of the amorphous carbon layers.[41, 44] The latter was found to have typically very high reactivity since it can be readily hydrogenated in flowing H_2 even at low temperature regimes.[44, 45, 49]

Fischer-Tropsch catalyst testing at the Center for Applied Energy Research at the University of Kentucky has demonstrated that the catalytic activity of an iron only catalyst typically declines in parallel with the fraction of iron carbide not converted to iron oxide.[50-52] This has led to the hypothesis that there may be an additional narrow interface layer between iron carbide and amorphous and/or graphitic carbon layers. If such a layer exists, its nature needs to be investigated to determine composition and ordering (crystalline versus amorphous) and the role it plays as a possible host for active sites. The controversy still remains over the existence and exact location of additional crystalline phases because of the immense challenges to analyze these catalyst materials during or after the FTS reaction process. A clear solution to this challenge would require in-situ reaction studies utilizing reaction chambers linked with high-resolution electron microscopy (EM) and spectroscopy applications. Although a small number of in-situ TEM facilities have been available for quite a while and now represent a growing field that focuses on phase changes under controlled conditions, the FTS samples of interest so far have not been applied in in-situ EM studies mainly because of the difficulties of imaging catalyst materials that are engulfed by FT-wax products. Instead, experiments that were designed to simulate the FTS reaction followed by controlled passivation and sampling provided a host of Fe- catalyst samples for a systematic identification of the morphology and crystallinity of the Fe-catalyst particles.[44] It also shed light onto the multifaceted phase transformations and zonation processes that proceeded during CO hydrogenation reactions in the presence of iron catalyst materials. Results were based on TEM, STEM and EELS investigations that distinguished three carbonacious species including reactive amorphous carbon, graphitic carbon, and carbide carbon based on the exact location and intensity of either the carbon $C\text{-}_{K\sigma}$ or $C\text{-}_{K\pi}$ edge and comparing the results within the energy loss spectra collected for pure phases.[44]

The objective of the current study is to examine the formation of crystalline iron carbides and oxides formed during slurry bed FTS reaction conditions and determine their spatial relationships with respect to the activity/ selectivity of the Fe-catalyst which needs to be investigated to be able to understand the role played by these phases in terms of catalyst performance. The complexity of the nano-zonation process may hold the key to further improve the Fe-FTS catalyst performance. At this time it is not known how the nano-zonation process matures during bona fide FTS. Therefore, our samples were derived from slurry phase reactors that will be analogous to normal commercial operations.

The current study demonstrates the presence of iron oxide-rich and carbide-rich nanozones and also the presence of multiple narrow nanometer thin carbon layers. Previously amorphous carbon and/or graphitic carbon layers have been shown to occur at the outer rim of the Fe-catalyst grains. Our study reveals that nanometer thin carbon zones also form at the interface between the iron carbide core and iron oxide layer. The outer oxide layer which is coated by an amorphous carbon zone is attributed to sample preparation for the EM study. Because of accessibility factors of reactive gas molecules, oxide and carbide "nano" zones and multiple nanozones of amorphous and crystalline carbon may be of great significance to the activity level of the FTS catalyst. Further studies will be aimed at understanding the mechanisms that control the build up and/or migration of these nano-zones during reaction conditions.

2. Experimental

The catalyst material is a precipitated un-promoted crystalline iron-oxide catalyst. Iron nitrate and concentrated ammonia solutions were added to a continuous stirred tank reactor maintained at pH of ~ 9.0 to effect precipitation. Samples for this investigation were withdrawn from the stirred tank reactor after activation in CO (catalyst was pretreated in CO at 270 ° C, 175 psig for 24 hours). The longest reaction time before a sample of the reactor slurry was withdrawn was 168 hours of synthesis and the catalyst was embedded in wax. The sample withdrawn at 168 hours of FTS was used for the EM study for this manuscript. Solvent extraction, using ortho-xylene, was carried out under an inert gas atmosphere to prepare the spent samples for EELS-STEM applications. In order to avoid aging/oxidation as much as possible, the catalyst samples were stored in inert gas atmosphere and prepared for the analytical characterization study using a glove box. Transfer of lacy-carbon-grid mounted samples into the electron microscope chamber was the only occasion the materials had to be exposed to air. In the current work we opted to use EELS-STEM and EELS quantitative spectrum imaging and profiling to investigate the formation and composition of what we refer to as nanozones in Fe-catalysts used for FTS. The procedures are discussed in detail in the next section.

3. Analytical Characterization Methods

The electron microscope used in this work was a JEOL 2010F equipped with a field emission gun, Oxford EDS detector with an Emispec EsVision computer control system, a Gatan CCD multiscan camera, Gatan imaging filter (GIF) model 2000 also known as an Electron Energy Loss Spectrometer (EELS), Gatan Digiscan II, Gatan Digital Micrograph with a full suite of EELS quantification software for Energy Filtered TEM (EFTEM) and spectrum imaging spectral analysis and mapping, and finally a Fischione High Angle Annular Dark Field (HAADF) detector for STEM imaging. The JEOL 2010F is a high resolution instrument with the ability to obtain a STEM probe of better than 2 Angstrom.[53, 54] The FEG provides high brightness so that EELS spectrum imaging and profiling can be performed using the high resolution probe. However this probe does not have sufficient brightness to allow Energy Dispersive Spectroscopy (EDS) spectrum profiling and imaging.

The HAADF dark field detector operated with a 60 [mrad] inner radius detects electrons that are for the most part scattered as a result of thermal and direct atomic scattering events with the elastically scattered diffraction electrons passing through the inner hole in the detector. The intensities in the dark field image produced can be interpreted directly in terms of the atomic species present and relative concentrations of differing atomic species. Intensity changes can also be interpreted in terms of sample thickness.[55-57]

EELS spot analysis can easily be performed in STEM imaging mode using Digital Micrograph and the Digiscan II. The advantage of this method is that it allows the rapid analysis of selected points on a sample with short acquisition times. This results in very low noise spectra allowing for in depth study of Energy-Loss Near-Edge Spectra (ELNES) equivalent to Near-edge X-ray Absorption Spectroscopy (NEXAFS) and other subtle effects in the EELS spectra that may occur during analysis.

EELS spectrum profiling can be performed with the combination of the Gatan Digiscan II and Digital Micrograph.[58] Here the electron probe moved linearly across a defined region with a set number of points. At each point an EELS spectrum is gathered with a user selected dwell time. Automatic sample drift correction is usually employed when gathering the data. After data acquisition Digital Micrograph has a power set of EELS analysis software that can analyze all the spectra simultaneously and produces quantitative graphs of the results.

Besides the standard absolute and relative elemental analyses capability, the user can employ the software and its scripting ability to develop an unlimited number of custom EELS analyses for other phenomena observed in EELS spectra.

EELS spectrum imaging expands the EELS spectrum profiling from one to two dimensions. The same powerful software analytic tools may be used on this data as with the spectrum profiling data. EELS spectrum imaging in two dimensions is more difficult due to the long acquisition times required, though they are considerably shorter than those required in EDS spectrum imaging. For example a 256 by 256 pixel image with a spectrum acquired at each pixel have a dwell time of 0.1 seconds would take 24 minutes not including the time taken for each drift correction. This amount of time requires the sample to be very stable in the microscope with minimal drift. Also because of the shorter dwell times used the spectra have much more noise making quantification less reliable. However spectrum imaging in two dimensions can provide invaluable information about the distribution of elements in a sample and their relative percentages allowing the mapping of compounds in materials. This is especially valuable for materials that have compositional zoning with nanoscale dimensions. The unique aspect of employing these methods to nanoscale materials versus other techniques such as NEXAFS is that the analysis can be performed with a 2 Angstrom resolution and is ideally suited for the nano-powdered Fe catalyst samples of interest to this study. This means that changes in composition and electronic structure can be studied on the atomic scale including catalyst particles and their active sites. The application of all these methods in combination along with other bulk methods such as XRD creates a very powerful set of tools to analyze structure and phase transitions in catalyst particles over time.

Given all of the analytical techniques available mentioned above, we followed a sequence of analyses so that the information gathered from one measurement provides direction leading to the next method applied. We typically start with low magnification followed by high resolution TEM imaging to quickly assess basic sample structure including morphology, size, zoning pattern, crystallinity, etc. This is followed by high resolution HAADF STEM imaging to further evaluate compositional changes and to act as a guide for EELS spot analysis. The spot analysis is followed by spectrum imaging to help determine the elemental distribution in the sample material to create a two dimensional map of each element of interest. From this information detailed spectrum profiles are acquired for the catalyst grains to help assess atomic and electronic structural information through quantitative analyses of the EELS spectra. The spectral changes observed in the profiles can be used as a guide for the acquisition of two dimensional spectrum images that can be used to generate two dimensional bonding maps.

4. Results and Discussion

In the current HR-TEM-study of the Fe-FTS catalyst sample, the majority of the lattice images were identified as either iron oxide (Fe_3O_4 and Fe_2O_3), or iron carbide (predominantly ε-Fe_2C and ε-Fe_5C_2). The lattice parameters for various iron carbide and oxides are shown in Table 1 and the relative intensities for the carbide reference compounds are shown in Figure 1. We observed a large

Table 1. Iron oxide and carbide phases

Name	Structure	PDF Card #	Lattice Parameters (Å)		
			a	b	c
Fe_3O_4	Cubic	19-629	8.396		
χ-$Fe_{2.5}C$	Monoclinic	36-1248	11.563	4.573	5.058
ϵ-Fe_2C	Orthorhombic	26-782	4.704	4.295	2.830
ϵ'-$Fe_{2.2}C$	Monoclinic	17-897	2.794	2.794	4.360
Fe_3C	Orthorhombic	35-772	5.0910	6.7434	4.526
FeC	Orthorhombic	3-411	4.30	2.50	6.70
Fe_7C_3	HCP	17-333	6.882		4.54

number of catalyst samples that represent different exposure to increasing length of synthesis time and the HR-TEM results in Figure 2 show the typical structural characteristics observed in all samples. Based on lattice d-spacings measured for crystallites occurring in the Fe-FTS samples at various locations including the core and inner or outer crystalline rim the corresponding iron phases are illustrated in Figure 2. The results suggest that the core of the Fe-FTS particles is composed of Fe_xC_y which is a mixture of preferentially monoclinic χ–$Fe_{2.5}C$ but other carbide phases including ε-Fe_2C and the hexagonally close-packed Fe_7C_3 were also observed in this order of abundance. There are also a lesser number of magnetite (Fe_3O_4) crystallites present in the vicinity of the core region. However, the core was clearly surrounded by a crystalline zone that was predominantly Fe_3O_4 and Figure 2 shows the location of magnetite (Fe_3O_4) occurring at the inner rim versus hexagonal hematite crystallites (Fe_2O_3) at the outer zone which is most likely an effect of sample exposure to air. The level of crystallinity of the outer zone did not appear to

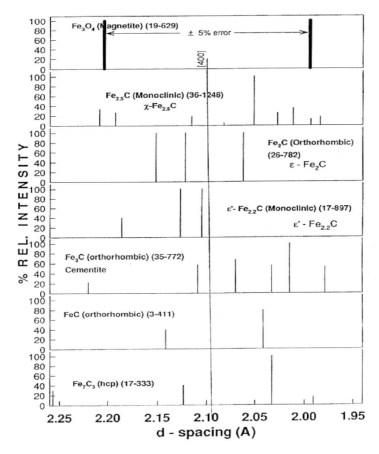

Figure 1. The relative peak intensities for iron carbides and iron oxides with corresponding d-lattice parameters and crystallinity information

vary based on extended exposure to the electron beam which suggests that the Fe_2O_3 crystallites were already present and fully developed prior to the TEM investigation. Generalization of a direct relationship between the bulk phases of an Fe-FTS catalyst (iron carbides and iron oxides) and its activity for FTS does not seem possible; however, the d-spacing data suggest that the current study was able to isolate the location of the oxide vs. carbide phases and was also able to illustrate the development of a process-induced nano-zonation within Fe-FTS catalyst grains which appear to be caused by phase transformations during slurry-bed reactor runs with the exception of the narrow Fe_2O_3 layer. The HR-TEM results indicate that the different zones host crystals that are on the order of 1-5 nm in diameter. The crystals are densely packed inside a particular zone;

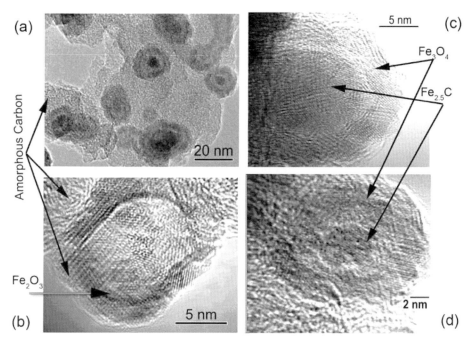

Figure 2. Illustration of the HR-TEM images for Fe-FTS catalyst grains. (a) shows a lower magnification TEM with several zoned catalyst grains embedded in an amorphous carbon matrix. (b), (c) and (d) show HR-TEM image of one of the grains shown in (a). The d-spacinngs were measured for the individual crystallites located in diffenrent zones and identified as indicated by the arrows

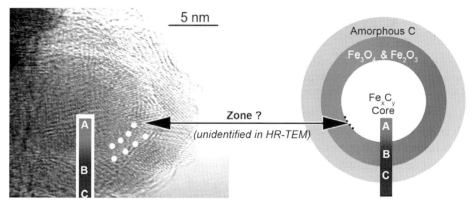

Figure 3. Illustration of a zoned Fe-FTS catalyst grain on the left with the corresponding nanozone model on the right side. Based on the d-spacings for crystals located in the core (A) and exterior zones (B and C) the model shows that the core is occupied by iron carbides Fe_xC_y (predominantly $Fe2.5C$) and is rimed by iron oxide layer (B) which is surrounded at the grain exterior by an amorphous carbon layer (C). The HR-TEM study suggests the presence of a narrow intermediate zone between A and B.

however, at the exterior rim of the catalyst grains the density of the zones diminishes and finally there is an amorphous carbon zone that occurs in close proximity to the oxide zone, but the amorphous carbon can also extend further away from the Fe-FTS grains like a lacy carbon matrix as can be seen in Figure 2a. The nanozones that are recognized by HR-TEM are schematically illustrated in Figure 3 and the zoned catalyst grains in Figure 2 were used to create the simplified nanozone model. This model has a carbide core followed by two outer layers which are iron oxide and amorphous carbon. The HR-TEM study did reveal an additional approximately 1.5 nm wide zone not shown in the model. The presence of this still unidentified Zone is suggested by the narrow band in the HR-TEM images that occurs between the crystalline carbide core and crystalline iron oxide Zone, but his narrow band appears to have a significantly reduced level of crystallinity and was not able to be identified by looking for characteristic lattice spacing information.

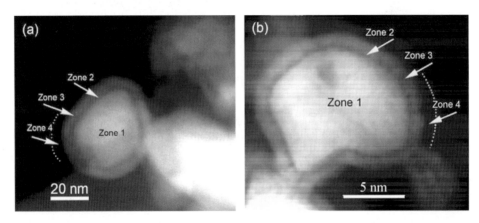

Figure 4. STEM images for two Fe-FTS catalyst grains. The nanozone areas (Zones 1-4) are marked by the arrows and the dotted lines show the location of the amorphous carbon layer which is not clearly visible in STEM mode.

In order to explore all nanozones in more detail this study also decided to take advantage of the High Angle Angular Dark Field (HAADF) STEM imaging technique where image contrast can be directly interpreted in terms of elemental and chemical composition variations due to the elimination of diffraction effects in image formation. Thus, the technique will allow the direct imaging of nanozones of differing phases and elemental compositions, which are recognized by variations in lightness and contrast. This type of imaging will not allow the interpretation of phase ordering (crystallinity vs. amorphous phase). However, Near Edge Energy Loss Spectroscopy (NEELS) can provide this information. As can be seen in Figures 4a and 4b the nanozone formation can be clearly observed with a very bright core, surrounded by a thin dark layer which is roughly 2 nm wide and is surrounded by a medium grey layer which is approximately four times as wide as the dark layer. Each layer represents a compositional change in the catalyst particle.

The composition of these nanozones can be easily determined in STEM mode by placing the electron probe exactly on a particular nanozone for EELS analysis. The advantage of using EELS in spot analysis is that the high resolution STEM probe (2 angstrom diameter) may be used. This provides a very precise spacial analysis and also allows us to analyze extremely narrow nanozones as in this case for the thin dark layer. An example of this work is provided in Figure 5 where the STEM-derived nanozone model is illustrated

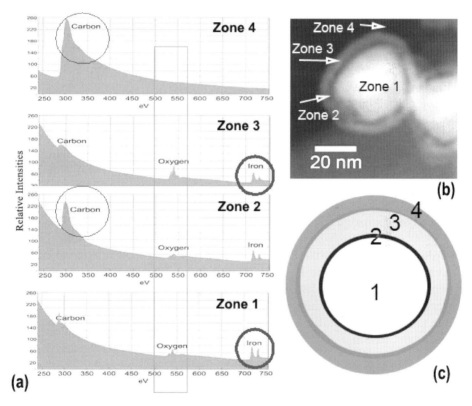

Figure 5. Illustration of EELS spot analyses for nanozones in Fe-FTS catalyst grain shown in the STEM image insert on the right.

based on the STEM image shown in Figure 5b. The outermost amorphous carbon zone (exterior rim) that was recognized earlier in the TEM study has poor contrast in STEM mode and is too dark to show up in the STEM image; however, the spot analysis using the high resolution probe can be placed on the exterior rim (Zone 4) and the composition can be identified as is shown in Figure 5a. This figure illustrates in sequence the spot analyses from the inner core (Zone 1) to the outer nanozones (Zones 2, 3 and 4). The EELS spectrum number 1 for Zone 1 (inner core) shows the presence of carbon, minor oxygen and iron with the $Fe-KL_{2,3}$ appearing with approximately the same strength (similar peak height). The EELS spectrum from Zone 2 which corresponds to the 2 nm thin dark layer shown in the STEM image of Figure 5b indicates a very high concentration of carbon and also the presence of small amounts of oxygen and iron as indicated in Figure 5a. The spectrum from Zone 3 indicates the presence of iron oxide and also has a minor carbon component. Note that the $Fe-K_{L2,3}$ ratio in the spectrum from Zone 3 has changed from that recorded in Zone 1. EELS spectra for Fe taken at the core (Zone 1) have peak intensities for $Fe-K_{L2}$ and $Fe-K_{L3}$ with similar height suggesting that iron is present predominantly as Fe^{2+}. The $Fe-K_{L3}$ peak occurs at an energy loss of approximately 710 eV and is the predominant peak versus the $Fe-K_{L2}$ peak which occurs at approximately at an energy loss of 722 eV. The relative strength of these peaks varies according to the density of unoccupied states of the iron phase. This can be used to determine the relative concentration of Fe^{3+} and Fe^{2+} present within each nanozone. This is very useful in determining possible phase transformations that occurred during the reaction or are a result of oxidation at the exterior rim of the catalyst grains. In Zone 3 in Figure 5a the $Fe-K_{L2}$ peak is observed to be significantly smaller compared with the $Fe-K_{L3}$, indicating that the iron phase is more oxidized compared with Zone 1 of the inner core (iron carbide core). Finally, the Zone 4 EELS spectrum reveals a pure carbon nanozone (Zone 4 was identified in TEM as amorphous carbon).

In order to study the morphology of the nanozones and their relation to catalytic activity, spectrum imaging has been incorporated. Spectrum imaging is used to determine the elemental distribution in the catalyst grains. The catalyst is first observed via STEM as shown in Figure 6a and the STEM image can act as a guide to create a 2D profile or map of the elements of interest (in this study we map carbon, iron and oxygen).

In spectrum imaging, an EELS spectrum is gathered at each pixel in a two dimensional array versus the one dimensional spectrum profile. The Gatan Digiscan system allows the quantification of spectrum images and also spectrum profiles such as line scans. This data can be used to generate images relating to composition and other properties of materials. Examples of this technology are illustrated in Figure 6. This figure shows the quantitative elemental spectrum images of a Fe-FTS catalyst particle where the intensity in each pixel in the image is directly related to the concentration of the element

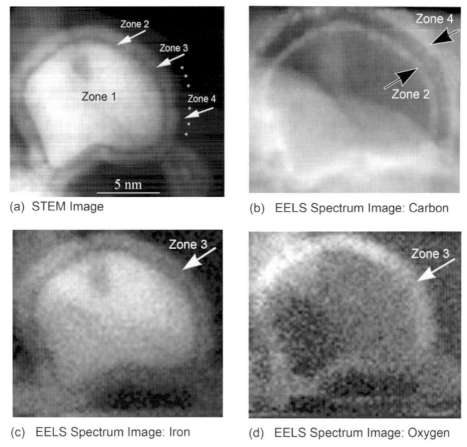

(a) STEM Image

(b) EELS Spectrum Image: Carbon

(c) EELS Spectrum Image: Iron

(d) EELS Spectrum Image: Oxygen

Figure 6. HAADF STEM image of a zoned catalyst particle. 6b. Carbon image, 6c. Iron image, 6d. Oxygen image, all areal density images generated from EELS spectrum image.

under investigation. Figure 6a is the actual STEM image showing the zoned catalyst grain. The spectrum imaging is able to identify and reveal the morphology and dimension of the nanozones. In Figure 6b carbon is imaged and this image clearly shows the two narrow nanozones of carbon. By noting the position of the dark 2 nm thin layer in the STEM image of Figure 6a and the corresponding inner carbon zone in Figure 6b it can be confirmed that there is an "interfacial carbon layer" that occurs between the iron carbide rich core (Zone 1) and the iron oxide rich Zone 3. The dark thin layer in the STEM image taken alone is not indicative of the presence of any material in this 2 nm wide dark region and only with the combination of the carbon spectrum image the true nature of Zone 2 could be identified. Zone 2 corresponds to the "unidentified zone" shown in the TEM-based nanozone model in Figure 3. In Figure 6b the carbon image taken in the carbide core (Zone 1) has less intensity than the two narrow carbon bands that surround the core (Zone 2 and Zone 4). This is due to the relative concentration of carbon in Zones 2 and 4 being very high. Figure 6c shows the relative iron concentration and clearly indicates how the high iron containing core is surrounded by a dark 2 nm thin layer which corresponds to the carbon of Zone 2. Outside of Zone 2 is a secondary iron layer corresponding to Zone 3 in the STEM image of Figure 6a. Finally, the spectrum image for oxygen in Figure 6d reveals only one single zone as indicated by the bright band in the spectrum image. This oxygen-rich band occurs in the same location as the iron-rich band (Zone 3) shown in Figure 6c. It is possible to create composite color images showing a variety of elemental distributions and other material properties in these grains simultaneously.

The spectrum profiles are acquired to help obtain atomic and electronic structural information through quantitative analyses of the EELS spectra across interfaces or, in our case, across nanozones. Profiles are used for this purpose because of the superior signal to noise ratio compared to spectrum imaging. The data are compiled in Figure 7. The catalyst particle of interest is shown in

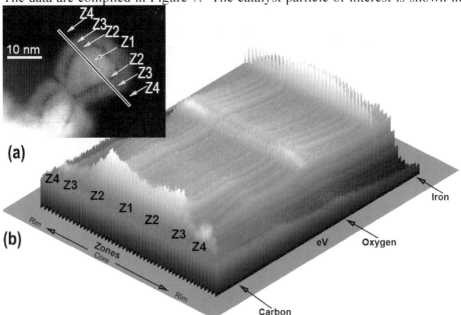

Figure 7. Surface Plot of an EELS spectrum profile across a catalytic particle shown in inset.

the STEM image in Figure 7a and has characteristic nanozones. The profile is indicated by a white linear trace that spans across the catalyst grain. The spectrum line profile follows this linear trace and was defined on the STEM image using Gatan Digital Micrograph along with other acquisition parameters such as dwell time and number of analysis points along the line. The Digiscan II will automatically acquire the data and also perform automatic drift correction. Figure 7b is a 3D surface plot of the spectrum profile for all the EELS spectra gathered in the linear profile. This image represents the background (pre-carbon) subtracted EELS data, and can serve as a guide to find the location of abrupt phase changes. The location of the zones and elemental peaks are marked in Figure 7b. The oxygen shows increased strength in Zone 3 indicated in Figure 7b as Z3. The carbon shows its greatest strength in the Zone 2 region as shown in Figure 7b as Z2 and the iron shows a large hump in the core region (Zone 1). In order to interpret this data in depth, background subtraction must be performed for each element on all the EELS spectra, appropriate quantification methods employed, and the spectra must be associated with the structure of the zones present in the STEM image. Ultimately, this information can be compared to theoretical calculations from differing models in order to understand each model's strengths and weaknesses.

Figure 8 shows three of the EELS spectra that were collected from the inner iron carbide core or Zone 1 of the catalyst particle. These three spectra were randomly chosen somewhere within the core region. Features to be noted in these spectra include the prominent carbon $C-K\pi$ peak (286 eV) versus the $C-K\sigma$ peak (296 eV). The strong $C-K\pi$ peak indicates an iron carbide phase in the core region. Also noted is the small amount of oxygen in the spectrum.

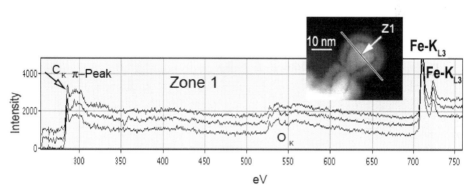

Figure 8. Three random spectra taken from the core region.

All of these features will change from nanozone to nanozone which can be viewed in the following set of Figures 9-11. For example, the strength of the $Fe-K_{L3}$ peak versus the $Fe-K_{L2}$ peak in the Fe-rich Zone 3 compared with that of the core region (Zone 1) spectra shown in Figures 8 and 10 illustrates how reduced iron is concentrated in the core while more oxidized iron occurs in the outer Zone 3. The small amount of oxygen detected in the core region is due to the electron probe passing through the oxide layer above and below the carbide core, but could also signify minor phase transformations within the core. It can be said that the core region of the Fe-FTS catalyst grains is oxygen deficient even after having been exposed to air for sample preparation etc. The prominent Fe-K_{L2} peak in EELS spectra collected from the core region in relation to the much smaller Fe-K_{L2} in EELS spectra collected for Zone 3 (Zone 3 defines an outer layer in the nanozone model) indicating a change in the valance state of the iron.

Figure 9 shows the EELS spectra that were collected from the narrow 1.5-2 nm wide dark layer that was identified earlier during spectrum imaging as a carbon-rich band labeled as Zone 2. Since Zone 2 intersects the spectrum profile at two locations, we separated three EELS spectra for the upper (Figure 9a) and three EELS spectra for the lower (Figure 9b) cross over area.

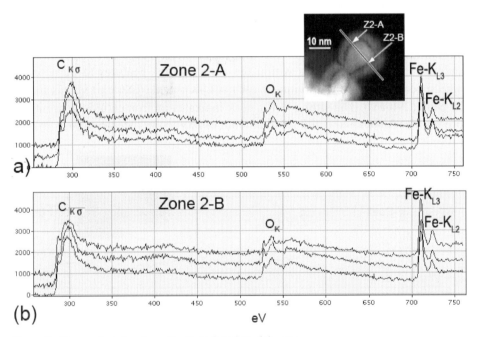

Figure 9. Six spectra taken from the Zone 2 carbon rich zone.

Our step size of .5 nm produced only three spectra from each of the two intersects. The Zone 2 width is then confirmed to be 1.5 to 2 nm wide. The results indicate that on both sides of the core the narrow dark Zone 2 has produced EELS spectra with a strong carbon edge with a greatly reduced carbon

C-Kπ (286 eV) peak strength, a slightly increased oxygen edge, and a slightly reduced strength in the iron edge. This indicates the carbon has changed its chemical state between Zone 1 (core region) and Zone 2 (dark narrow band outside the core region). The Fe-K_{L2} peak has also lost strength in the EELS spectra taken from Zone 2 indicating the iron has a different valence state in Zone 2 compared with Zone 1, implying some oxidation.

Figure 10 shows the EELS spectra that were collected from the iron oxide-rich band that surrounds the inner carbon band known as Zone 3. Since Zone 3 intersects the spectrum profile at two locations, we separated three EELS spectra for the upper (Figure 10a) and three EELS spectra for the lower (Figure 10b) cross over area in the same fashion as was done earlier for the carbon-rich band. The results indicate that Zone 3 has produced EELS spectra with a strong oxygen peak and also strong Fe-$K_{L2,3}$ peaks, while the carbon peak has significantly reduced strength compared with Zone 2. The results are very similar for EELS spectra collected on both sides of the particle center. Looking at the Fe-K_{L2} peak it can be seen that it has lost considerable strength compared to the Zone 1 collected Fe-K_{L2} spectra. In Zone 3 the principle composition is iron and oxygen.

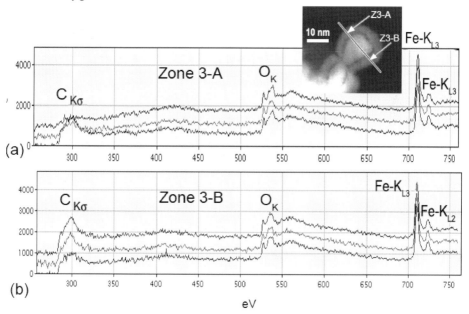

Figure 10. Six spectra taken from the Zone 3 iron oxide zone.

Figure 11 shows the EELS spectra that were collected from the exterior rim of the catalyst grain, also known as Zone 4 in this study. Again, this is the nanozone that is difficult to image in STEM mode which is based on the low density in the exterior region and, therefore, Zone 4 appears at the outside of the catalyst grain shown in the STEM insert in Figure 11. This Zone 4 was earlier interpreted with HR-TEM as an amorphous carbon layer based on characteristic soot-like structures and the lack of crystalline features. Also, the EELS spot analysis results shown in Figure 5 already confirmed the presence of amorphous carbon in Zone 4. The results obtained by the spectrum line profile in Figure 11 further confirm that the exterior rim is enriched in carbon that is neither carbide carbon nor graphitic carbon (the lack of a carbon C-Kπ (286 eV) peak indicates the carbon is in an amorphous state). The carbon was nearly 100 % in the three spectra shown in the intersection in Figure 11. The intersection of the EELS spectrum profile shown in Figure 7 with Zone 4 on one side of the grain showed only a very weak carbon peak. This indicates that the outer zone 4 carbon rim is either not uniform or was destroyed during sample preparation. Also, where the profile did intersect the amorphous carbon, it shows a very thin layer of only 1.5 to 2 nm. It is important to point out that large amounts of amorphous carbon were found further away from the catalyst grain itself.

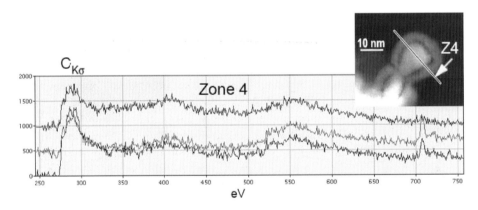

Figure 11. Three spectra taken from the outer Zone 4 carbon zone.

All of the EELS spectra were quantified with Gatan's Digital Micrograph using standard routines. The carbon and iron area density ratio were computed from the area densities of carbon and iron. This ratio will give peaks in carbon rich or iron deficient regions. Figure 12 shows the C over Fe graph and provides information on the location of the inner and outer carbon bands (Zone 2 and Zone 4). The inner carbon Zone 2 appears to extend into the carbide core Zone 1 inwards and into Zone 3 outwards, but this could be due to the curvature of the carbon zone (Zone 2) around the carbide core and the geometry in which the electron probe passes through it. Also note that in the core region (Zone 1) the

C/Fe ratio is close to 0.5. This could correspond to several of the carbide phases illustrated in Figure 1 and also in Table 1, but it is remarkable that the C/Fe ratio can be measured at all on the nanometer scale and changes monitored for nanozoned catalyst grains like the one shown in Figure 12.

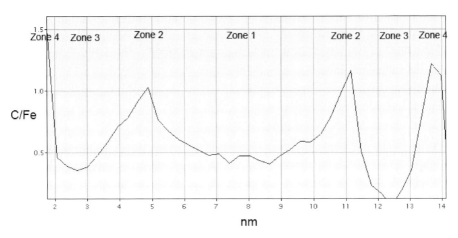

Figure 12. Carbon over Iron profile.

Figure 13 shows the O over Fe graph computed from area densities and provides information on the location of the iron oxide rich area (Zone 3). The

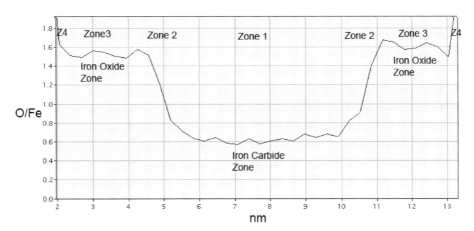

Figure 13. Oxygen over Iron profile.

gradual transition from the core to oxide could also be a geometrical curvature effect, especially since the HR-TEM and EELS image spectra show a more abrupt change. Also note that in the oxide zone the ratio of oxygen to iron

120

shown in Figure 13 is approximately 1.5. This is what would be expected for Fe_2O_3. The non-zero ratio in the core region is due to the beam passing through the oxide layer above and below the core.

The Gatan Digital Micrograph software system allows for the definition of regions of interest that can then be manipulated and useful information can be extracted. In this case two regions of interest were defined with widths of 4 eV centered on the $C\text{-}K_\pi$ peak and the $C\text{-}K_\sigma$ peak. These regions were integrated and the ratio of the strength of the $C\text{-}K_\sigma$ peak over the $C\text{-}K_\pi$ peak was computed from the spectrum profile. The resulting peaks will then correspond to either non-carbide phase carbon or amorphous carbon. Figure 14 graphs the $C\text{-}K_\sigma$ over the $C\text{-}K_\pi$ peak strengths. The carbide region which has a very large $C\text{-}K_\pi$ peak shows a strong dip. The Zone 2 carbon inner layer shows a strong prominence corresponding to carbon in a non-carbide form. Finally the outer Zone 4 carbon layer shows a peak corresponding to the amorphous carbon state.

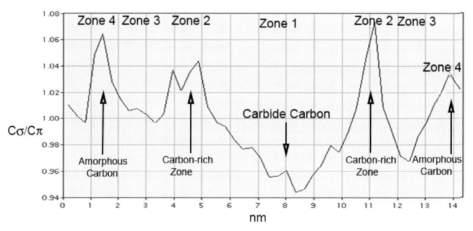

Figure 14. Ratio of Carbon sigma peak over Carbon π peak profile

The region of interest described above can also be applied to the $Fe\text{-}K_{L2,3}$ peaks and ratios computed then will show the variation in the valance state of iron along the profile. Figure 15a shows the Fe III over Fe II signal and provides information on the location of the reduced iron versus oxidized iron in the different nanozones. Because the $Fe\text{-}K_{L2,3}$ peaks give a relatively strong signal in these samples this method could also be applied to the spectrum image obtained for iron. Thus an image of iron reduction in two dimensions can be generated as shown in Figure 15b. The Zone 3 oxide ring is highlighted due to the diminished strength of the $Fe\text{-}K_{L2}$ peak. Figure 15b shows the results for the Image of FeIII/FeII.

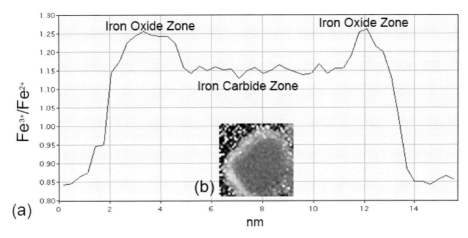

Figure 15a. Ratio of Fe-K_{L3} peak over Fe-K_{L2} peak profile Fe-K_{L3}/Fe-K_{L2}, 15b Two dimensional image of iron reduction or valence state.

5. Summary and Conclusions

Iron Fischer-Tropsch (Fe-FTS) catalyst particles were studied using high resolution transmission electron microscopy (HR-TEM) and electron energy loss spectroscopy (EELS) as well as energy filtered TEM (EFTEM) with spectrum imaging spectral analysis and mapping capabilities. These studies have independently demonstrated the formation of various nanozones within iron catalyst particles. By this study's classification schema, nanozones comprised of nm-wide layers that host a dense array of either iron carbide or iron oxide crystallites and multiple carbon-rich nano-layers that occur both on the exterior of the Fe-FTS catalyst grains and also further inwards. The precursor catalyst material that underwent phase transformations that led to the formation of the reported nanozones during the FT-experiments was a precipitated unsupported iron oxide and the materials under investigation were derived from slurry-bed reactor runs that were sampled after exposure to increasing length of synthesis time. The presence of a carbon-rich layer closer to the core was a surprise, while the amorphous carbon zone that occurs at the rim of the catalyst grains had already been known. What was also discovered in this study is the fact that the carbon-rich layer located closer to the core in fact has carbon EELS signatures that are different from those of the carbide carbon that is concentrated in the core region and also those of the exterior amorphous carbon layer. This suggests that the inner carbon nanozone either was derived from the carbide core after some phase transformations of the iron carbide crystallites that caused a spatial repositioning of extra carbon outside of the iron

122

carbide nanozone, or that the carbon was independently deposited onto the inner core during the reactor runs. Furthermore, the inner carbon zone is surrounded by an iron oxide-rich nanozone that has both Fe_3O_4 and Fe_2O_3 crystallites present which was determined in the HR-TEM study and also confirmed by EELS measurements. The oxidation of the iron phases in this zone was most likely caused by the exposure of catalyst particles to air during sample preparation for the EM study. This could also result in the deposition of carbon if the only reaction of oxygen was to produce Fe_3O_4 and Fe_2O_3. It is therefore suggested to perform independent future experiments using reactors that allow *in-situ* measurements using STEM-EELS and EFTEM for determining the nanozones that are solely formed during the FT-reaction and to exclude any phase transformation caused during sample preparation. This will allow us to identify if active sites are associated with the inner carbon nanozone and what role the outer carbon nanozone plays for deactivation.

Acknowledgement

This work was supported by US DOE contract number DE-AC22-94PC94055 and the Commonwealth of Kentucky.

References

1. Anderson, R. B., *The Fischer-Tropsch Synthesis*. 1984; p 301 pp.
2. Shultz, J. L.; Hofer, L. J. E.; Cohn, E. M.; Stein, K. C.; Anderson, R. B., *Bulletin - United States, Bureau of Mines* **1959**, No. 578, 139 pp.
3. Schultz, J. F.; Abelson, M.; Shaw, L.; Anderson, R. B., *Journal of Industrial and Engineering Chemistry (Washington, D. C.)* **1957**, 49, 2055-60.
4. Dalai, A. K.; Davis, B., *Abstracts of Papers, 229th ACS National Meeting, San Diego, CA, United States, March 13-17*, **2005**, PETR-120.
5. Davis, B. H., *Topics in Catalysis* **2005**, 32(3-4), 143-168.
6. Davis, B. H.; Dalai, A. K., *Abstracts of Papers, 227th ACS National Meeting, Anaheim, CA, United States, March 28-April 1*, **2004**, PETR-083.
7. Davis, B. H.; Miller, S. J. Slurry-phase Fischer-Tropsch reaction over potassium-promoted iron catalysts for production of olefin-rich diesel fuel and lubricating base oil feedstocks. US Patent 6602922, 2003.
8. Dry, M. E.; Hoogendoorn, J. C., *Catalysis Reviews - Science and Engineering* **1981**, 23(2), 265-78.
9. Li, J.; Jacobs, G.; Das, T.; Zhang, Y.; Davis, B., *Applied Catalysis, A: General* **2002**, 236(1-2), 67-76.
10. Shi, B.; Davis, B. H., *Preprints - American Chemical Society, Division of Petroleum Chemistry* **1999**, 44(1), 106-110.
11. Storch, H. H.; Golumbic, N.; Anderson, R. B., *The Fischer-Tropsch and Related Syntheses*. 1951; p 603 pp.
12. Schulz, H.; Nie, Z.; Ousmanov, F., *Catalysis Today* **2002**, 71(3-4), 351-360.
13. Schulz, H., *Topics in Catalysis* **2003**, 26(1-4), 73-85.

14. Schulz, H., *Preprints of Symposia - American Chemical Society, Division of Fuel Chemistry* **2003,** 48(2), 782.
15. Schulz, H., *Preprints of Symposia - American Chemical Society, Division of Fuel Chemistry* **2003,** 48(2), 557-558.
16. Jacobs, G.; Chaney, J. A.; Patterson, P. M.; Das, T. K.; Maillot, J. C.; Davis, B. H., *Journal of Synchrotron Radiation* **2004,** 11(5), 414-422.
17. Schaberg, P. W.; Zarling, D. D.; Waytulonis, R. W.; Kittelson, D. B., *Society of Automotive Engineers, [Special Publication] SP* **2002,** SP-1724, (Diesel Fuel Performance and Additives), 55-67.
18. Tobias, H. J.; Beving, D. E.; Ziemann, P. J.; Sakurai, H.; Zuk, M.; McMurry, P. H.; Zarling, D.; Waytulonis, R.; Kittelson, D. B., *Environmental Science and Technology* **2001,** 35(11), 2233-2243.
19. Calis, H. P. A.; Haan, J. P.; Boerrigter, H.; Van der Drift, A.; Peppink, G.; Van den Broek, R.; Faaij, A. P. C.; Venderbosch, R. H., *Pyrolysis and Gasification of Biomass and Waste, Proceedings of an Expert Meeting, Strasbourg, France, Sept. 30-Oct. 1, 2002* **2003,** 403-417.
20. Boerrigter, H.; Uil, H. d.; Calis, H.-P., *Pyrolysis and Gasification of Biomass and Waste, Proceedings of an Expert Meeting, Strasbourg, France, Sept. 30-Oct. 1, 2002* **2003,** 371-383.
21. Davis, B. H.; Luo, M.; Devilliers, D., *Abstracts of Papers, 227th ACS National Meeting, Anaheim, CA, United States, March 28-April 1, 2004* **2004,** PETR-080.
22. Jacobs, G.; Chaney, J. A.; Patterson, P. M.; Das, T. K.; Davis, B. H., *Applied Catalysis, A: General* **2004,** 264(2), 203-212.
23. Jacobs, G.; Patterson, P. M.; Chaney, J. A.; Conner, W.; Das, T. K.; Luo, M.; Davis, B. H., *Preprints - American Chemical Society, Division of Petroleum Chemistry* **2004,** 49(2), 186-191.
24. O'Brien, R. J.; Davis, B. H., *Catalysis Letters* **2004,** 94(1-2), 1-6.
25. Shi, B.; Keogh, R. A.; Davis, B. H., *Journal of Molecular Catalysis A: Chemical* **2005,** 234(1-2), 85-97.
26. Bartholomew, C. H., *American Institute of Chemical Engineers, [Spring National Meeting], New Orleans, LA, United States, Mar. 30-Apr. 3,***2003,** 2884-2897.
27. Jacobs, G.; Chaudhari, K.; Sparks, D.; Zhang, Y.; Shi, B.; Spicer, R.; Das, T. K.; Li, J.; Davis, B. H., *Fuel* **2003,** 82(10), 1251-1260.
28. Wu, B.; Bai, L.; Xiang, H.; Li, Y.-W.; Zhang, Z.; Zhong, B., *Fuel* **2003,** 83(2), 205-212.
29. O'Reilly, K. T.; Michael, E. M.; O'Rear, D. J. Inhibition of biological degradation in Fischer-Tropsch products. US Patent 6,924,404, August 2, 2005.
30. Motal, R. J.; O'Rear, D. J. Protection of Fischer-Tropsch catalysts from traces of sulfur. US Patent 6,682,711, January 27, 2004.
31. Liu, Z.-T.; Li, Y.-W.; Zhou, J.-L.; Zhang, B.-J.; Feng, D.-M., *Fuel Science & Technology International* **1995,** 13(5), 559-67.
32. Luo, M.; O'Brien, R. J.; Bao, S.; Davis, B. H., *Applied Catalysis, A: General* **2003,** 239(1-2), 111-120.
33. Wang, Y.; Tonkovich, A. L.; Mazanec, T.; Daly, F. P.; Vanderwiel, D.; Hu, J.; Cao, C.; Kibby, C.; Li, X. S.; Briscoe, M. D.; Gano, N.; Chin, Y.-H. Fischer-Tropsch synthesis for the manufacture of C>=5 aliphatic hydrocarbons using microchannel technology and a novel catalyst and microchannel reactor. US Patent Application 2004-766297, 2005.
34. Bukur, D. B.; Lang, X., *Industrial & Engineering Chemistry Research* **1999,** 38(9), 3270-3275.
35. Li, S.; O'Brien, R. J.; Meitzner, G. D.; Hamdeh, H.; Davis, B. H.; Iglesia, E., *Applied Catalysis, A: General* **2001,** 219(1-2), 215-222.
36. Jin, Y.; Mansker, L.; Datye, A. K., *Preprints - American Chemical Society, Division of Petroleum Chemistry* **1999,** 44(1), 97-99.

124

37. Davis, B. H.; Miller, S. J. Process for the production of highly branched Fischer-Tropsch products and potassium promoted iron catalyst. US Patent Application 2002-80148, 2003.
38. Li, S.; Li, A.; Krishnamoorthy, S.; Iglesia, E., *Catalysis Letters* **2001,** 77(4), 197-205.
39. Liu, Z.-T.; Zhou, J.-L.; Zhang, Z.-X.; Zhang, B.-J., *Journal of Natural Gas Chemistry* **1995,** 4(1), 66-76.
40. Graham, U. M.; Dozier, A.; Srinivasan, R.; Thomas, M.; Davis, B. H., *Preprints - American Chemical Society, Division of Petroleum Chemistry* **2005,** 50(2), 178-181.
41. Jin, Y.; Datye, A. K., *Journal of Catalysis* **2000,** 196(1), 8-17.
42. Datye, A. K.; Jin, Y.; Mansker, L.; Motjope, R. T.; Dlamini, T. H.; Coville, N. J., *Studies in Surface Science and Catalysis* **2000,** 130B, (International Congress on Catalysis, 2000, Pt. B), 1139-1144.
43. Li, S.; Krishnamoorthy, S.; Li, A.; Meitzner, G. D.; Iglesia, E., *Journal of Catalysis* **2002,** 206(2), 202-217.
44. Jin, Y.; Xu, H.; Datye, A. K., *Microscopy and Microanalysis* **2006,** 12(2), 124-134.
45. Datye, A. K.; Shroff, M. D.; Harrington, M. S.; Sault, A. G.; Jackson, N. B., *Studies in Surface Science and Catalysis* **1997,** 107, (Natural Gas Conversion IV), 169-174.
46. Kuhrs, C.; Arita, Y.; Weiss, W.; Ranke, W.; Schlogl, R., *Topics in Catalysis* **2001,** 14(1-4), 111-123.
47. Xu, J.; Bartholomew, C. H.; Sudweeks, J.; Eggett, D. L., *Topics in Catalysis* **2003,** 26(1-4), 55-71.
48. Bukur, D. B.; Sivaraj, C., *Applied Catalysis, A: General* **2002,** 231(1-2), 201-214.
49. Jackson, N. B.; Datye, A. K.; Mansker, L.; O'Brien, R. J.; Davis, B. H., *Studies in Surface Science and Catalysis* **1997,** 111, (Catalyst Deactivation 1997), 501-516.
50. Sirimanothan, N.; Hamdeh, H. H.; Zhang, Y.; Davis, B. H., *Catalysis Letters* **2002,** 82(3-4), 181-191.
51. O'Brien, R. J.; Xu, L.; Milburn, D. R.; Li, Y.-X.; Klabunde, K. J.; Davis, B. H., *Topics in Catalysis* **1995,** 2, (1-4, Fischer-Tropsch and Methanol Synthesis), 1-15.
52. Rao, K. R. P. M.; Huggins, F. E.; Huffman, G. P.; O'Brien, R. J.; Gormley, R. J.; Davis, B. H., *Preprints of Papers - American Chemical Society, Division of Fuel Chemistry* **1995,** 40(1), 153-7.
53. Browning, N. D.; Wallis, D. J.; Nellist, P. D.; Pennycook, S. J., *Micron* **1997,** 28(5), 333-348.
54. Egerton, R. F.; Editor, *Electron Energy-Loss Spectroscopy in the Electron Microscope, Second Edition.* 1996; p 480 pp , (approx).
55. Xu, P.; Kirkland, E. J.; Silcox, J.; Keyse, R., *Ultramicroscopy* **1990,** 32(2), 93-102.
56. Bleloch, A.; Lupini, A., *Materials Today (Oxford, United Kingdom)* **2004,** 7(12), 42-48.
57. Pennycook, S. J.; Jesson, D. E., *Physical Review Letters* **1990,** 64(8), 938-41.
58. Rizzi, A., *Fresenius' Journal of Analytical Chemistry* **1997,** 358(1-2), 15-24.

Fischer-Tropsch Synthesis, Catalysts and Catalysis
B.H. Davis and M.L. Occelli (Editors)

Effect Of Mo Loading And Support Type On Hydrocarbons And Oxygenates Produced Over Fe-Mo-Cu-K Catalysts Supported On Activated Carbons

Wenping Ma, Edwin L. Kugler, and Dady B. Dadyburjor

Department of Chemical Engineering, West Virginia University, Morgantown, WV 26506-6102

1. Abstract

The effects of Mo loading and type of activated-carbon (AC) support on activity, stability, and selectivities to hydrocarbons and alcohols during Fischer-Tropsch synthesis over Fe-Mo-Cu-K/AC catalysts are studied under the conditions of 310-320 °C, 2.2 MPa, 3 Nl-gcat/h and $H_2/CO = 0.9$ in a fixed-bed reactor. Mo loadings of 0, 6 and 12 weight percent were used to determine the effect of Mo loading, while four different carbon supports, i.e. peat, generic wood, pecan-shell and walnut-shell derived ACs, were used to determine the effect of carbon support. Addition of Mo promoter to the Fe-Cu-K catalyst supported on peat AC leads to improved catalyst stability and improved selectivities to methane and oxygenates. The 6% Mo catalyst is more active and produces more hydrocarbons and oxygenates than the 12% Mo catalyst. Carbon-support type affects both catalyst activity and selectivity. Catalysts supported on pecan and peat ACs are the most active and produce the largest amount of hydrocarbons; the catalyst supported on pecan AC also shows the best C_{5+} selectivity. The catalyst supported on walnut-shell AC shows high C_{5+} selectivity too, but it is not as active. The catalyst using a wood AC support is the least active, with the smallest C_{5+} hydrocarbon selectivity (least chain-growth probability) but the highest oxygenate selectivity. In all cases, hydrocarbons upto ~C_{34} and alcohols upto ~C_5, are detected, and total oxygenates make up less than 6 wt% of total products. It is hypothesized that the Mo promoter prevents active sites from agglomeration, and that textural properties, such as BET and pore volume of carbon supports, might be similarly related to catalytic performance.

Keywords: Fischer-Tropsch Synthesis, Molybdenum, Iron, Activated Carbon, C_1-C_{34} hydrocarbons

2. Introduction

Fischer-Tropsch synthesis (FTS) has been widely studied during the past 80 years due to its significance in indirectly converting coal/natural gas to transportation fuels. However, one of big challenges is how to control hydrocarbon chain growth during the FTS reaction [1]. It has been proposed that proper process conditions or some kinds of porous supports can be used to restrict chain propagation [1, 2].

In recent years, studies over molecular sieves such as MCM-41 and SBA-15 indicate that these meso-porous materials are very promising for primary production of diesel-range hydrocarbons by FTS [3-7]. Co supported on these materials was reported to produce hydrocarbons up to C_{28} with high selectivity to diesel-range hydrocarbons at typical FTS conditions. Activated carbon (AC) has been found to be another promising catalyst support for restricting hydrocarbon chain length [8-12]. Limited earlier studies with AC-supported iron, cobalt and molybdenum-nickel catalysts [8-13] indicate that chain length could be limited below C_{20} over the activated-carbon support.

In this paper, we report reaction results of FTS over Fe-Mo-Cu-K catalysts in a fixed-bed reactor. Two Mo loadings (6 and 12%) and four types of carbon supports are used to study the effects of Mo loading and carbon support type on activity, stability, productivities and selectivities of both hydrocarbons and oxygenates. The interaction between Mo and Fe and different textural properties of the AC supports are discussed as possible factors related to the catalytic behavior.

3. Experimental

3.1. Carbon support treatment

Four different types of AC were used as supports of the catalysts, one from peat, two from specific woods (pecan and walnut) and the last as a generic "wood". AC from peat was supplied by Sigma-Aldrich; AC from "wood" by Norit, whereas ACs from pecan and from walnut were provided by US Department of Agriculture. The four ACs were washed in hot distilled water and then calcined at 500°C for 2 h by flowing N_2. All AC materials were ground to 20-40 mesh before impregnation.

BET surface and pore volume data of the four carbon supports are shown in Table 1. Pecan-based AC has the largest surface area and the largest total pore size. Wood-based AC has the highest fraction of microporous surface area and the highest fraction of micropore volume, but the largest average pore diameter.

Table 1 BET measurement results of activated carbon supports.

AC Type	Total SA	Micropore SA	$f_{\mu S}$	Total PV	Micropore Vol.	$f_{\mu V}$	Ave. pore dia.
	$[m^2/g]$	$[m^2/g]$		$[cm^3/g]$	$[cm^3/g]$		$[\text{Å}]$
Peat	606	455	0.75	0.48	0.22	0.46	59
Wood	970	914	0.94	0.58	0.44	0.76	66
Pecan	1016	736	0.72	0.72	0.36	0.50	51
Walnut	892	659	0.74	0.50	0.32	0.64	39

$f_{\mu S}$ - fraction of surface area present in the micropores;
$f_{\mu V}$ - fraction of pore volume present in the micropores

Table 2 summarizes EDS results. While pecan- and walnut-based ACs show only C and O, the wood-based AC shows Ca and S in addition, and the peat-based AC shows a number of trace elements in addition to those.

Table 2. Elements observed by EDS analysis, in order of abundance.

AC Type	Elements
Peat	C, O, Mg, Al, Si, Ca, Fe, S
Wood	C, O, Ca, S
Pecan	C, O
Walnut	C, O

Figure 1 shows SEM images that allow a qualitative comparison of the pore morphologies of the support materials. The peat-based AC has slit-like pores and the wood-based AC has pan-like pores, while pecan- and walnut-based ACs have honeycomb-like pores.

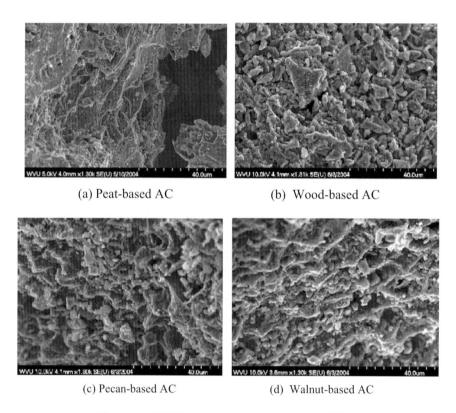

(a) Peat-based AC   (b) Wood-based AC

(c) Pecan-based AC   (d) Walnut-based AC

Figure 1. SEM images from four types of AC.

3.2. Catalyst synthesis

AC-supported Fe-Cu-K catalysts with or without Mo were prepared using sequential incipient wetness impregnation. If Mo was required, then an appropriate amount of aqueous ammonium molybdate solution was first impregnated on the AC support. The material was then dried in air at 90-100 °C overnight. Aqueous ferric nitrate and cupric nitrate solution corresponding to final iron and copper contents of 15.7 wt% and 0.8 wt%, respectively, on the catalyst was impregnated on the support or the prepared Mo/C sample, again followed by drying in air at the same temperature overnight. Potassium nitrate solution (corresponding to 0.9 wt% K) was the last to be put onto the sample, again followed by drying in air at the same temperature overnight.

3.3. Reactor testing

Catalysts were tested in a computer-controlled fixed-bed reactor system. A detailed description of the reactor system has been reported previously [12]. The amount of catalyst used in each test was nominally 1.0 g. The catalyst was reduced by flowing H_2 at 400 $^{\circ}$C, 0.5 MPa, and 3 Nl/g-cat/h for 12 h. The reaction conditions were 310-320 $^{\circ}$C, 300 psig, 3 Nl/g-cat/h, and H_2/CO = 0.9. The runs were carried out for 72-396 hours.

3.4. Product analysis

Inlet and outlet gases were analyzed on-line by TCD and FID in a HP-5890 GC equipped with packed and capillary columns. The liquid organic phase (C_4-C_{34} hydrocarbons) was analyzed by FID and a MXT-5 capillary column in a Varian 3400 GC. The liquid water phase was analyzed by FID and a Porapak-Q packed column in the same Varian 3400 GC. The alcohol components in the water phase were identified by matching retention times of known standards. The water-phase composition was quantified using tert-amyl alcohol as an internal standard.

4. Results and Discussion

4.1. Effect of Mo loading

Supported Fe-Cu-K catalysts were prepared with 0, 6 and 12 wt % Mo to evaluate the effects of Mo loading. In addition to Mo, the catalyst contained 15.7% Fe, 0.8% Cu and 0.9% K. The support used was activated carbon from peat.

4.1.1. Effect on catalyst activity and stability

The effect of Mo loading on CO conversion is shown in Figure 2. On the catalyst without Mo, the initial CO conversion is high, approaching 97%. However, the catalyst deactivated rapidly and the reaction was stopped after 72 h of time on stream at 320°C. A catalyst containing 6% Mo shows the opposite

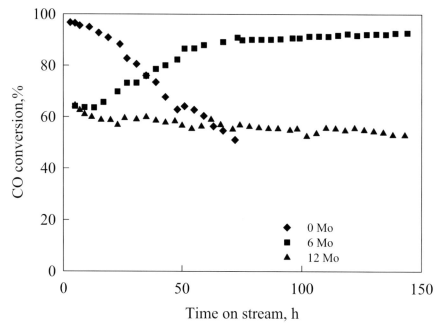

Figure 2. Effect of Mo loading on CO conversion. (320 °C, 300 psig, 3 Nl/gcat/h, $H_2/CO = 0.9$)

effect. Over this catalyst, the CO conversion increases from 60 to 90% over the first 72 hours and then remains stable for the next 72 hours. Finally, in the case of the catalyst with 12% Mo, the CO conversion is initially the same as for 6% Mo, but then decreases slowly to 53% conversion at 143 h. (The Mo-containing catalysts were continuously tested at 310-320°C for up to 396 h, but the results after 144 h are not discussed here.)

Figure 2 suggests two important effects of Mo. First, the Mo-containing catalysts either are stable or show a very slow deactivation rate compared to the Mo-free catalyst, suggesting that the Mo promoter improves catalyst stability. Because Mo promoter has been reported to be able to incorporate at the iron surface [13], the effectiveness of the Mo promoter can be ascribed to Mo segregation of iron active sites and prohibition of iron crystal agglomeration. Second, the 6% Mo catalyst is much more active than the 12% Mo catalyst, which clearly suggests loading excess Mo significantly decreases catalyst activity.

4.1.2. Effect on hydrocarbon selectivity

Changes of CH_4 and C_{5+} selectivity with time on stream are shown in Figure 3. The catalyst with 0% Mo produces 6-8% CH_4 and 48-53 % C_{5+} hydrocarbons during 72 hours of testing at 320 °C. Catalysts with Mo show 11-14% CH_4 and 38-51 % C_{5+} hydrocarbon selectivities at the same temperature. The addition of Mo into Fe-Cu-K/AC improves CH_4 production and inhibits the production of higher-molecular-weight hydrocarbons. This is probably due to an increase in the dissociation of CO [14].

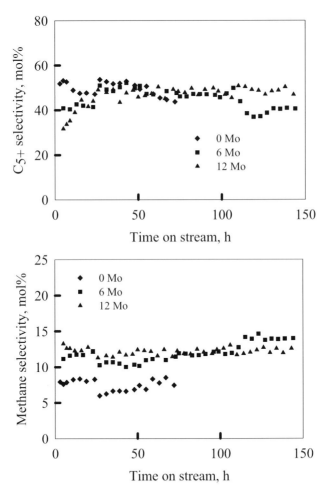

Figure 3. Effect of Mo loading on selectivities of (top) CH_4 and (bottom) C_{5+} (320 °C, 300 psig, 3 Nl/gcat/h, H_2/CO = 0.9).

4.1.3. Effect on productivities of hydrocarbon and oxygenate

The effect of Mo loading on Space Time Yields (STYs) of hydrocarbon and oxygenate is shown in Figure 4. The data reported in the figure correspond to times on stream between 48 and 72 h. The 6% Mo catalyst exhibits the highest productivities of hydrocarbon (490 g/kg-cat/h) and oxygenate (12.7 g/kg-cat/h) at 320 °C. This is consistent with the stability and the high activity of the catalyst. However, values of oxygenate selectivity (Figure 5) for all

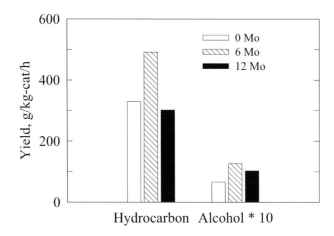

Figure 4. Effect of Mo loading on hydrocarbon and oxygenate yields (320 °C, 300 psig, 3 Nl/gcat/h, $H_2/CO = 0.9$; 48-72 hours).

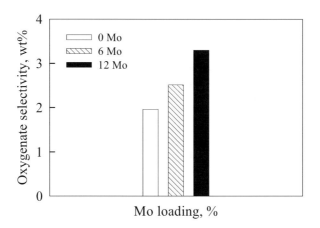

Figure 5.Effect of Mo loading on oxygenate selectivity at 320 °C, 300 psig, 3 Nl/g-cat/h, $H_2/CO = 0.9$; 48-72 h. (Oxygenate selectivity, % = 100 × $r_{oxygenate}/\{ r_{oxygenate} + r_{hydrocarbon}\}$)

three catalysts are less than 3.3 wt%, but selectivity increases with Mo loading. These results indicate that hydrocarbons make up the bulk of the products, and Mo promoter improves oxygenate selectivity.

4.1.4. Effect on alcohol distribution

The distribution of individual C_1-C_5 straight alcohols is shown in Figure 6. Small amounts of other oxygenates, such as acetaldehyde, acetone and C_3-C_5 branch alcohols, were also quantified, but are not shown here. As can be seen from the figure, ethanol is the dominant oxygenate for all three catalysts, followed by C_3-OH, C_1-OH, C_4-OH and C_5-OH. This qualitative sequence of alcohol selectivity was also reported by Kugler et al.[11] over Mo-Ni-K/AC alcohol catalysts.

The distribution of alcohols changes somewhat with the Mo loading. Methanol increases, and ethanol decreases, as the Mo loading is increased, with little or no discernable change in the (smaller) amounts of the higher alcohols. This result is consistent with the finding by Kip et al.[14] that MoO_3 improves methanol selectivity.

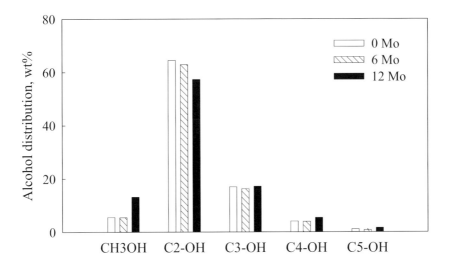

Figure 6. Effect of Mo loading on alcohol distribution (320 °C, 300 psig, 3 Nl/g-cat/h, $H_2/CO = 0.9$; 48-72 h).

134

4.2. Effect of AC Type

4.2.1. Effect on catalyst activity and stability

For this series of experiments, the catalyst composition was fixed at 6.0% Mo, 15.7% Fe, 0.8% Cu and 0.9% K. The catalysts using the peat and pecan-shell ACs were tested at both 310 and 320 °C, whereas the catalysts supported on wood and walnut-shell ACs were tested at 320 °C only. Other process conditions remained the same during the test periods. Only the test results at 320 °C are discussed in this paper.

Figure 7 shows the change of CO conversion with time on stream for the four different AC supports. Catalysts with peat and pecan AC supports reach a similar constant, high CO conversion activity after approximately 70 h. Catalysts with walnut and wood AC supports reach nearly stable CO conversions as well, but the values are much lower. All four activities decrease initially with time on stream; one levels off, one returns slightly before leveling off, both below their initial values, and the other two return to their initial high values. It can be concluded, based on the steady state values of CO conversion, that the catalyst using wood AC support is the least active, while the catalysts using peat and pecan-shell AC supports are the most active.

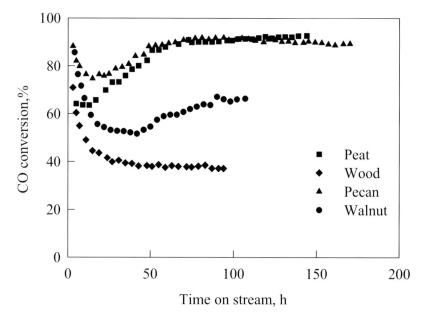

Figure 7. Effect of carbon support type on CO conversion (320 °C, 300 psig, 3 Nl/g-cat/h, H$_2$/CO = 0.9).

The activity trend is not correlated with values of total BET surface area or total pore volume of the four carbon supports (Table 1). However, the peat and pecan-shell ACs, which have the lowest fraction of micropore volume (highest fraction of mesopores), exhibit higher CO conversion, while the wood-AC, which has the highest fraction of micropores (lowest fraction of mesopores), exhibits the lowest CO conversion. This suggests that the activity of catalyst is associated with the fraction of mesopores in the carbon support.

4.2.2. Effect on hydrocarbon selectivity

Changes of CH_4 and C_{5+} selectivity with time on stream for the four AC supports are shown in Figure 8. Methane selectivities over catalysts with the walnut- and pecan-shell AC supports are generally the highest. Methane selectivity is the lowest for the peat-AC support. C_{5+} selectivities over catalysts with pecan- and walnut-shell-AC supports are highest. The catalyst with the wood-AC support shows the poorest C_{5+} selectivity, at least during the first 94 h of testing. Note that the wood AC has the largest absolute amount of micropores, as well as the largest fraction of micropores, both in terms of surface area and volume. Hence, the existence of mesopores seems to correlate well with C_{5+} selectivities.

4.2.3. Effect on productivity of hydrocarbon and oxygenate

Figure 9 illustrates the yields of hydrocarbon and oxygenate at 320°C over the four different carbon supports. Hydrocarbon yields over catalysts with peat- and pecan-shell-AC supports are nearly the same, and are the largest among all four carbon supports. Hydrocarbon yield is the least over the catalyst with the wood-AC support. In addition, the catalysts with wood- and peat- AC supports show the largest oxygenate yields. However, oxygenate selectivities on all four carbon supports are rather small, less than 6% (Figure 10). The oxygenate selectivity over the catalyst with wood-AC support is 2-3 times higher than that of other three supports. Thus, it can be concluded that the peat- and pecan-shell ACs are the best support choices for hydrocarbon production, while wood AC is a good support for oxygenate production.

4.2.4. Effect on overall hydrocarbon distribution

Overall hydrocarbon distributions on the four carbon supports are shown in Figure 11. The hydrocarbon mole fractions comply with the Anderson-Schulz-Flory (ASF) distribution. Note that hydrocarbons of carbon number up to 34 are detected for all carbon supports. However, the chain-growth probability (α) for catalyst with wood-AC support is 0.65 (at least for carbon numbers of up to 20), smaller than that of the catalysts using the other three AC supports (~0.71).

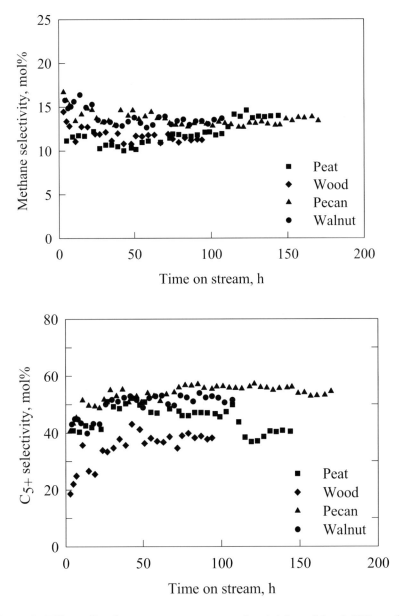

Figure 8. Effect of carbon support type on selectivities of (top) CH_4 and (bottom) C_{5+} (320 °C, 300 psig, 3 Nl/g-cat/h, $H_2/CO = 0.9$).

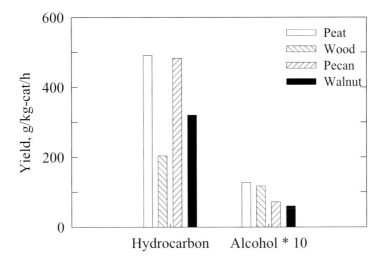

Figure 9. Effect of carbon support type on hydrocarbon and oxygenate yields (320 °C, 300 psig, 3 Nl/g-cat/h, $H_2/CO = 0.9$; 48-72 h).

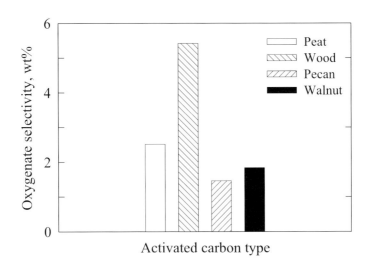

Figure 10. Effect of carbon support type on oxygenate selectivity at 320 °C, 300 psig, 3 Nl/g-cat/h, $H_2/CO = 0.9$ and 48-72 h. (Oxygenate Selectivity, % = 100 x $r_{oxygenate} / \{r_{oxygenate} + r_{hydrocarbon}\}$)

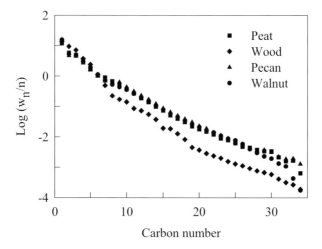

Figure 11. Overall hydrocarbon distribution over catalysts with different carbon supports (320 °C, 300 psig, 3 Nl/g-cat/h, $H_2/CO = 0.9$; 48-72 h).

4.2.5. Effect on Alcohol Distribution

Figure 12 shows the distribution of alcohols over catalysts having each of the four types of AC supports. Ethanol is the dominant alcohol, as in Figure 6. The order of the selectivities for the other alcohols does not change with carbon support type. Peat and pecan-shell supports lead to the largest production of ethanol, whereas catalysts with wood- or walnut-AC supports lead to larger amounts of the higher-molecular-weight alcohols.

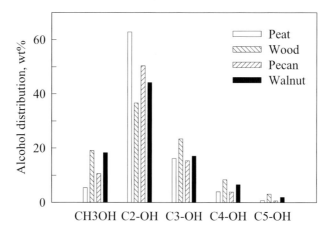

Figure 12. Effect of carbon support type on alcohol distribution (320 °C, 300 psig, 3 Nl/g-cat/h, $H_2/CO = 0.9$; 48-72 h).

5. Conclusions

The effects of Mo loading and carbon support type on the catalytic performance (activity, selectivity and stability) of AC supported Fe-Mo-Cu-K catalysts are studied in this paper. In all cases, hydrocarbons up to C_{34} and alcohols up to C_5 are detected, regardless of Mo loading or source of AC support. Oxygenate compounds produced during FTS on all carbon supported catalysts are less than 6% of the total oxygenates and hydrocarbons. In all cases, ethanol is the dominant alcohol, followed by C_3-OH, methanol, C_4-OH and C_5-OH.

The outcome of the addition of Mo promoter to the Fe-Cu-K/AC catalyst is to improve catalyst stability and oxygenate selectivity, but this also increases CH_4 selectivity. Of the catalysts used, the best performance with respect to activity and yields of hydrocarbons and oxygenates is found on the catalyst with 6% Mo loading. Increasing the Mo loading results in an increase in the oxygenate selectivity. The improvement of catalyst stability has been attributed to Mo inhibiting iron particle agglomeration during the FTS reaction.

The type of AC support used affects catalytic performance. Catalysts with AC derived from pecan-shell or peat exhibit the highest activity and the highest yield of hydrocarbons. Those with pecan- and walnut-shell-AC supports give the best C_{5+} selectivity. The generic wood-derived support is the least active and it produces the smallest amounts of C_{5+} hydrocarbons. The wood- and peat-derived AC supports typically give rise to the largest values of oxygenate productivities, with the wood-based support having the highest values of oxygenate selectivity. The total activity and C_{5+} selectivity depend on the number of mesopores on the carbon support.

6. Acknowledgements

This study was supported by the U. S. Department of Energy under Cooperative Agreement DE-AC22-99FT40540 with the Consortium for Fossil Fuel Science (CFFS). We are grateful to Professor Wayne E. Marshall from the Agriculture Research Center, United States Department of Agriculture, for providing us two AC samples and to Mr. James Wright, Mr. Steve Carpenter and Dr. Liviu Magean in Chemical Engineering at WVU for help with BET and SEM measurements.

References

[1] M.E. Dry, Appl. Catal., 276 (2004) 1.

[2] A.A. Nikolopoulos, S.K. Gangwal, J.J. Spivey, Stud. Surf. Sci. Catal. 136 (2001) 351.

[3] D.H. Yin, W.H. Li, W, S. Yang, H.W Xiang, Y.H. Sun, B. Zhong, S.Y. Peng, Microporous Mesoporous Mater., 47 (2001) 15.

[4] A.Y. Khodakov, A. Griboval-Constant, R. Bechara, V.L. Zholobenko, J. Catal., 206 (2002) 230.

[5] M. Agustín, L. Carlos, M. Francisco, D. Isabel, J. Catal. 220 (2003) 486.

[6] Y. Ohtsuka, Y. Takahashi, M. Noguchi, T. Arai, S. Takasaki, N. Tsubouchi, Y. Wang, Catal. Today 89 (2004) 419.

[7] J. Panpranot, J.G.Jr. Goodwin, A. Sayari, J. Catal. 211 (2002) 530.

[8] H.J. Jung, M.A. Vannice, L.N. Mulay, R.M. Stanfield, W.N. Delgass, J. Catal. 76 (1982) 208.

[9] W.P. Ma, Y.J. Ding, J. Yang, X. Liu, L.W. Lin, React. Kinet. Catal. Lett. 84 (2005) 11.

[10] W.P. Ma, Y.J. Ding, L.W. Lin, Ind. Eng. Chem. Res. 43 (2004) 2391.

[11] E.L. Kugler, L. Feng, X. Li, D.B. Dadyburjor, Stud. Surf. Sci. Catal. 130A (2000) 299.

[12] X.G. Li, L.J. Feng, Z.Y. Liu, B. Zhong, D.B. Dadyburjor, E.L. Kugler, Ind. Eng. Chem. Res. 37 (1998) 3853.

[13] J.M. Zhao, Z. Feng, F.E. Huggins, N. Shah, G.P. Huffman, I. Wender, J. Catal. 148 (1994) 194.

[14] B.J. Kip, E.G.F. Hermans, J.H.M.C. Van Wolput, N.M.A. Hermans, J. Van Grondelle, R. Prins, Appl. Catal. 35 (1987) 109.

Fischer-Tropsch Synthesis, Catalysts and Catalysis
B.H. Davis and M.L. Occelli (Editors)
141

Study of carbon monoxide hydrogenation over supported Au catalysts

Yanjun Zhao,[a] Arthur Mpela,[a,b] Dan I. Enache,[a] Stuart H. Taylor,[a] Diane Hildebrandt,[b] David Glasser,[b] Graham J. Hutchings,[a*] Martin P. Atkins,[c] Michael S. Scurrell[b,d]

[a] School of Chemistry, Cardiff University, Main Building, Park Place, CF10 3AT, U.K.
[b] School of Process and Materials Engineering, University of Witwatersrand, Private Bag 3, Wits 2050, Johannesburg, South Africa
[c] BP Chemicals, Sunbury-on-Thames, Bd B Desk 37, Middlesex, TW16 7LL, U.K.
[d] Molecular Science Institute, School of Chemistry, University of Witwatersrand, Johanesbug, South Africa
[*] corresponding author – e-mail: hutch@cf.ac.uk

1. Abstract

Recently, there has been a marked increase in the interest shown in catalysis by gold. It is now recognised that gold has unique properties as a catalyst for many reactions with pre-eminence in the oxidation of carbon monoxide. However, it is also known that supported gold catalysts can be used for other reactions involving carbon monoxide, for example the water gas shift reaction. Supported gold catalysts have also been shown to be effective for hydrogenation reactions. This paper reports the possible use of gold as a catalyst for the hydrogenation of carbon monoxide. In particular, we describe the preparation and characterisation of Au/ZnO and Au/Fe$_2$O$_3$ as catalysts for CO hydrogenation and for the synthesis of alcohols in particular. Alcohols including methanol, ethanol, 1-propanol, 2-propanol and 1-butanol have been successfully synthesised at 300 °C at a pressure of 25 bar over supported Au catalysts.

1. Introduction

Presently there is a great deal of interest in the use of gold as both heterogeneous and homogeneous catalysts. For many years gold was perceived as a relatively inert catalyst material; however the discovery in the 1980s that finely supported divided nanoparticles of gold could act as catalysts for reactions at low temperatures has stimulated considerable research effort on gold catalysts. Bond and co-workers [1] were amongst the first to demonstrate that very small gold particles supported on silica could give interesting catalytic performance for hydrogenation of butadiene. Subsequently, we showed that Au/ZnO could be used for selective hydrogenation of α,β-unsaturated aldehydes [2]. Haruta and co-workers discovered that supported Au catalysts are very active for low temperature CO oxidation [3]. In addition, there have been extensive studies on the oxidation of carbon monoxide, which have been reviewed [4,5]. Haruta and co-workers [6] have reported that gold supported on ZnO and Fe_2O_3 could be used as a catalyst for carbon monoxide hydrogenation, and a small amount of methanol was observed.

In this paper we extend this earlier study and explore the preparation and characterisation of Au/ZnO and Au/Fe$_2$O$_3$ as catalysts for the hydrogenation of carbon monoxide, particularly for the synthesis of alcohols. Mixed alcohols are widely used as fuel additives in the petroleum industry. The introduction of mixed alcohols into the fuels decreases the unwanted CO, NO_x and hydrocarbon emissions. It is for this reason that we wished to explore supported Au catalysts, which could function as CO hydrogenation catalysts and we now report our initial results.

2. Experimental

2.1. Catalyst preparation

5wt%Au/ZnO catalysts used in this study were prepared by co-precipitation from $HAuCl_4 \cdot 3H_2O$ (Johnson-Matthey) and $Zn(NO_3)_2 \cdot 6H_2O$ (ACROS). An aqueous mixture of the precursors ($HAuCl_4$ 0.002M and Zn $(NO_3)_2 \cdot 6H_2O$, 0.1M) was introduced at the rate of 7.5 ml/min into an aqueous solution of 1M Na_2CO_3 (pH 9-11.5) under vigorous stirring (~600 rpm) for 90-120 min. The precipitation temperature was maintained at 70-80°C. The co-precipitated sample obtained was aged for about 24 h, filtered, washed several times with warm distilled water, and then dried. The powder obtained was calcined at 400°C in air. ZnO catalyst was prepared in a similar way without adding the gold source (HAuCl$_4$·3H$_2$O).

5wt%Au/Fe$_2$O$_3$ catalyst was used as obtained, as a standard catalyst provided by the World Gold Council [7]. Fe$_2$O$_3$ catalyst was prepared by a precipitation method. An aqueous solution of Fe nitrate (Fe(NO$_3$)$_3$·9H$_2$O, 0.25M) was heated to80 °C. Na$_2$CO$_3$ solution (0.25M) was added dropwise to the nitrate solution until a pH of 8.2 was reached. The precipitate was washed with distilled water and then dried. The dried brown colour catalyst was calcined in air at 400 °C for 6 h.

For the Au/ZrO$_2$ catalyst, ZrO(NO$_3$)$_2$ (35ml, 0.1M) solution was added dropwise to HAuCl$_4$·3H$_2$O (10ml, 0.025M) followed by the ammonia 4.98N (d:0.88) solution. The coprecipitate was filtered, washed with warm water, dried at 80 °C and calcined at 300 °C for 6h. The Au/SiO$_2$ catalyst was prepared by mixing gold solution (HAuCl$_4$·3H$_2$O 0.025M) with silica solution made by dissolving SiO$_2$ powder (2.18g) in distilled water (100ml). The resulting solution was evaporated. The precursor obtained was calcined at 300 °C under ammonia vapour for 6h.

2.2. Catalyst characterisation

Catalysts were characterised by powder X-ray diffraction using an Enraf Nonius PSD120 diffractometer with a monochromatic CuK$_\alpha$ source operated at 40 keV and 30 mA. Phases were identified by matching experimental patterns to the JCPDS powder diffraction file. BET surface areas were determined from the nitrogen adsorption isotherm, using Micromeritics Gemini equipment. Raman spectroscopy characterisation was made using a Renishaw Ramascope spectrometer.

2.3. Catalyst testing procedure

The catalysts used in our study were evaluated in a stainless steel tubular reactor (0.25 in external diameter). The catalyst (ca. 0.2 gram) was pre-treated with 1% H$_2$/N$_2$ (flow rate 10 ml/min) for 1 h at 250°C under atmospheric pressure. Then syngas (CO/H2/N2=47.5/47.5/5, BOC UK) was introduced to the reactor and the pressure was increased to 25 bar by using a back-pressure regulator (TESCOM 26-1700). The temperature of reaction (300°C) was adjusted and maintained with a Cole Parmer controller. The gas phase products were analysed using an online gas chromatograph (Varian GC 3800). Concentrations of carbon monoxide, carbon dioxide and nitrogen were analysed by a thermal conductivity detector and the other organic compounds such as hydrocarbons and oxygenates were determined by a flame ionisation detector. Conversion was determined using nitrogen as internal standard. The liquid products were collected using a trap that was kept at room temperature and under catalytic test pressure. The liquid products were identified by GCMS (Perkin Elmer, TurboMass). Oxygenates were quantified by a second gas chromatograph (Chrompack) equipped with capillary column (CP-Sil 8CB 30m, 0.32mm, 1µm) and FID detector.

3. Results and discussion

3.1. Characterisation of the catalysts

The XRD pattern of the 5%Au/ZnO and 5%Au/Fe$_2$O$_3$ catalyst present only the diffraction lines characteristic to the support ZnO and Fe$_2$O$_3$ respectively (Figure 1). The amount of Au on the sample should be sufficient to provide an X-ray diffraction pattern of crystalline Au particles if they had been sufficiently large. The observed X-ray pattern may suggest that the Au particle is very small in the investigated catalysts.

The laser Raman spectra of the ZnO support and 5%Au/ZnO catalyst are presented in Figure 2. Unfortunately, significant fluorescence is observed for the ZnO support. For the Au/ZnO sample, the fluorescence is decreased and new bands at 3224 and 3472 cm^{-1} were observed and these are assigned to hydroxyl groups, which, as they are absent in the ZnO support, may be associated with the interface between the ZnO and the Au nanocrystal. The measured BET surface areas are 35 m^2/g and 41 m^2/g for Fe$_2$O$_3$ and 5%Au/Fe$_2$O$_3$ catalysts respectively.

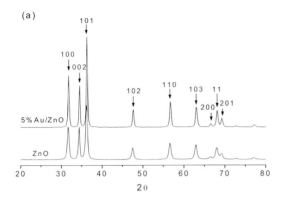

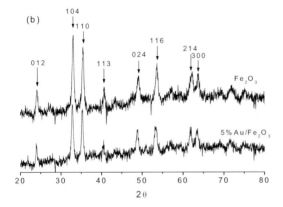

Figure 1. XRD diagram of (a) 5% Au/ZnO catalyst and ZnO support; (b) 5% Au/Fe$_2$O$_3$ catayst and Fe$_2$O$_3$ support.

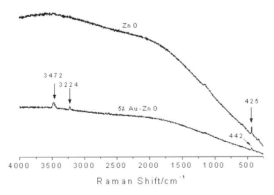

Figure 2. Raman spectra of ZnO and 5% Au/ZnO catalysts.

3.2. CO hydrogenation

3.2.1 Au/Fe₂O₃

Two sets of experiments of CO hydrogenation were carried out to examine the effect of Au on the Fisher Tropsch synthesis with Fe_2O_3 as support material. Figure 3 shows the on-line analysis data when Fe_2O_3 was used alone and when Au was supported on Fe_2O_3, i.e. these data are for the gas phase products only.

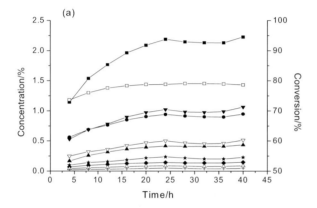

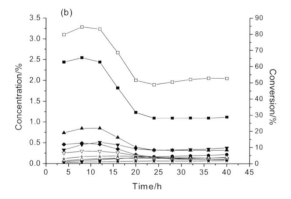

Figure 3. Gas phase online data of catalyst (a) Fe_2O_3 (b) 5% Au/Fe_2O_3 by operating the reactor at 300°C and 25 bar; (!) CH_4 concentration; (7) C_2H_6 concentration; (,) C_2H_4 concentration; (Λ) C_3H_8 concentration; (B) C_3H_6 concentration; (8) C_4H_{10} concentration; (X) C_4H_8 concentration; (ξ) C_5H_{10} concentration; ($\square$) C_5H_{12} concentration; ($\forall$) CO conversion .

Several features of the results for Au/Fe$_2$O$_3$ are noteworthy when compared with Fe$_2$O$_3$ alone. First, for Au/Fe$_2$O$_3$ the conversion increases slightly in the first 10 h and then decreases rapidly until after 30 h when it reaches a steady state with a conversion of ca. 55%. This is significantly different from the catalytic behaviour of Fe$_2$O$_3$. In the catalytic test for Fe$_2$O$_3$, a smooth but small increase in conversion was observed at the first 15 h and then the reaction reached a steady state for the rest of the running time, giving a conversion of ca. 80%. Second, the concentration of alkanes for bulk Au/Fe$_2$O$_3$ and Fe$_2$O$_3$ follows quite similar trends as the conversion. However, a different pattern was observed for alkene generation for Au/Fe$_2$O$_3$ which increased with time on stream. These results may indicate that for the Au/ Fe$_2$O$_3$ catalysts, the conversion and alkane synthesis rates are correlated.

Detailed analysis data of the collected liquid is shown in Table 1. For the above two sets of experiments, both hydrocarbons and alcohols were produced. Two phases were collected in the trap, one of which, Phase I, was water rich. It was noted that the addition of Au to the Fe$_2$O$_3$ catalyst did not give a distinct effect on the liquid product distribution. The percentage of alcohol produced is slightly lower for 5% Au/Fe$_2$O$_3$ than that for Fe$_2$O$_3$ alone, and vice versa for hydrocarbons. A total distribution for the products combining liquid and gas phase products is shown in Table 2.

Table 1. Liquid Phase Analysis of Catalyst (a) Fe$_2$O$_3$ (b) 5% Au/Fe$_2$O$_3$ by Operating the Reactor at 300°C and 25 bar.

Phase I (wt.%)	Fe$_2$O$_3$	5%Au/Fe$_2$O$_3$	Phase II (wt.%)	Fe$_2$O$_3$	5%Au/Fe$_2$O$_3$
Methanol	0.34	0.74	C$_5$	1.56	2.6
Ethanol	3.9	1.5	C$_6$	2.09	6.5
2-propanol	0.08	Trace	C$_7$	9.6	8.8
1-propanol	0.58	0.41	C$_8$	10.2	9.3
2-butanol	0.18	0.16	C$_9$	9.44	8.5
water	20.3	17.2	C$_{10}$	7.7	6.7
			>C$_{10}$	34.1	37.6

Table 2. Total Products Analysis of Catalyst (a) Fe_2O_3 (b) 5% Au/Fe_2O_3 by Operating the Reactor at 300°C and 25 bar.

Hydrocarbons (wt.%)	Fe_2O_3	5%Au/Fe_2O_3	Alcohols (wt.%)	Fe_2O_3	5%Au/Fe_2O_3
CO_2	73.2	73.5	Methanol	0.06	013
CH_4	2.0	2.0	Ethanol	0.7	0.3
$C_2 - C_5$	8.0	6.9	2-propanol	0.01	Trace
$C_6 - C_{10}$	6.7	7.2	1-propanol	0.1	0.07
$>C_{10}$	5.8	6.8	1-butanol	0.03	0.03
			water	3.5	3.12

3.2.2 Au/ZnO catalyst

Two sets of gas phase on-line data for the catalytic reaction of CO hydrogenation over ZnO and Au supported on ZnO is shown in Figure 4(a) and (b) respectively. An average conversion of ca. 20% was achieved with only ZnO powder as the catalyst in Figure 4(a), which is much higher than the ca. 6% conversion observed with 5%Au/ZnO as the catalyst. Meanwhile, the percentage of hydrocarbons decreased from ca. 0.36% to ca. 0.26% for methane and from a maximum of *ca.* 0.17% to a maximum of *ca.* 0.07% for other hydrocarbons. The decreased conversion could be partially attributed to the decreased production of hydrocarbons.

Table 3 presents the liquid phase results in the CO hydrogenation over ZnO and 5%Au/ZnO. Small quantities of liquid products were collected for both cases. For ZnO, the products include mixed alcohols, C_5 to C_{23} hydrocarbons and wax. The 5%Au/ZnO produces the mixed alcohol products only. The calculation of alcohol and hydrocarbon phases was carried out separately for the purpose of a direct comparison. The observed results suggest that the presence of Au suppresses the activity of the catalyst for the production of hydrocarbons, and also it shifts the selectivity towards higher alcohols.

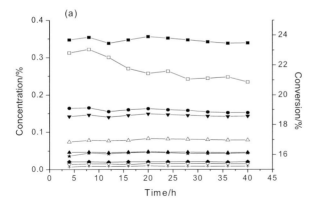

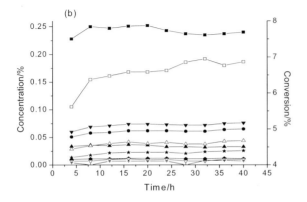

Figure 4. Gas phase online data of catalyst (a) ZnO (b) 5%Au/ZnO by operating the reactor at 300 °C and 25 bar; (!) CH$_4$ concentration; (7) C$_2$H$_6$ concentration; (,) C$_2$H$_4$ concentration; (Λ) C$_3$H$_8$ concentration; (B) C$_3$H$_6$ concentration; (8) C$_4$H$_8$ concentration; (X) C$_4$H$_{10}$ concentration; (ξ) C$_5$H$_{10}$ concentration; (□) C$_5$H$_{12}$ concentration; (∀) CO conversion.

Table 3. Liquid Phase Analysis of Catalyst (a) ZnO (b) 5% Au/ZnO by Operating the Reactor at 300^3C and 25 bar.

Phase I (wt.%)	ZnO	5% AuZnO	Phase II (wt.%)	ZnO	5% AuZnO
Methanol	75.8	62.9	C_5	7.4	0
Ethanol	17.9	26.2	C_6	0.5	0
2-propanol	1.13	0.7	C_7	2.9	0
1-propanol	3.8	7.5	C_8	7.4	0
a-butanol	1.3	2.7	C_9	7.7	0
			C_{10}	51.7	0
			$>C_{10}$	34.1	0

It is known that ZnO alone is a methanol synthesis catalyst [8,9,10], however, the impurities (for example, alkaline residues) introduced to the catalyst during the preparation accelerate side reactions including higher alcohol synthesis and hydrocarbon synthesis [9]. In this paper, the support and the gold catalysts were prepared in a similar way, which enables us to investigate the function of gold over the supported catalysts. The results above show that the catalytic behaviour is improved with the addition of Au to the catalyst, since the Au/ZnO selectivity produces alcohols as products and there is a distinct increase in the selectivity for higher alcohols for Au/ZnO when compared with ZnO alone.

3.2.3 Gold over other supported catalysts

An Au/SiO_2 catalyst was also tested but no observable conversion or products were found. The CO hydrogenation over Au/ZrO_2 produced only CH_4, C_2H_6 and C_3H_8 in our current system. It was reported by Baiker [11] in 1993 that methanol was synthesized over Au/ZrO_2 from hydrogenation of CO and CO_2. Hence, our initial results show some differences in this respect, since we observed no activity for methanol formation with $Au/ZrO2$ under our reaction conditions.

4. Conclusions

Supported gold catalysts were studied for CO hydrogenation to investigate the possibility of using gold as a higher alcohol synthesis catalyst. The initial results, particularly for the Au/ZnO catalyst, are interesting and suggest that Au could play a role in the synthesis of mixed alcohols: [i] by suppressing the side hydrocarbon synthesis reaction and [ii] shifting the product selectivity towards higher alcohols.

- **References**

1. Sermon, P.A.; Bond, G.C.; Wells, P.B.; *J. Chem. Soc., Faraday Trans. 1,* **1979,** 75, 385.
2. Bailie, J. E.; Hutchings, G. J.; *Chem. Commun.,***1999**, 2151.
3. Haruta, M.; Kobayashi, T.; Sano, H.; Yamada, N.; *Chem. Lett.*, **1987**, 4, 405.
4. Bond, G.C.; Thompson, D.T.; *Catal. Rev.-Sci. Eng.*, **1999**, 41, 319.
5. Bond, G.C.; Thompson, D.T.; *Gold Bull.*, **2000**, 33 41.
6. Sakurai, H.; Haruta, M., Applied Catalysis, A: General, **1995**, 127 (1-2), 93.
7. http://www.gold.org/discover/sci_indu/gold_catalysts/refcat.html.
8. Wilmer, H.; Kurtz, M.; Klementiev, K.V.; Tkachenko, O.P.; Grunert, W.; Hinrichsen, O.; Birkner, A.; Rabe, S.; Merz, K.; Driess, M.; Woll, C. and Muhler, M.; Phys.Chem.Chem.Phys., **2003**, 5, 4736.
9. Bridger, G.W.; Spencer, M.S.; 'Methanol Synthesis' in Catalysis Handbook ed. by Twigg, V.M.**1989.**
10. Hoflund, G.B. and Epling, W.S.; Catalysis letters **1997**, 45, 135.
11. Baiker, A.; Kilo, M.; Maciejewski, S.; Menzi, S. and Wokaun, A.; Srud. Surf.Sci.Catal., **1993**, 75, 1257.

Fischer-Tropsch Synthesis, Catalysts and Catalysis
B.H. Davis and M.L. Occelli (Editors)
© 2007 Elsevier B.V. All rights reserved.

ZrFe INTERMETALLIDES FOR FISCHER-TROPSCH SYNTHESIS: PURE AND ENCAPSULATED INTO ALUMINA-CONTAINING MATRICES

S.F.Tikhov*[1], A.E.Kuz'min[2], Yu.N.Bespalko[1], V.I.Kurkin[2], V.A.Sadykov[1], E.I.Bogolepova[2], S.V.Tsybulya[1], A.V.Kalinkin[1], V.P.Mordovin, A.N.Salanov, V.I. Zaikovskii[1], A.A.Shavorsky[1]

[1]*Boreskov Institute of Catalysis SD RAS, 5 Lavrentieva St., 630090, Novosibirsk*
[2] *Topchiev Institute of Petrochemical Synthesis, RAS, 9 Leninskii Av., 117912, Moscow, RUSSIA*

Abstract

Performance of the bulk hydrogenated ZrFe intermetallides both pure and encapsulated into alumina-containing matrix (Al_2O_3/Al, Al_2O_3) was studied in catalysis of Fischer-Tropsch synthesis. Their structural, textural and surface properties were characterized by combination of such methods as XRD, SEM, TEM, nitrogen adsorption-desorption isotherms and XPS, and impact of these properties on catalytic activity and selectivity was analyzed. The highest activity per the surface Fe atom ($\sim 5 \cdot 10^{-19}$ CH_x/at.Fe·h) was obtained for pure active component, while the highest activity per the unit of volume (~ 168 g C_{5+}/l·h) was revealed for a composite catalyst at 300°C, 3 MPa and space velocity ~ 7000 h^{-1}.

1. Introduction

In the fixed-bed Fischer-Tropsch synthesis (FTS), polymetallic systems based upon bulk hydrogenated ZrFe alloys are known to be among the most active iron-containing catalysts. However, they have such drawbacks as fragility and a low coking stability [1, 2], which could be overcome by their encapsulation in an oxide or cermet matrix. Earlier, some data concerning activity of ZrFe alloys, both pure and incapsulated into Al_2O_3/Al matrix were published [3, 4]. This paper presents results of studies aimed at synthesis and characterization of hydrogenated ZrFe alloys, both pure and encapsulated into alumina-containing matrix, as FTS catalysts. Their catalytic performance data are analyzed with a due regard for their bulk and surface chemical composition, structural and textural properties. For comparison, the data obtained for Fe/ZrO_2 catalysts prepared through impregnation of zirconia are considered [5].

2. Experimental

The catalysts with different Zr/Fe atomic ratios (~2:1, 1:1 and 1:2) were prepared by alloying Zr and Fe at 2200-2400°C under Ar followed by hydrogenation of those alloys at a room temperature and hydrogen pressure of 5 MPa. Approximate stoichiometry of hydrides measured both gravimetrically and by the chemical analysis was found to be $Zr_{2.6}FeH_{1.5}$, $ZrFeH_{0.5}$, $ZrFe_2H_0$. respectively. Hydrogenated active components (AC, fraction 0.5-0.25 mm) were mixed with the aluminum (PA-HP and PA-4 grade) or amorphous aluminum hydroxide powder and subjected to hydrothermal treatment (HTT) at 100°C for 4 hours in a special die immersed into boiling water. This die was specially constructed to ensure a free access of water to the loaded powder and an easy removal of formed hydrogen from its internal volume. HTT procedure results in self-pressing of blends into ceramic or metallo-ceramic monoliths [3,4]. After HTT, the sound monolith was removed from the die followed by its drying and calcination at 350°C and 540°C. Encapsulation in alumina was also carried out through molding of blended AC and alumina.

The phase composition was analyzed with an URD-63 diffractometer using CuK_α radiation. The total pore volume was estimated from the values of true and apparent densities of granulated cermets [3]. The details of microtexture were characterized by using the nitrogen adsorption isotherms measured at 77 K on an ASAP-2400 Micromeritics instrument. The crushing strength was measured using a PK-2-1 machine. The XPS spectra were recorded with a VGESCA-3 spectrometer using MgK_α radiation. Binding energy (BE) of C_{1s} line 284.8 eV was used for calibration. The deconvolution procedure was carried out using peaks profiles approximation by Gaussian function and Shirley background subtraction [6]. The surface concentration of Fe atoms (Θ) was estimated using next formula:

$$\Theta = S \cdot \frac{a}{1+a} \cdot L,$$

where S – specific surface area (m^2/g), a = Fe/Zr atomic ratio according to XPS data, L – number of surface atoms in the monolayer (~10^{19} at/m^2).

SEM analysis was carried out on a LEO-1430 machine supplied with an X-Ray microprobe spectrometer RÖNTEC (beam diameter ca. 1μm). TEM micrographs were obtained with a JEM-2010 instrument with the lattice resolution 1.4 Å and accelerating voltage 200 kV. Local elemental analysis was performed with EDX method on an Energy-dispersive X-ray Phoenix Spectrometer equipped with the Si(Li) detector (the energy resolution not worse than 130 eV).

The catalyst performance in FTS was evaluated for hydrogenated AC and composite catalysts (fraction 2-3 mm) in a flow type reactor at 300-310°C, 3 MPa, H_2:CO ≈2:1, GHSV 6500-7400 h^{-1} with GC analysis of reagents and products.

3. Results and discussions

3.1. Bulk intermetallides

As is seen from the Fig. 1, the activity in FTS per the mass unit of ZrFe intermetallides passes through the maximum with variation of the bulk Zr:Fe ratio. Independently of the total level of activity, selectivity towards different types of reaction products remains almost unchanged being high for methane and low for C_{10+} hydrocarbons formation (Fig. 2). The average parameter α of the Anderson-Schulz-Flory distribution is in the range of 0.6-0.7. These selectivity values are typical for intermetallic iron-zirconium catalysts (including those modified by some other elements, such as Mn and Ce) tested in similar conditions. Under the same reaction conditions, such traditional FTS catalyst as fused iron promoted by K, SiO_2, Al_2O_3 etc shows a lower methane selectivity (21% wt.) and a higher – to heavy (especially, C_5-C_{10} fraction) hydrocarbons (31% wt. vs. 20-22 % wt) [2]. Taking into account a higher selectivity of ZrO_2 –supported iron-containing catalysts to high-molecular products [5,7], conclusion about the negative effect of Fe-containing FTS catalysts modification by the metal zirconium can be made. A key role of the metal–oxide interface for the case of zirconia-supported catalysts can thus be inferred.

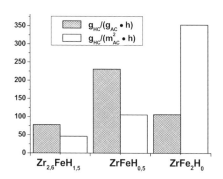

Fig.1 The total hydrocarbon productivity of the hydrogenated Zr-Fe intermetallides in Fischer-Tropsch synthesis with varied Fe:Zr ratio.

Specific activity of catalysts in FTS related to their specific surface area (SSA) increases

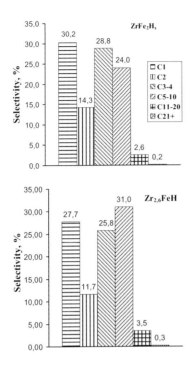

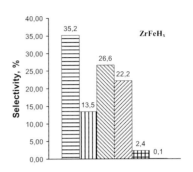

Figure 2. Selectivities to different hydrocarbon fractions of hydrogenated Zr-Fe intermetallides in FTS, wt.%.

Table 1. Surface properties of ZrFe intermetallides

Sample	XPS data (0-20Å) BE, eV		Fe/Zr	SSA, m^2/g	$\Theta \cdot 10^{-16}$, at.Fe/g
	Fe $2p_{3/2}$	Zr $3d_{5/2}$			
$ZrFe_2$	709.6, 710.9	180.3, 182.1	2.2	-	-
$ZrFe_2H_x$	709.6, 711.0	180.5, 182.2	2.6	0.03	22
ZrFe	709.5, 711.2	180.9, 182.6	0.5	-	-
$ZrFeH_x$	709.6, 710.8	181.1, 182.6	0.4	0.22	63
$ZrFeH_x$ (H_2O, O_2)	709.6, 710.9	181.2, 182.8	0.2	0.29	48
$Zr_{2,6}Fe$	709.5, 710.9	181.8	~ 0	-	-
$Zr_{2,6}FeH_x$	709.5, 710.9	180.2, 182.2	0.2	0,17	15

monotonically with the bulk iron content (Fig. 1), correlating thus with the surface iron concentration by XPS (Table 1). On the other hand, specific surface area of AC goes through the maximum with Fe content (Table 1), which results in the non-monotonous variation of catalytic activity related per the mass unit of AC. This trend can be tentatively explained by decline of the intermetallide particles fragility at a high iron content, thus decreasing the crack density for Zr:Fe=1:2 sample.

It should be mentioned that the total output of CH_x related to the surface Fe atom (~5×10^{-19} gCH$_x$/at.Fe×h) estimated for the initial catalysts composition remains almost constant for three types of AC. This suggests that for these catalysts the nature of active sites is identical. Apparently, the actual surface concentration of iron could change significantly in the course of samples activation and interaction with the reaction media due to segregation of iron and carbides formation [8, 9, 10]. However, the effect of these processes is not expected to be significant even when the bulk phase composition varies. For example, according to XRD (Fig. 3), pure intermetallides are mainly comprised of tetragonal Zr_2Fe phase [11] and $ZrFe_2$ phase, the latter one of a hexagonal structure for ~2:1 composition, or of a cubic structure for 1:1 and 1:2 samples [12, 13] The amount of the first phase increases, while that of the second one decreases with the Fe content. For Zr:Fe=1:2 sample, a cubic β-Zr phase is also observed. The phase composition of all hydrogenated intermetallides is characterized by the presence of a cubic $ZrFe_2$ phase along with ε-$ZrH_{1.95}$ phase [14] (for $Zr_{2.6}FeH_{1.5}$ and $ZrFeH_{0.5}$ samples); and by earlier unknown hydride

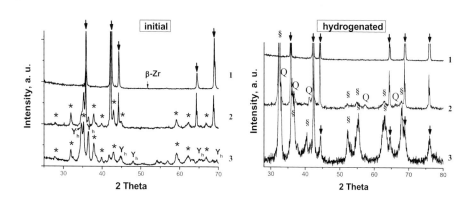

Fig.3. X-ray diffraction patterns of initial and hydrogenated intermetallides: 1 – $ZrFe_2$, 2 - ZrFe, 3 – $Zr_{2.6}Fe$, where Q - $ZrFeH_x$, ↓ - $ZrFe_2$ (cub.), * - Zr_2Fe (tetr.), §- ε-$ZrH_{1.95}$, Y_h - $ZrFe_2$ (hexag.).

phase (designated here as ZrFeH) for ZrFeH$_{0.5}$ sample. According to TEM and EDX, in the accrete particles of the initial "ZrFeH$_{0.5}$" phase, separate single-phase domains with the atomic ratio Zr:Fe $\approx$1:1 (probably – "uknown" phase) (Fig.4a) [4, 15] along with multiphase regions comprised of Zr-enriched matrix (probably – ZrH$_{1.95}$) with iron -enriched (Zr:Fe $\approx$ 1:2) inclusions are revealed. Though the surface Fe/Zr ratio as well as the surface concentration of Fe varies with the bulk composition (Table 1), the specific activity related to the unit of weight varies rather moderately with the bulk Zr content, passing through the maximum at Fe/Zr = 1 (Fig. 1).

Comparison of the most active ZrFeH$_{0.5}$ sample with Fe/ZrO$_2$ catalyst prepared via impregnation (SSA ~207 m^2/g) [5] revealed a much higher surface concentration of Fe per the mass unit of the latter catalyst (~10^{18} vs. ~800×10^{18} at.Fe/g), while Fe concentrations per the surface unit were almost the same (~4 ×10^{18} at.Fe/m^2). The specific activity of "impregnated" catalyst related per the surface Fe atom was found to be lower by ~ three orders of magnitude (~1·10^{-22} gCH$_x$/at.Fe×h). Taking into account significant difference of GHSV in [5] (500 h^{-1}) and in this work along with some variation in other conditions (300°C, 2.5 MPa in this work versus), a higher efficiency of the

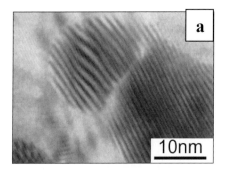

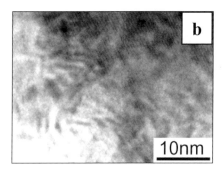

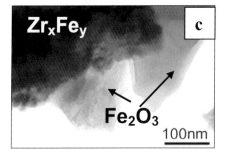

Fig.4. TEM micrographs of the particles "ZrFeH$_x$": initial (a); after calcinations at 350°C (b); after calcinations at 550°C (c) [14].

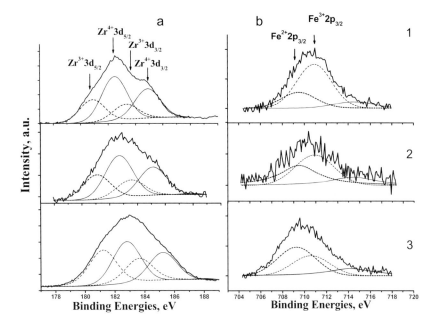

Fig.5. XPS spectra of Zr:Fe=1:1 intermetallide: 1 – initial, 2 - hydrogenated, 3 – after hydrothermal treatment (H_2O, O_2), followed by calcinations at 350°C.

surface active center in $ZrFeH_{0.5}$ sample at the same GHSV can be inferred. The same phenomena has been observed for Ni-Zr and Co-Zr hydrogenated intermetallides in hydrogenation reactions [16].

A high specific activity of hydrogenated intermetallides can be explained by specificity of their surface structure and composition which are quite different from their bulk characteristics. The surface layer of all initial zirconium intermetallides is oxidized due to contact with the oxygen of air [16]. Similar to Fe/ZrO_2 catalysts [5], for our samples, main XPS peaks at ~711 eV ($Fe2p_{3/2}$ peak corresponding to Fe(III) state [17]) and at ~182.2-182.6 eV ($Zr3d_{5/2}$ peak corresponding to Zr(IV) [18]) were observed (Table 1). In addition, after deconvolution of the complex spectra (Fig.5), peaks corresponding to Fe(II) (~709.5 eV with a satellite at 715-716 eV) and Zr(III) (~181.5 eV) [17, 18] are clearly visible. The latter feature is especially unusual being earlier observed only for ZrY sputtered on the steel surface and mildly oxidized [18]. This reduced state probably facilitates active carbide formation as well as activation of hydrogen. Hence, a high activity of ZrFe intermetallides in general and especially that of 1:1 sample can be assigned to a complex structural arrangement of the intermetallide particles where several phases coexist including hydride phase, a surface zirconium oxide (stable in hydrogen up to 800-1000°C [16]) and a highly dispersed surface iron oxide (forms active carbide phase in the reaction media).

3.2. Encapsulated ZrFeH$_x$ intermetallides

Encapsulation of hydrogenated intermetallides (0.25-0.5 mm fraction) in alumina- containing matrix via HTT (vide supra) increases activity (by 4-5 times) and C$_{5+}$ output (by 4-6 times) (Fig.6) as compared to pure AC (2-3 mm fraction). Hence, encapsulation permits to improve the AC effectiveness in the fixed-bed layer presumably due to increase of dispersion without increasing the pressure drop, thus overcoming a drawback inherent for small-sized granules in such a layer.

Main factors determining variation of the AC properties after encapsulation are as follows:
a) grinding, leading to decrease of AC fraction size;
b) oxidation in the course of calcination;
c) interaction with HTT products.

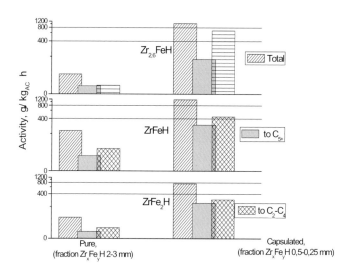

Figure 6. Effect of encapsulation of ZrFe intermetallides with varied Fe:Zr ratio on their activity to different FTS products.

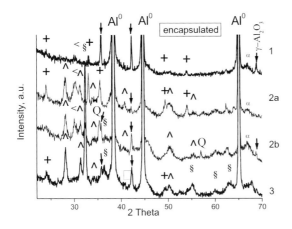

Fig. 7. X-ray diffraction patterns of encapsulated intermetallides:

1 – $ZrFe_2H_0$, 2a – $ZrFeH_{0.5}$ before SFT, 2b – $ZrFeH_{0.5}$ after SFT, 3 – $Zr_{2.6}FeH_{1.5}$, where Q - $ZrFeH_x$, $\downarrow$ - $ZrFe_2$ (cub.), + - Fe_2O_3, §- ε-$ZrH_{1.95}$, < - ZrO_2 (tetr), ^ - ZrO_2(monocl.), a – γ-Al_2O_3.

Variation of synthesis parameters (temperature, type of aluminum powder, fraction of AC etc) results in variation of the composite structure, texture as well as the surface state of AC.

According to XRD data, after encapsulation and calcination at 540°C, such phases as Al^0, γ-Al_2O_3, α-Fe_2O_3 and ZrO_2 (monoclinic and tetragonal) are observed (Fig.7). This agrees with TEM and EDX data detecting an iron oxide phase and Zr-enriched phase. In addition, the phase with Zr:Fe ratio ~1:2 was revealed by EDX as well (Fig.4c).

After calcination of HTT product at 350°C, a large broadening of XRD peaks of almost all components but Al^0 was observed, apparently caused by further oxidation of composites. For particles of calcined samples, "nucleation" of iron- enriched and Zr -enriched regions observed for the initial sample by TEM was detected as well (Fig.4b). This is accompanied by a large increase of microstrains density. So, apparently, calcination at a high temperature further oxidizes AC causing phase segregation (Fig.4c). Calcination at intermediate temperatures leads to more random distribution of Zr and Fe in AC. The activity of AC molded in alumina without any calcination is lower than that of a mildly calcined sample (Table 2).

In agreement with the structural data, thermal analysis revealed a large exoeffect with a mass increment at ~380°C (Fig. 8) due to a bulk oxidation of AC [19]. At lower temperatures, only surface oxidation occurs. Hence, a highest activity of mildly calcined catalyst (Table 2) can be explained by a maximum dispersion of iron in partially oxidized zirconium-containing matrix. For those weakly oxidized samples, an "unknown" ZrFeH phase is restored in

Table 2. Some properties of "$ZrFeH_{0.5}$" (0.25-0.5 mm) encapsulated in different matrices

Characteristics	"$ZrFeH_{0.5}$" in different matrices				
	in Al_2O_3/Al, PA-HP	in Al_2O_3/Al, PA-HP	in Al_2O_3/Al, PA-4	in Al_2O_3 molded	in Al_2O_3 from $Al(OH)x$
Content, wt. %	40	40	25.5	50	55
$T_{calcination}$, ^{0}C	350	550	550	-	350
Total conversion of CO, %	46.7(13.3)*	24.4	21.1	52.7	35.9
Conversion of CO to CO_2, %	10.1(0.7)	2.6	1.5	14.1	8.6
Total output of hydrocarbons, $g/kg_{cat}\cdot h$ $g/kg_{AC}\cdot h$	632(85.8) 1580(-)	381 953	300 1176	524 1048	513 933
Output of hydrocarbons to C_{5+}, $g/kg_{cat}\cdot h$	157(36.8)	99	82	88	104
V_{total}, ml/g	0.22(-)	0.64	0.31	0.24	0.64
SSA, m^2/g	42(15)	42	11	74	40
$V_{under\ 2000Å}$, ml/g	0.050(0.012)	0.051	0.019	0.133	0.110
$D_{average}$, under 2000 Å, Å	39(57)	49	53	72	111

*In brackets the data for impregnated catalyst.

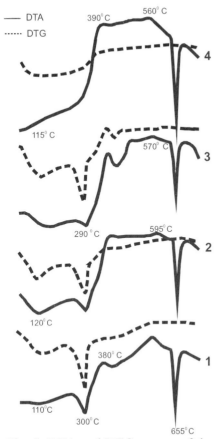

Fig. 8. DTA and DTG curves of the HTT products for various blends of the AC fractions with aluminum (ZrFeH$_x$:Al$_{PAHP}$ = 1:1): 1) 0.25-0.50, 2) 0.20-0.25, 3) 0.06-0.06, and 4) ≤0.04 mm.

SFT tests (Fig.7, curve 2b), thus suggesting some reversibility in formation of this phase in reducing conditions.

For composites calcined at 350°, the total efficiency first declines as the AC particle size decreases increasing again for the finest particles (Fig. 9). The efficiency with respect to C$_{5+}$ hydrocarbons is lower for the finest AC than that for the most coarse fraction. For this series of samples, selectivity to various hydrocarbon fractions varied slightly and irregularly. Specific catalytic activity in FTS is usually expected to grow with decreasing the AC particle size below ~0.1 mm due to decreasing diffusion limitations for the reactants (first of all, CO) transport in the catalyst micro-and mesopores filled with condensed hydrocarbons [20]. Hence, for composite catalysts, performance of even coarse fractions seems not to be affected by the mass-transfer limitations.

X-ray data revealed no considerable effect of grinding on the phase composition of AC [19]. The total pore volume of composite catalysts varies only slightly, while their micropore and mesopore volumes and the specific surface area changes non-monotonically (Tables 1, 2) which will be discussed later. For the finest fraction of AC, specific surface area decreases markedly, while the micropore volume for fractions less than 0.25 mm is negligible (Table 3). Probably, the number of cracks per particle decreases for fine fractions of AC, so grinding leads only to further disappearance of cracks due to percussion riveting. However, it has no effect on the specific activity of AC (related per gram) (Fig.9)

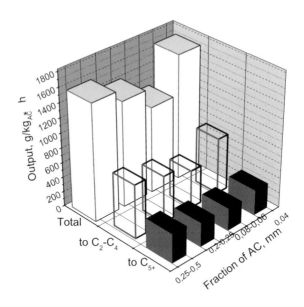

Fig. 9. The influence of the AC dispersion in Al_2O_3/Al matrix on its activity in FTS (Al^0 PA-HP, $T_{calc.}=350°C$, $ZrFeH_{0.5}$).

Understanding the reason of a weak effect of AC dispersion on its specific activity is possible through the detailed analysis of the impact of encapsulation procedure on the properties of AC. Firstly, the products of the hydrothermal oxidation of aluminum can decorate the surface of AC (Fig.10b) or even cover it by the layer of porous aluminum hydroxides (or alumina after calcination) (Fig.10c) which is clearly visible from comparison of SEM data for the AC surface without HTT products (Fig. 10d). This hypothesis is supported by the XPS data demonstrating very low surface content of Zr and Fe in the composite catalyst as compared to its bulk content (Al:Zr:Fe=~1:0.007:0.008). Similarly, a higher catalytic activity was revealed for AC encapsulated in Al_2O_3/Al matrix prepared from PA-4 aluminum with a lower reactivity with respect to steam [15] and, hence, a lower content of aluminum hydroxide in composite. Encapsulation of active component carried out by using more reactive PA-HP aluminum produced a stronger cermet (Table 2) with a lower activity. Comparing specific activities of "$ZrFeH_{0.5}$"/Al_2O_3/Al and "$ZrFeH_{0.5}$"/Al_2O_3 (prepared with aluminum hydroxide) calcined at the same

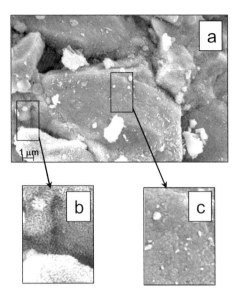

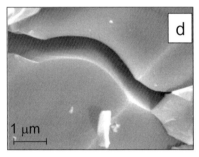

Figure 10. SEM micrographs of $ZrFeH_x/Al_2O_3/Al$ catalyst prepared through HTO and calcination at 350°C (a), $ZrFeH_{0.5}$ after the same treatment (d), inserts (b) and (c) – surface of encapsulated AC ((b) – with porous alumina, (c) – without porous alumina.

(350°C) temperature also revealed the negative effect of the HTT products on the performance of encapsulated AC (Table 2).

Secondly, the partial oxidation can affect the specific surface area and the surface concentration of iron in AC (Table 1). On the stage of calcination, it depends upon the temperature as well as the AC fraction size. As follows from the thermal analysis data [19] (Fig.8), the strongest effect of the AC fractional composition is observed for oxidation of composite obtained via HTT. For three samples with a coarse AC fraction, both DTA and DTG curves, though being somewhat different in shape, have a pronounced common feature, namely, a sharp peak at 290-300°C due to decomposition of crystalline bayerite. This peak is absent in the encapsulated catalyst containing AC fraction <0.04 mm. Presumably, dispersed AC affects aluminum oxidation and formation of aluminum hydroxides. The reactivity of aluminum and, hence, the alumina content in the composite decreases with decreasing the particle size. Smaller hydroxide content in the HTT product means a greater amount of remaining aluminum, which is indeed revealed by more intense endothermic peak at 655°C (Fig. 10, curves 4) corresponding to Al^o melting.

Therefore, during HTT, there is a strong interaction between the components of emerging cermets. The products of aluminum HTT can "decorate" the AC surface through the chemical interaction between Al and iron hydroxides [21] as well as by formation of iron-aluminum mixed oxides/hydroxides. An aluminum hydroxide "coating" may be formed, which can give a porous alumina layer on the AC surface after the thermal treatment. This coating can obstruct both the access of reactants and removal of products. Both effects would decrease the specific (per the unit surface area) activity of AC and the composite catalysts as a whole if the particle size decreases from 0.25-0.50 to 0.06-0.08 mm. On the other hand, a finer AC affects the reactivity of aluminum during HTT and, thus, decreases the total amount of hydroxocompounds in the pores of resulting composite. This reduces the "decoration" effects and increases the catalytic activity of the catalyst fraction <0.04 mm. Hence, the encapsulation procedure results in a complex (negative and positive) effects leading to the surface and bulk modification of AC which can mask the effect of AC dispersion.

4. Pore structure and peculiarities of catalytic properties

Hydrogenated AC powders (0.25-0.5 mm fraction) are characterized by a low SSA (Table 1) as well as a small (<0.01 cm^3/g) pore volume due to presence of pores with diameter >400 Å. Special experiments did not reveal a substantial effect of HTO and calcination on the porosity of intermetallides [4]. For finer AC fraction, the pore volume of micro- and mesopores is negligible (Table 3).

The pore structure of composite catalysts has a polymodal character. According to the nitrogen adsorption isotherm data, it has very narrow (in the range of 40-60 Å) pore size distribution (~70-120 Å for the alumina matrix), a relatively high SSA and the micropore volume (Table 2). The pore structure characteristics of composites are close to those for pure Al_2O_3/Al or alumina matrix (Fig. 11). Possibly, in spite of alumina deposition on the surface, a high activity of AC is provided by a high permeability of this porous layer towards reagents and FTS products. Another factor determining a high activity is a developed ultra-macroporosity. Qualitatively, the pores with diameters up to tens of μm are visible in SEM images of composite "$ZrFeH_{0.5}$"/Al_2O_3/Al catalyst (Fig. 12). Quantitatively, the presence of these pores follows from the total pore volume exceeding that of pores with sizes below 2000 Å (Table 2). These ultramacropores formed between particles of AC, aluminum or alumina in the initial blends provide an easy transfer of gaseous substances between the interior of granules and the gas phase, thus helping to achieve a high performance of catalysts (compare Table 1 and Table 2).

Comparison of the textural characteristics of catalysts with their catalytic performance in Fischer-Tropsch synthesis revealed that for alumina- containing composites there is a clear tendency for decreasing selectivity to C_{5+} products and increasing that to methane with the increase of the micropore and mesopore volume and SSA (Fig. 13). Some deviation can be explained by the

Table 3. Effect of the Particle Size of the AC on the Mechanical and Textural Properties of the Composite Catalysts $ZrFeH_{0.5}/Al_2O_3/Al$ (Al^0_{PA-HP}, $T_{calc.}=350°C$).

Characteristics	AC fraction, mm			
	0.25-0.5	0.2-0.25	0.056-0.08	<0.04
Crushing strength, MPa	2.1	1.4	0.9	2.1
Total pore volume, cm³/g	0.22	0.28	0.31	0.23
Specific surface area, m²/g:				
$ZrFeH_x/Al_2O_3/Al$	42	49	25	36
$ZrFeH_x$	0.22	~0.12	~0.21	~0.04
Micro- and mesopore volume, cm³/g:				
$ZrFeH_x/Al_2O_3/Al$	0.050	0.036	0.023	0.032
$ZrFeH_x$	0.012	<0.001	<0.001	<0.001
Pore diameter (up to 2000 Å), Å :				
$ZrFeH_x/Al_2O_3/Al$	39	30	37	35
$ZrFeH_x$	440	-	-	-

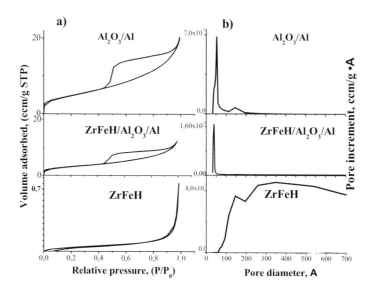

Fig. 11. Textural properties of matrix, $ZrFeH_x$ and encapsulated $ZrFeH_x$: a) isotherms, b) pore size distribution.

Fig.12. SEM micrographs of (a) the Al$_2$O$_3$/Al matrix, (c) original ZrFeH$_x$ particles (<0.04 mm), and (b) the composite catalyst ZrFeH$_x$/Al$_2$O$_3$/Al (the batch composition is ZrFeH$_x$/Al PAHP = 1 : 1) [18].

effect of the average pore diameter which could differ for composites with the same integral pore volume. Observed dependence between selectivity and SSA could be explained by variation of the concentration of active sites (surface concentration of iron) as well as by the impact of diffusion characteristics of porous alumina, both factors are known to be of a great importance for the FTS selectivity. Detailed analysis of these factors has been earlier presented for cobalt- containing FTS catalysts [22]. Probably, the surface of alumina strongly affects the FTS characteristics through readsorption of reaction intermediates leading to chain termination, while readsorption on the active sites situated on the surface of AC leads to the chain growth. Another factors of importance could be filling of micropores by water and liquid hydrocarbons formed in FTS or a non-uniform distribution of active sites inside the composite catalysts. In all cases, the microporosity of alumina has a negative effect on the chain growth reaction.

The same negative impact of microporosity on the C$_{5+}$ hydrocarbons selectivity was earlier observed by Hongwei et al for Fe/ZrO$_2$ catalyst [5] (cf. Fig. 14 where the data of [5] are presented). The authors discussed these data in terms of the AC particle size quite traditional for porous impregnated catalysts. However, in our case, intermetallide particles in composites have much bigger sizes greatly exceeding typical sizes of FTS products molecules. Hence,

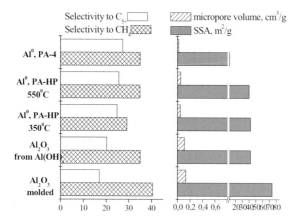

Fig.13. Influence of micropore structure on the selectivity of composite catalysts based upon hydrogenated ZrFe intermetallides.

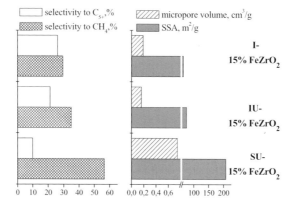

Fig.14. Influence of micropore structure on the selectivity of the impregnated Fe/ZrO$_2$ catalysts (derived from [5]).

observed trends can be explained by the indirect effect of the porous oxide matrix described earlier. Similarly, for pure ZrFe intermetallides, non-monotonous variation of the C$_{5+}$ and methane selectivity with Fe:Zr ratio (Table 1) can also be explained by the effect of the mesopores volume and SSA [4]

To improve selectivity by decreasing microporosity and, perhaps, affecting the active sites distribution, the composite "ZrFeH$_{0.5}$"/Al$_2$O$_3$/Al catalyst was impregnated by water solution of Fe(NO$_3$)$_3$ followed by drying and calcination. This procedure really changed the micropore structure decreasing SSA and the pore volume, while increasing the average pore diameter (Table 2). This suggests that deposition of iron oxides primary occurs within alumina ultramicropores. As a result, the selectivity to CH$_4$ decreases while C$_{5+}$ selectivity reaches the highest value for this type of catalysts (Table 2). Unfortunately, the total activity was found to be lower, possibly due to deposition of less active bulk iron oxides on the surface of AC. Indeed, according to XPS, the ratio of components on the surface was changed from Al:Zr:Fe =~1:0.007:0.008 before deposition to ~1:0.04:0.1 after deposition. Nevertheless, this model experiment illustrates one of the possible ways to further improve the performance of composite catalysts.

5. Advantages of the macroporous composite catalysts

Activity of the unit of a catalytic layer volume is very important characteristics of the catalyst, which depends on the activity per the weight unit as well as on apparent density of the layer. For bulk ZrFe intermetallides (fraction 2-3 mm), the apparent density (~3 g/cm^3) is much higher than that for composite catalysts (~1 g/cm^3). However, enhancement of AC specific activity due to increase of dispersion permits to obtain activity of the composite catalysts volume unit comparable with that of pure AC with the same size of granules (Fig. 15).

Granulated Fe/ZrO$_2$ catalysts obtained via impregnation of zirconia support have an apparent density of the fixed bed catalyst layer about 1 g/cm^3 [5]. For these catalysts, the rate of liquid hydrocarbons production varies from ~6 to 15 g$_{CHx}$/l hr. Obviously, extrapolation of these data obtained for 20-40 μm

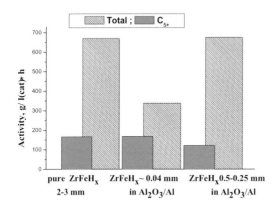

Figure 15. The output of different FTS products per volume of bulk ZrFeH$_{0.5}$ and composite catalysts with varied dispersion of AC [14].

granules at GHSV=500 h^{-1} [5] to our conditions (~7000 h^{-1} and 2-3 mm granules) would give the rate of liquid hydrocarbons production below that for composite catalysts.

As described above, the most effective composite catalysts demonstrate the specific volume activity close to that for bulk intermetallides. This is apparently caused by specificity of the composites structure and texture, though variation of the state of AC surface caused by encapsulation could be important as well. As for the texture effects, a significant decrease of the encapsulated AC particle size versus that for pure AC minimizes the effect of the internal diffusion on the total reaction rate. Similarly, while condensed hydrocarbons usually fill meso- or micropores of traditional FTS catalysts, they are nearly absent in macropores of cermets, thus minimizing the diffusion resistance of the inner space of matrix

The quantitative abalysis of the relative effectiveness of composite catalyst can be carried out following next assumptions:

a) for iron catalysts, at low or intermediate conversions, the rate of FTS can be described by the first-order equation:

$$dC_{CO}/dt = kC_{CO}$$

б) conditions within a catalyst pellet are close to isothermal ones. This is supported by the fact that catalysts are comprised of metals instead of oxides and, hence, the pellet is characterized by a rather high thermal conductivity.

The volumetric hydrocarbon productivity for bulk hydrogenated $ZrFeH_{0.5}$ intermetallide and the most active composite samples within the 40 % ZrFeH/PA-HP series is nearly identical (~670-690 g/l×h, Fig. 15), which is equal to 48-49 mole CO/l×h. For the isothermal plug-flow reactor at GHSV≈ 7000 h^{-1}, this yields the efficient FTS rate constant k≈420 h^{-1}. An apparent density of the composite containing 40 % wt. of AC is approximately 3 times lower than that of the bulk intermetallide. Therefore, a volume AC fraction is about 1/7 and k value, referred to it, is, respectively, ~2900 h^{-1}.

Using the first-order kinetic equation, it is possible to apply well-known expression for the calculation of particle effectiveness factor η [20, 23]

$$\eta = \frac{3\phi cth(3\phi)-1}{3\phi^2}$$

where Thiele modulus ϕ as a function of the intrinsic rate constant k_{int}, the particle radius R_p and the effective H_2 diffusion coefficient D_{eff} is expressed by equation $\phi = (2k_{int}R_p^2/D_{eff. H2})^{1/2}$. In turn, $D_{eff. H2} = D_{H2}\varepsilon/\tau \approx 0.36$ cm^2/h, since at 300 °C D_{H2} for condensed hydrocarbon media ≈0.95 cm^2/h, AC pore tortuosity τ is taken as 2.6 [24].

Values of AC porosity ε given in Table 4 were estimated by relation: $\varepsilon=\rho_{AC}V_{micro}$, where ZrFeH$_x$ particle density $\rho_{AC}\approx6$ g/cm^3, V_{micro} – AC pore volume, cm^3/g, see Table 3. Using the data for bulk ZrFeH (fraction 2-3 mm), one can calculate the value of $k_{int}\approx50\ 000$ h^{-1} and effectiveness factor $\eta\approx$ 0.0082 (Table 4). For composite with the incapsulated particles (AC fraction 0.25-0.5 mm), η increases up to 0.042. Using the proportion $k_i/k_j=\eta_i/\eta_j$ for any pair of pellets i and j with different sizes for a considered composite, it is easy to find that the value of observed k is ≈ 2560 h^{-1}. These results are in a good agreement with the experimental data, namely, the increase of FTS rate and specific AC productivity of hydrocarbons for 5-6 times after encapsulation.

As for smaller AC particles, the situation is different due to their lower porosity. One can find that for the composite with the AC fraction 0.2-0.25 mm both actual and calculated productivity are lower than those for 0.25-0.5 mm fraction: the first – for nearly 1.2 times (Fig.9), the second – for 1.7 times (Table 4). So, diminished porosity causes a strong influence on the activity but the surface blocking is negligible. However, for finer (0.06-0.08 mm) AC fraction, calculated k is 2 times higher than that for 0.25-0.5 mm fraction, but the actual activity is $\approx$1.4 times lower. This implies that for this fraction size, HTT products blocking phenomenon plays an important role. For the composite with a finest fraction <0.04 mm (much lower SSA, and, respectively, 4-5 times lower value of k_{int}), both calculated and actual values of activity are nearly the

Table 4. Effectiveness factors and Thiele modulus for various Fe-containing catalysts

Catalyst	ε	η	ϕ	R_p, cm	Ref.
ZrFeH bulk	0.07	0.0082	122	0.125	This study
ZrFeH/PA-HP	0.07	0.05	19.5	0.02	This study
-"-	0.006	0.0297	33.3	0.01	This study
-"-	0.012	0.102	9.4	4×10^{-3}	This study
-"-	0.0025	0.19* 0.047**	5.02*	2×10^{-3}	This study
Fe fused	0.445	0.43	1.95	0.024	[20]
-"-	-"-	0.73	0.85	0.011	[20]
-"-	-"-	0.95	0.32	3.9×10^{-3}	[20]
Fe-Cu-K	0.51	0.12	~20	0.2	[24]
-"-	-"-	0.42	~6	0.05	[24]
Fe precipitated	-	~0.2	-	0,125	[25]

* Here $k_{int}\approx12\ 000$ h^{-1}
** Value referred to $k_{int}\approx50\ 000$ h^{-1}

same. This is explained by a small degree of AC surface blocking due to a specificity of aluminum oxidation under HTT (vide supra).

The values of Thiele modulus calculated for studied samples (Table 4) are much higher (values of η much lower) as compared with those determined earlier for various Fe-containing catalysts [20, 24, 25]. Thus, in [20] the same equation for η calculation was used as in this work. However, the dependence of Thiele modulus on the particle size was different due to a lower and varied porosity of the AC particles in the present work. Another relation between φ and η was applied in [24]; however, in that work more complex kinetic model for the less active catalyst was used.

A low degree of the cermet matrix transport macropores filling by condensed hydrocarbons is apparently caused by the following factors:

a) The specific macropore volume of composite catalyst is considerably higher than the specific pore volume for granules of bulk intermetallide. For example, for $ZrFeH_x$ this volume is equal to 0.01 cm^3/g, while for the composite catalyst it is 0.2 cm^3/g. A certain balance between formed and removed hydrocarbons is achieved, and the amount of product condensate collected within pores remains the same for a given kind of the catalyst. Therefore, at the same conditions (i.e. generation of the same amount of high-molecular hydrocarbons per the unit of time during FTS) in the flow reactor, a degree of pore filling in the case of macroporous composite can be essential smaller (in particular, if the level of high-molecular condensed product formation is low as in the case of Fe-Zr intermetallides at 300 °C);

б) In addition, for meso- and micropores, the liquid/gas phase equilibrium can be shifted to a liquid phase due to the capillary effect. According to Kelvin equation

$$p/p_0 = \exp(2\sigma v/rRT)$$

the ratio of the vapor saturation pressure of considered substance (with a surface tension σ and a liquid molar volume v) for a capillary and for the free surface at a temperature T is inverse-exponentially interrelated with the curvature radius of a surface. The latter is determined by the radius of a capillary and a degree of wetting of the particle material. Thus, the smaller is the pore size, the lower is the saturation pressure of condensed hydrocarbon, and, respectively, the larger is the amount of condensate.

It is interesting to note that in the modeling of FTS in the catalyst pellets [20, 22, 24], only particles with pores completely filled by condensed products are always considered. Such an idea about the pellet state has caused well-known direction of FTS technology development, namely "egg-shell" catalysts [26], where the active substance is situated mainly in the near-surface layers of a support. Here, the internal pellet volume is not used at all. As follows from the aforesaid, the composite catalysts realize the opposite approach to the pellet design: AC is distributed more or less uniformly within the bulk of the granule, and the unrestricted reagents (products) supply (removal) to it is provided due to a special method of the matrix organization around AC. Obviously, both approaches have their own advantages and disadvantages; their detailed analysis is the subject of further studies

6. Conclusion

The variation of Zr:Fe ratio in hydrogenated intermetallides permits to obtain AC with the maximum activity for 1:1 sample due to formation of mixed "ZrFeH" phase; such intermetallides revealed extremely high FTS activity per the surface Fe atom. Calcination at moderate temperatures improves the uniformity of Zr-Fe atomic distribution and, hence, activity in FTS. At higher temperatures AC is almost fully oxidized and decomposed on different iron - enriched and Zr- enriched components which leads to deactivation of AC. These processes also occur in the course of AC encapsulation in the alumina/aluminum matrix. The latter, due to a high amount of ultramacropores, provides a good permeability of reagents and reaction products in FTS. However, the microporous alumina negatively affects the selectivity to longer hydrocarbons. Further improvement of composite catalysts is possible through the increase of the amount of active phase in AC as well as by diminishing microporosity with more random distribution of active sites in alumina. The results obtained can be applied to other types of composite catalysts including cobalt-containing ones. Composite metalloceramic catalysts are capable to combine advantages of granulated catalysts for the fixed bed (simplicity of loading/unloading and products separation, absence of mechanical erosion) and fine catalysts for fluidized (two-phase and three-phase slurry) beds (significantly higher activity per the mass unit). The composite catalysts based upon Al_2O_3/Al matrix can be shaped as granules, honeycombs, and thick coatings on inner or outer sides of metallic tubes improving heat conductivity which is important for FTS process.

References

1. A.Ya .Rozovsky, Kinet.Catal. 40 (1999) 358.
2. L.A.Vytnova, V.P.Mordovin, G.A.Kliger, E.I.Bogolepova, V.I.Kurkin, A.N.Shuykin, E.V.Marchevskaya, E.V.Slivinky,. Russ.J.Petrochemistry, 42(2002) 111.

3. S.F.Tikhov, V.A.Sadykov, Yu.V.Potapova, A.N.Salanov, S.V.Tsybulya, G.N.Kustova, G.S.Litvak, V.I.Zaikovskii,. S.N.Pavlova, A.S.Ivanova,

A.Ya.Rozovskii, G.I.Lin, V.V.Lunin, V.N.Ananyin, V.V.Belyaev, Stud.Surf.Sci.Catal., 118 (1998)797.

4. S.F.Tikhov, V.I.Kurkin, V.A.Sadykov, E.V.Slivinsky, Yu.N.Dyatlova, A.E.Kuz'min, E.I.Bogolepova, S.V.Tsybulya, A.V.Kalinkin, V.B.Fenelonov, V.P.Mordovin, Stud.Surf.Sci.Catal., 147(2004) 337.

5. Z.Hongwei, Z.Bing, P.Shaoyi, W.Dong, F.Wenhao, J.Molec. Catal. (China) 9(1995)13.

6. D.Briggs, M.P. Seach. Practical Surface Analysis by Auger and X-Ray Photoelectron Spectroscopy, John Wiley & Sons Ltd., New York, 1983.

7. K.Chen, Y.Fan, Z.Hu, Q.Yan, Catal. Lett., 36(1996) 139

8. B.T.Davis, Catal.Today 84(2003)83.

9. S.Li, S.Krishnamoorthy, A.Li, G.D.Meitzner, E.Iglesia, J.Catal 206(2002)202.

10. J.Xu, C.H.Bartholomew, J.Sudweeks, D.L.Eggertt, Topics Catal 26(2003)55.

11. JCPDS 25-04220

12. JCPDS 26-0809.

13. JCPDS 18-0666.

14. JCPDS 236-1339.

15. S.F.Tikhov, Yu.N.Dyatlova, A.E.Kuzmin, V.I.Kurkin, V.A.Sadykov, E.V.Slivinsky, E.I.Bogolepova, S.V.Tsybulya, A.V.Kalinkin, V.B.Fenelonov, V.P.Mordovin, A.N.Salanov, V.I.Zaikovskii, Prep. Pap.-Am.Chem.Soc., Div.Petr.Chem. 50(2005)188.

16. V.V.Lunin, A.Z. Khan, J.Molec.Catal., 25 (1984) 317.

17. A.V.Kalinkin, V.I.Savchenko, A.V.Pashis, Catal.Lett. 59(1999)115.

18. G.M.Ingo, J.Am.Chem.Soc., 74(1991)381.

19. A.E.Kuz'min, Yu.N.Dyatlova, S.F.Tikhov, V.I.Kurkin, V.A.Sadykov, E.V.Slivinskii, E.I.Bogolepova, S.V.Tsybulya, V.B.Fenelonov, V.P.Mordovin, G.S.Litvak, Kinet.Catal., 46(2005) (accepted).

20. W.H.Zimmerman, J.A.Rossin, D.B.Bukur, Ind. Eng. Chem. Res., 28(1989) 406.

21. A.I.Rat'ko, V.E.Romanenkov, E.V.Bolotnikova, Zh.V.Krupen'kina, Kinet. Catal., 45(2004)162.

22. E.I.Iglesia, S.C.Reyes, R.J.Madon, S.L.Soled, Adv.Catal. 39(1993) 221.

23. Zhorov Yu. M. Modeling of Oil Refining and Petrochemistry Physicochemical Processes (Russ. Ed), Moscow, Khimia, 1978 [in Russian].

24. Y.-N.Wang, Y.-Y.Xu, H.-W.Xiang, Y.-W.Li, B.-J.Zhang, Ind. Eng. Chem. Res., 40(2001) 4324.

25. A.Jess, R.Popp, K.Hedden, Erdol Erdgas Kohle, 113(1997) 531.

26. E.Iglesia, S.L.Soled, J.E.Baumgartner, S.C.Reyes, J. Catal., 152(1995) 137.

178

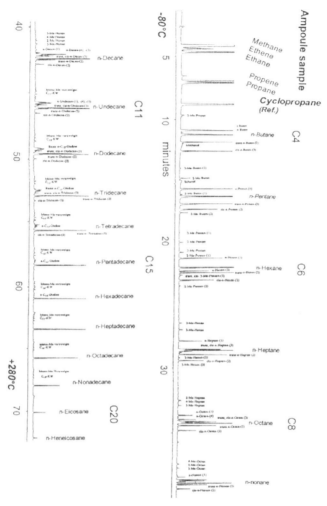

Fig. 1: Chromatogram of a Fischer-Tropsch synthesis product [20]
Ampoule sample, taken from the hot (220 °C) gas/vapour product stream after a hot trap at 200°C
and after depressurizing. The ampoule is sealed and stored for later analysis. Sampling duration is
less than 0.1 second.
FT-synthesis: Cobalt catalyst, steady state, reaction temperature 190 °C, total pressure 5 bar.
Gas chromatography: Hewlett Packard 5880, adapted for ampoule samples and pre-column
hydrogenation, FID 280 °C.
GC-conditions: Fused silica capillary, 100 m, 0,25 mm, cross linked methyl-silicon, 0.5 µm film
Sample introduction: Nitrogen flow 150 ml/min (NTP) for 2.5 minutes at 250°C after ampoule
breaking. Split 1:20 to 1:600 depending on sample size, ampoule volume 0.5 – 5 cm^3
Carrier gas: Hydrogen 1,6 ml/min (NTP)
Temperature: - 80 °C for 3,75 min; up to - 35 °C with 15 °C/min, 4 minutes at - 35 °C; up to - 5
°C with 2.5 °C/min, 2.5 min at -5°C; up to 280°c with 5°C, 10 min at 280 °C.
Reference compound for quantitative analysis: Cyclopropane.

The multiplicity of ordered product composition provides information about the events on the catalytic sites and about the nature of the sites. The following features of compare are used in this article:

Methanation in relation to the Fischer Tropsch synthesis
Olefin/paraffin selectivity
Olefin isomerization
Chain length distribution
Chain branching
Self-organization of the catalyst.

The discussions and conclusions refer to kinetic schemes, probabilities of growth, branching and desorption, secondary olefin reactions, the episodes of formation of the FT-regime and catalyst reassembling, the active sites, the "true" FT-catalyst and the basic FT-principle.

A key technique of this work is *time resolved analysis of product composition,* by means of **ampoule sampling** and adapted gas chromatography [2,3,4,5]. A reference chromatogram is presented in **Fig. 1**. It comprises the organic compounds C_1 to C_{21}. Using GC together with pre-column hydrogenation, as shown for the C_7-secction in **Fig. 2,** yields a chromatogram of only paraffinic compounds which allows calculation of branching probability, using the kinetic model of "non trivial surface polymerization"[6,7,8,9,10].

Temporal changes of structure and composition of the iron catalyst have been studied with XRD-, Moesssbauer- and XPS-techniques [11]. Temporal changes of the surface structure of cobalt during FT-synthesis have been observed by tunneling electron microscopy [12].

180

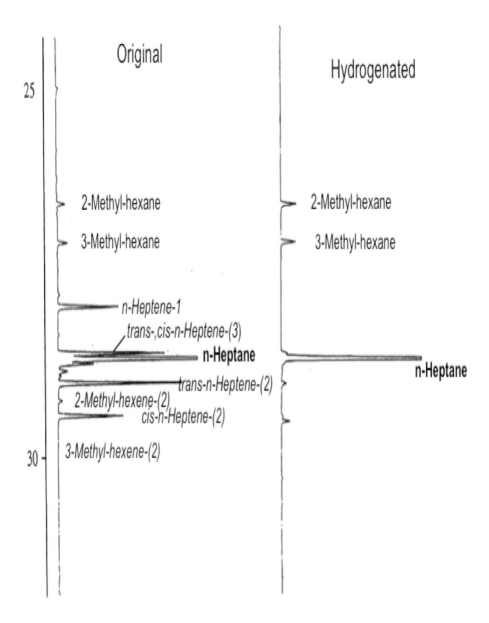

Fig. 2: C$_7$- section of the FT-product-chromatogram-
original (left) and hydrogenated (right) [20]

Pre-column hydrogenation: Catalyst 3 ml of 0.35 wt-% Pt on fused silica particles (dp 0.2 -0.25 mm). Temperature of hydrogenation: 250 °C, 3 minutes. Further legend as in Fig.1.

Methanation

Methane formation in FT-synthesis is an issue of theoretical and technical interest, CH_4 being a "true" FT-product and an independent methanation product as well. Methane formation is thermodynamically favored against the formation of higher hydrocarbons. It follows that CH_4-formation is suppressed in FT-synthesis.

With alkalized iron catalysts, the FT-regime can be amazingly stable against excessive methane formation, even at high temperature (ca 350 °C) and low CO partial pressure [11,13,14] and also in the episodes of self-organization, when the metallic iron is being converted to iron carbide [11]. Consequently it is concluded that the iron-FT-sites are static entities [15] and methanation sites are not present. Thus iron catalysts allow the industrial production of low molecular weight hydrocarbons by increasing the reaction temperature (without excessive methane formation).

With cobalt catalysts, the FT-regime is sensitive against changes of partial pressures (mainly CO and H_2) and temperature [9,10,16]. Methane formation is restricted at low temperature (180-220 °C), sufficiently high CO partial pressure and through self-organization of the FT-regime (restructuring of the catalyst by surface segregation) during the early episodes of synthesis. Otherwise, the methane formation precedes excessively (on unspecific metallic sites) [16]. Cobalt is well suited for low temperature FT-synthesis to produce diesel fuel, waxes and lubricants together with gasoline.

Then - as evident from Fig.7 - the comparison of Fe- and Co- catalysts in methane selectivity (on the basis of growth probability as a function of carbon number) results in only limited methane formation on iron (only growth sites), but on the Co-catalysts the methane formation tends to be excessive (low value of p_g at Nc =1).

The kinetic scheme for methanation in **Fig. 3** shows as the slow step in the sequence of hydrogen addition to carbon, to be the final (irreversible) associative reaction between H and CH_3. This has been reported also for methanation on ruthenium [17].

As indicated in the kinetic scheme in Fig. 3, and resulting from Fig. 7 (left), CH_4-formation on growth sites is slow, e.g., only one third as fast as CH_2-addition for growth.

Thermodynamic Trend: Methanation much favored against FT-Synthesis

	With Iron (K-promoted)	**With Cobalt**
Lowest CH4-Selectivity	ca. 5 C%	ca. 5C%
Increasing Temperature	little change	strong increase
Decreasing pCO	little change	strong increase
Increasing pH2O		decrease
Initial time-on-stream	little change	decrease
Methanation sites	static/merely present	dynamic

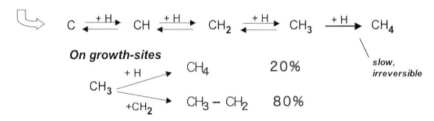

On methanation sites

CO-Dissociation

$$C \xrightleftharpoons{+H} CH \xrightleftharpoons{+H} CH_2 \xrightleftharpoons{+H} CH_3 \xrightarrow{+H} CH_4$$

slow, irreversible

On growth-sites

$$CH_3 \begin{array}{c} \xrightarrow{+H} CH_4 \quad 20\% \\ \xrightarrow{+CH_2} CH_3 - CH_2 \quad 80\% \end{array}$$

Fig. 3; Methanation in FT-synthesis

The olefin/paraffin selectivity

With iron- and particularly with cobalt catalysts, the paraffin to olefin ratio of the produced hydrocarbons can vary strongly, as depending on several parameters. It has been shown (e.g. by co-feeding of olefins) that both, olefins and paraffins, are primary FT-products, obtained in a molar ratio of olefins to paraffins of about 4:1 to 3:1 [9,10,18]. This primary ratio is almost the same with iron and with cobalt catalysts. It is not much depending on reaction conditions.

The primary olefin-selectivity can be obscured by secondary reactions [9,10,18,19] through

- Olefin double bond shift
- Olefin hydrogenation
- Olefin re-adsorption on growth sites.

These reactions are carbon number sensitive (**Figs. 4 and 5**), which eases their identification. Due to these reactions, α-olefins cannot be produced on cobalt catalysts.

As seen in **Fig. 4 (right),** with a **cobalt** catalyst at the steady state (t_{exp} = 9,600 min), the curve shape of olefin content as a function of carbon number shows a low value at Nc = 2 (reflecting the high reactivity of ethene), the highest value at Nc = 3 (lowest reactivity for secondary reactions) and then a steady decline with increasing carbon number, as expected because of with carbon-number-favored-adsorption of the olefins [20, 21].

In the early episodes of synthesis (t_{exp}=12 and 30.3 minutes) the olefin content is low, as caused by incomplete cobalt surface segregation and more pronounced secondary reactions. Surprisingly, here the curve shape is different; specifically the value at C_3 is now lower than that at C_4. This is to be understood from the simultaneous double bond shift, together with the rule of easier hydrogenation of terminal than of the internal olefin double bonds: The double bond shift at C_3 only converts two identical α-olefins into each other

$$CH_3 - CH = CH_2 \quad \Leftrightarrow \quad CH_2 = CH - CH_3$$

whereas at Nc = 4 (and higher values of Nc) less reactive internal olefins (mainly β-olefins) are formed.

For the **iron** catalyst, the curves of olefin content in dependence of carbon number (**Fig.4**) do not exhibit the typical shape at secondary olefin

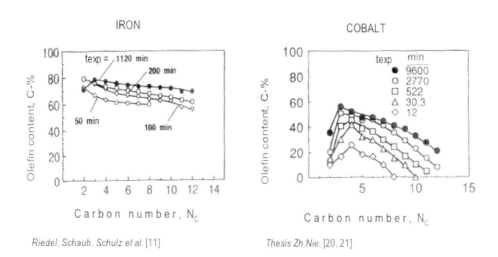

Iron (100Fe-13Al-11Cu-25K); 250°C, 10 bar, H_2/CO = 2.3:1
Cobalt (97Co-12Zr-100Aerosil); 190°C, 5 bar, H_2/CO = 2:1

Fig. 4: Olefin selectivity in FT-synthesis on iron and cobalt as a function of carbon number and time

hydrogenation (except the a little low value at Nc = 2 of the product at t_{exp} = 1120 min) indicating primary olefin-selectivity, characterized by (almost) horizontal lines - these only smoothly declining with carbon number.

These lines are at a lower level at short times (t_{exp}), indicating the alternative desorption as α-olefin or as paraffin to change in direction of preferably α-olefin desorption, referring to increasing suppression of paraffin desorption during catalyst self-assembling.

The moderate decline of olefin content at increasing carbon number could be caused by increasing α-olefin re-adsorption on growth sites.

With (alkalized) iron catalysts, the secondary olefin reactions (except some re-adsorption on growth sites) can be (almost) completely absent and high yields of α-olefins are recovered [11, 18]. Thus iron catalysts are used for Fischer-Tropsch olefin production [13].

Olefin isomerization and -hydrogenation on cobalt are thought to perform at "on-plane sites" of the metal, which are not active for chain growth [19, 22]. With alkalized iron there may "no" metallic surface anymore exist, to allow for olefin secondary reactions (see below).

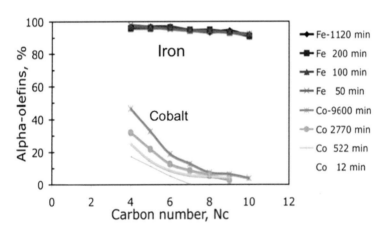

Iron (100Fe-13Al-11Cu-25K); 250°C, 10 bar, H_2/CO = 2.3:1 *Riedel, Schaub, Schulz et al.* [11]
Cobalt (97Co-12Zr-100Aerosil); 190°C, 5 bar, H_2/CO = 2:1 *Thesis Zh.Nie,* [20, 21]

Fig. 5: Olefin isomerization
Contents of α-olefins among the straight chain olefins as a function of carbon number and time (t_{exp}) with iron and cobalt as catalyst

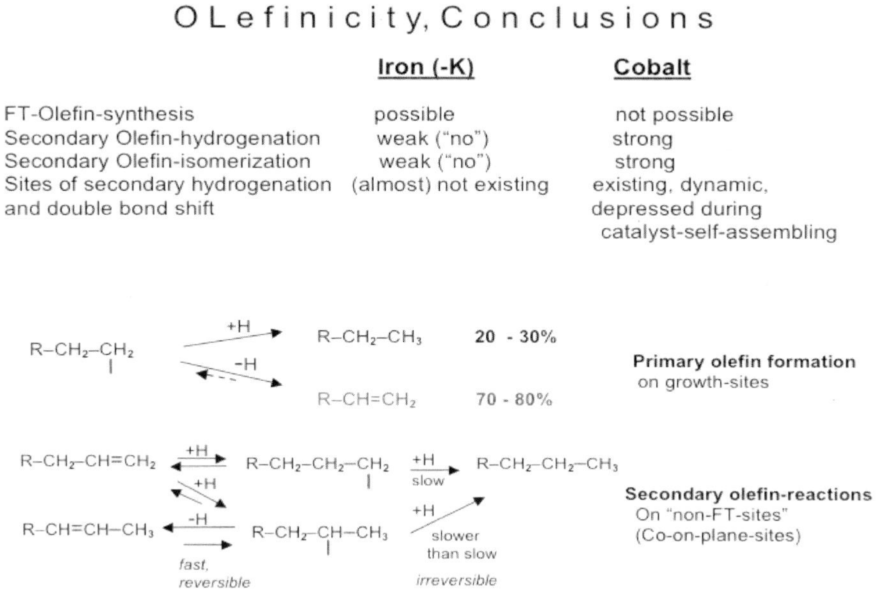

OLefinicity, Conclusions

	Iron (-K)	Cobalt
FT-Olefin-synthesis	possible	not possible
Secondary Olefin-hydrogenation	weak ("no")	strong
Secondary Olefin-isomerization	weak ("no")	strong
Sites of secondary hydrogenation and double bond shift	(almost) not existing	existing, dynamic, depressed during catalyst-self-assembling

Fig. 6: FT-olefin formation principles on iron and cobalt as the catalysts

Olefin double bond shift can be characterized by the percentage of terminal olefins among the straight chain olefins as a function of carbon number. When comparing iron- with cobalt catalysts on this basis (**Fig.5**) there is (almost) no double bond shift noticed with the iron catalyst (ca 95 - 98 % α-olefins at any time t_{exp}), whereas with cobalt, double bond shift is extensive: only a small portion of the primary α-olefins remains. Double bond shift declines during FT-catalyst assembling.

The proposed kinetic schemes of primary olefin- and paraffin formation and of olefin secondary hydrogenation and isomerization are presented in **Fig.6** together with comparative conclusions for iron and cobalt as the catalysts.

Chain growth

At a first glance, cobalt and iron are similar FT-catalysts: with both, waxes, diesel fuel, gasoline and LPG can be obtained. Mainly the reaction temperature controls the probability of chain prolongation p_g. High values of p_g are noticed at low temperature (ca. 190 – 230°C). This corresponds to the general trend for the hydrocarbon-intermediates: the increase of desorption probability with increasing temperature [16].

186

Main different behaviors of iron and cobalt catalysts are the following:
- Increase of p_g with increasing partial pressure of carbon monoxide with cobalt - but not with iron [9, 10, 16].
- Increase of p_g during self-organization with cobalt - but not with iron).

These differences indicate a different nature of the growth sites on cobalt and on iron.

Fig. 7 shows the growth probability as a function of carbon number during catalyst self-assembling for FT-synthesis on iron and cobalt. A horizontal line would indicate ideal polymerization (constant p_g). The observed deviations and the changes with time (t_{exp}) have to be explained:

The value at $N_C = 1$ refers to methane selectivity. With iron, the value of p_g at $N_C = 1$ is about the same as at $N_C = 2$ and $N_C = 3$. This means there is "no" further methane formation than on the growth-sites. With cobalt, the value of p_g at $N_C = 1$ is much lower than at $N_C = 2$ and $N_C = 3$, as relating to extensive "extra methane"-formation (presumably on non-specific on-plane-sites).
The average level of p_g with iron (ca 0.7 in the actual case) does not change much with time (t_{exp}). This means that the nature of the FT-growth sites on iron does not change in the formation period. But with cobalt, the FT-sites develop their nature with time: the average growth probability increases drastically from ca 0,6 at $t_{exp} = 10$ min to 0.8 at $t_{exp} = 7470$ min.

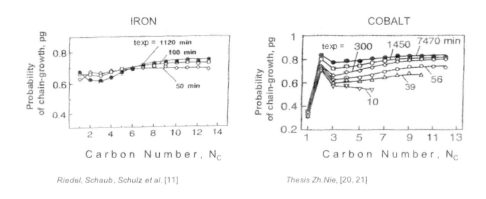

Iron (100Fe-13Al-11Cu-25K); 250°C, 10 bar, $H_2/CO = 2.3:1$
Cobalt (97Co-12Zr-100Aerosil); 190°C, 5 bar, $H_2/CO = 2:1$

Fig. 7: Chain length Distribution
Chain growth probability as a function of carbon number and time (t_{exp}) for iron and cobalt catalysts

	Iron (K)	Cobalt
pg-change with time, texp	marginal	increase
pg-change with pCO	marginal	increase
Methanation	marginal	extensive
The FT-regime	static	dynamic

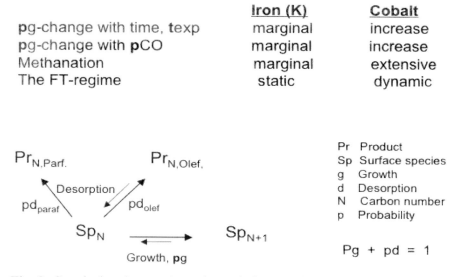

Pr   Product
Sp   Surface species
g   Growth
d   Desorption
N   Carbon number
p   Probability

Pg + pd = 1

Fig. 8: Conclusions/comparison about chain growth on iron and cobalt catalysts

With iron, the curve shape changes smoothly with time: The values at Nc = 2 and Nc = 3 decrease and the values in the range Nc = 8 to Nc = 12 increase. This curve shape has been explained and kinetically modelled as reflecting the (carbon-number-depending) re-adsorption of α-olefins on growth sites, which is favoured as time on stream (t_{exp}) increases [9,10].

The conclusions about chain growth on iron and cobalt catalysts are summarized in **Fig. 8** together with the kinetic scheme.

Chain branching

With both - iron and cobalt as catalysts - the Fischer-Tropsch product consists mainly of aliphatic chains with a few methyl side groups. The branching reaction can be assumed more demanding in space than linear growth of chains and thereby constrained.

Fig. 9 shows curves of the branching probability in dependence of carbon number for different times t_{exp}. Principle differences are noticed for iron and cobalt catalysts. On alkalized iron catalysts -as known from many experiments - the spatial constraints for the growth reaction appear to be high, with little temporal change or influences of CO- and H_2 partial pressures.

With both kinds of catalysts, the **first value of p_{br}** (at Nc = 3) is generally low. Specific spatial constraints are to be visualized as the reason:

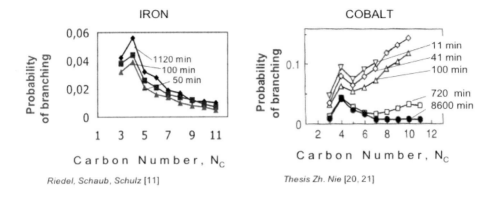

Riedel, Schaub, Schulz [11]   Thesis Zh. Nie [20, 21]

Iron (100Fe-13Al-11Cu-25K); 250°C, 10 bar, $H_2/CO = 2.3:1$
Cobalt (100Co-11Zr-0,15Pt-100Aerosil); 190°C, 5 bar, $H_2/CO = 2:1$

Fig. 9: Chain branching
Branching probability (p_{br}) as a function of carbon number Nc (of the species to be branched) and time (t_{exp}) in Fischer-Tropsch synthesis with iron and cobalt as the catalysts

Desorption directly after branching can be assumed strongly constrained because of a much more spatially demanding transition state as desorption after linear growth. The assumption of "slow desorption after branching" has been introduced by Wojciechowsky et al [23]. Of course, this argument holds for both, iron- and cobalt catalysts, and the phenomenon is persistent with both sorts of catalysts.

From Nc = 2 to higher values of Nc, branching probability principally decreases, as expected because of with chain-length-of-the-species increasing spatial constraints at the site. This is a strong effect. For instance, with iron at steady state (t_{exp} = 1120 min in Fig. 9), p_{br} decreases by a factor of about ten, from Nc = 4 to Nc = 10.

Remarkably, there is an opposite **trend with time** for iron and cobalt: With iron, branching becomes more probable with time t_{exp}, indicating *decreasing* spatial constraints during catalyst assembling. With cobalt, branching probability decreases with time t_{exp}, indicating spatial constraints to *increase* during catalyst restructuring.

With **cobalt, (Fig. 9, right), initially** (at short time t_{exp}) the curve shape is different from that at steady state: The probability of branching p_{br} increases with carbon number, from Nc = 5 onwards. This indicates a new mechanism of branching to operate, relating to spatial constraints being initially lower than at

Branching, Conclusions

Spatial constraints on the FT-sites control branching!
- Strong decrease of **p**br with length of chain (Nc)
- Desorption after branching is suppressed
  (low value of **p**br at Nc=3)

On **iron** spatial constraints are of *static* nature,
On **cobalt** spatial constraints are of *dynamic* nature.

Fig. 10: Conclusions about chain branching on iron and cobalt catalysts

steady state. The explanation could be re-adsorption of the α-olefins on growth sites, now possible also at the second carbon atom and not only at the first. Growth then causes a methyl branch on the chain. This branching reaction, as being a secondary reaction of the α-olefins, would be favoured by increasing length of the chain – a trend known from secondary olefin hydrogenation and double bond shift (see above).

On cobalt catalysts, branching declines with increasing partial pressure of CO, as reflecting the dynamic nature of the sites. A short conclusion/comparison about branching on iron- and cobalt catalysts is presented in **Fig. 10**.

Water gas shift reaction

The water gas shift reaction is fast on alkalized iron catalysts, but merely proceeds on cobalt catalysts. Synthesis gases from high temperature coal- or heavy oil gasification with high CO-content can directly be used for FT-synthesis on iron. The respective formal stoichiometry of the reaction can be written as:

$$2\,CO\ +\ 1\,H_2\ =\ 1\,CH_2\ +\ 1\,CO_2\ .$$

On the other hand, gases containing little CO but much CO_2 (e.g. the gas produced from bio-mass) could profit from the reverse water gas shift reaction as according to the equation:

$$CO_2\ +\ 3H_2\ =\ CH_2\ +\ 2H_2O$$

For synthesis gas with the right stoichiometric composition (molar H_2/CO-ratio of about 2.1) as can be obtained by auto-thermal reforming of methane, the water gas shift reaction would not be useful, and cobalt catalysts (with no, or almost no water gas shift activity) are suited.

$$CO\ +\ 2H_2\ =\ CH_2\ +\ H_2O$$

It is assumed that the water gas shift activity is related to the alkalized oxidic part of the iron catalyst. Hydrogen - activated on the oxidic sites - could support the hydrogenation of carbon on the carbidic sites [11] to form the "CH_2 monomer".

Again, iron and cobalt are different and each may be the best choice for a distinct task.

Response to changes in kinetic parameters

Temperature. With alkalized iron, the dominating effect of increasing the reaction temperature is reducing the chain length of the product-molecules, whereas their distribution on compound classes (olefins or paraffins etc.) is not much changed [11, 15]. This is seen as reflecting the static nature of FT-sites on iron catalysts [15].

With cobalt, the situation is different: In addition to reducing the average product-molecule size, the methanation reaction and the secondary olefin reactions of hydrogenation and isomerization are enhanced by increasing the reaction temperature, indicating the release of their suppression [16, 19] (and the dynamic nature of the regime).

CO partial pressure (H_2/CO-ratio): With iron, the CO- and the H_2-partial pressure have no strong influence on the product distribution [19, 24]. It can be concluded that with iron catalysts (the adsorbed) CO is not much involved in defining the nature of the growth-site.

With cobalt as the catalyst, CO-chemisorption rules the nature of the active sites by controlling surface segregation and thus even "construction" of the true FT-catalyst [9, 10, 16].

Self-organization

With both - iron and cobalt catalysts - the steady state of FT-synthesis is attained during several episodes of time o steam [11, 19, 22], **Fig. 11**. In the early German FT-literature this time was termed "Formierung" (formation). But in those days the experimental tools for its further study did not exist.

Starting the run with a precipitated, reduced, **alkalized iron** catalyst, this is initially not active (**Fig. 11**). FT-activity is developed through formation of iron carbide and enrichment of alkali on the surface (**Figs. 12, 13, 14**). It reaches the steady state when all the metallic iron has reacted (**Figs. 12, 13**), and no metallic iron can be any more detected. It follows, *metallic iron is not needed for FT-synthesis* and also then not present for olefin reactions and CO-methanation (in contrast to FT-synthesis on Co-, Ni- and Ru-catalysts).

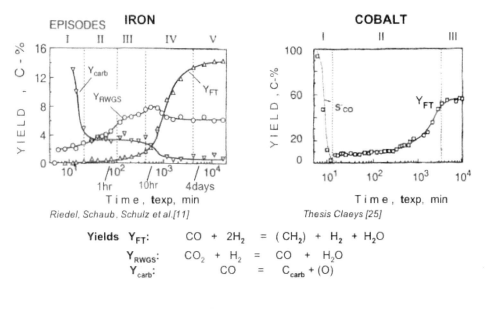

$$Yields\ Y_{FT}:\quad CO + 2H_2 = (CH_2) + H_2 + H_2O$$
$$Y_{RWGS}:\quad CO_2 + H_2 = CO + H_2O$$
$$Y_{carb}:\quad CO = C_{carb} + (O)$$

Iron (100Fe-13Al-11Cu-9K); 250°C, 10 bar, $H_2/CO_2 = 3:1$

Cobalt (100Co-15ZrO$_2$-0,66Ru-100Aerosil); 190°C, 10bar, $H_2/CO = 2:1$

Fig. 11: Activity in dependence of time
The yields of the FT-reaction (Y_{FT}), the reverse water gas shift reaction (Y_{RWGS}) and carbiding (Y_{CARB}) as a function of time (t_{exp}) with iron (left) and cobalt (right) as catalysts.
The diagram for FT on iron (left) refers to an experiment with a H_2/CO_2 synthesis gas. This is advantageous for a clearer separation of the episodes of self-organization

192

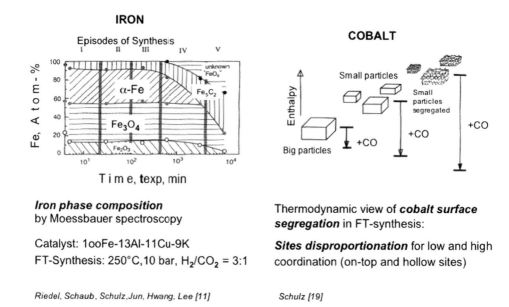

IRON

Episodes of Synthesis
I   II   III   IV   V

Iron phase composition
by Moessbauer spectroscopy

Catalyst: 1ooFe-13Al-11Cu-9K
FT-Synthesis: 250°C,10 bar, H_2/CO_2 = 3:1

Riedel, Schaub, Schulz,Jun, Hwang, Lee [11]

COBALT

Thermodynamic view of *cobalt surface segregation* in FT-synthesis:

Sites disproportionation for low and high coordination (on-top and hollow sites)

Schulz [19]

Fig. 12: Self-assembling of the FT-catalyst with iron (left) and cobalt (right)

Starting the run with a precipitated, promoted, supported and reduced **cobalt catalyst**, there is a low initial activity, which increases by a factor of 3 to 5 with time (**Fig.11-right**). Methane selectivity decreases, and olefin reactions are suppressed (see above). This has been explained by creation of activity via segregation of the metal-crystallite surface planes and thereby sites disproportionation for high coordination (hollow sites, in hole sites) and low coordination (on top sites), the on top sites being active for chain growth (insertion reaction) and the hollow sites active for CO-dissociation [19 22]; the on plane sites being deactivated for methanation by the strong CO-adsorption. Such FT-regime must be dynamic because the controlling (even corrosive) CO-chemisorption will be reversible and depending on reaction conditions.

It is surprising, how - in spite of all these differences - the sophisticated conversion of CO and H_2 to chains of aliphatic hydrocarbons is feasible with both kinds of catalysts.

Segregation of the planes of cobalt metal crystallites means increase of the number of surface atoms without change of particle diameter, the driving force being an increase of the number of sites for CO-adsorption. This restructuring of surface planes also concerns disproportionation of the sites

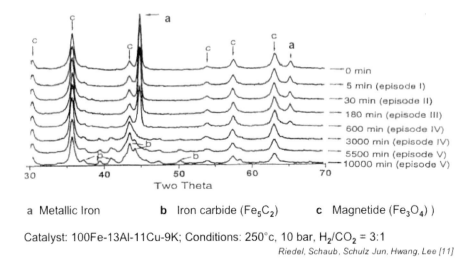

a Metallic Iron   b Iron carbide (Fe₅C₂)   c Magnetide (Fe₃O₄))

Catalyst: 100Fe-13Al-11Cu-9K; Conditions: 250°c, 10 bar, H_2/CO_2 = 3:1

Riedel, Schaub, Schulz Jun, Hwang, Lee [11]

Fig. 13: Iron phase changes during catalyst re-assembling
XRD-spectra of iron catalyst samples from different run lengths

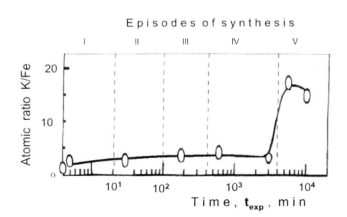

XPS of catalyst samples after different FT-run lengths
100Fe-13Al-11Cu-9K;   250 °C, 10 bar, H_2/CO_2 = 3:1

Riedel, Schaub, Schulz Jun, Hwang, Lee [11]

Fig. 14: Potassium accumulation on the surface during iron catalyst reassembling

to such of high and low coordination. The sites of low coordination allow several ligands to be positioned and accordingly allow coordination chemistry on the site, e.g. the insertion reaction. Interestingly, it had been understood early that FT-synthesis on cobalt (Ni, Ru) performs at reaction conditions where the reaction of the metal with CO to the metal carbonyl becomes thermodynamically possible [1]. This also explains the transient passage from heterogeneous FT-synthesis to homogeneous olefin-hydroformylation on cobalt at decreasing temperature and increasing CO partial pressure.

Site-specific activity (TOF), Fig. 15

FT-activity may be calculated per catalyst volume or weight. Calculating a site-specific activity (TOF) appears not possible and theoretically not justified in Fischer-Tropsch synthesis, the TOF-concept being not applicable. The active sites of FT-synthesis develop (are built) during the episodes of self-organization and can neither in number nor in nature be the same as those determined by chemisorption with the reduced fresh catalyst.

FT-sites on iron-catalysts:

- Are located on the surface of iron carbide (not on α-iron)
- Alkali constitutes part of the sites.
- Chemi-sorption with the reduced fresh catalyst refers not to the sites
  of the "working" catalyst
- Self-assemling provides a distinct surface from different precursors.
  This might appear like "structure insensitivity"

FT-sites on cobalt-catalysts

The segregated cobalt surface will be rich in FT-sites:
- On-top-sites for chain growth
- Hollow-sites for CH_2-from-CO-formation
(The remaining on-plane-sites will be widely occupied by CO- and CH_3-species)

Multi-functionality of the FT-catalyst surface is needed for simultaneous
- CO-dissociation
- H_2-dissociation
- Hydrogenation of C, CH, CH_2, CH_3, O, OH,
- Insertion, e.g.: $CH_3 + CH_2 \longrightarrow CH_3-CH_2$
and mobility of the monomer

Fig.15: Conclusions about the active sites on the reorganized iron- and cobalt catalysts

In addition, the mechanism combines basic reactions, which presumably perform on different kinds of sites of the catalyst. The problem of FT-sites characterization could be addressed further with methods of surface science for attaining new results on an atomic level. Proposing FT-synthesis to be "structure insensitive" appears erroneous and misleading.

Promoting FT-catalysts

With cobalt, the original "standard FT-catalyst" of Fischer with Meyer and Koch, had the composition (by weight) $100Co-18ThO_2-100Kieselguhr$ [26,27]. The co-precipitated beautifully shaped particles of Kieselguhr provided the right porosity. The role of the (un-reducible) thoria is not evident, but thoria presumably provided high cobalt dispersion and -stability during reduction, segregation of the surface, synthesis and even regeneration. Today, zirconia and other materials have replaced thoria. The principles for "designing" the FT-catalyst remain the same. Stabilizing the cobalt dispersion appears difficult because of the dynamic nature of the structure of the surface under FT-reaction -conditions.

FT-iron catalysts are promoted with alkali, which enriches on the surface during synthesis. The amount of alkali must be adjusted carefully. Too high alkali addition causes accumulation of carbon on the catalyst [28].

Further catalyst additions can be Cu for easier reduction and e.g. MnO, ZnO, alumina and others for improved performance. Stabilizing the dispersion of the working catalyst appears important.

Interestingly, a minimum "nano size" particle diameter seems necessary for both iron- [29] and cobalt- [30] FT-catalysts.

The common FT-principle for iron- and cobalt catalysts

Even as the working iron- and cobalt catalysts are seen to be different in composition and structure, there should be a common principle to rule FT-synthesis (**Fig. 16**). The principle has been proposed as "frustration of desorption" [19, 24]. More specifically, frustration concerns desorption of the alkyls. Associative desorption - to obtain a paraffin molecule - then is more retarded than dissociative desorption to obtain an α-olefin-molecule. The retarded desorption of the alkyl chains allows for their reaction with a monomer, leading to chain growth.

	Iron(K)	Cobalt
Extensive *methanation*	no	at increasing temperature and decreasing pCO
FT-*self-organization*	at static selectivity	at dynamic selectivity
Catalyst-*self-assembling*	carbide formation	surface segregation
Alkali	essential	no
Product distribution (on Nc-Frctns.)	s i m i l a r	
FT-*Monomers*	CH_2	CH_2 (CO, C_2H_4)
WGS-activity	yes	no
Olefin reactions:		
-Hydrogenation	no (little)	extensive
-Isomerization	no (little)	extensive
Branching reaction	static	dynamic

The *FT-principle*? ?

Frustration of alkyl desorption (to allow for chain growth)

Fig. 16: Comparing FT-synthesis on iron and cobalt catalysts

Frustration is established in a different way with iron- and cobalt catalysts. With cobalt, it performs dynamically through reversibly adsorbed CO. With iron, CO is of little influence on selectivity. The carbide FT-sites, strongly interacting with the alkali, are quite static and invariant in their chemistry.

Understanding the principles of FT-synthesis on the different catalysts shall help improving commercial processes of Fischer-Tropsch synthesis.

Conclusions

The plain idea of a common FT-reaction mechanism for iron- as well as cobalt catalysts is not supported by the basic comparison in this article. The essential common mechanistic feature is specified as "frustrated alkyl desorption", but other mechanistic aspects of FT on Co or Fe are different:

- the mode of self-organization of the FT regime,
- the structure and composition of the catalytic surface,
- the kinetic response to changes in reaction parameters as T, p_{CO} and p_{H2}
- and the occurrence of secondary reactions.

The idea to address FT-synthesis as a "structure insensitive" reaction appears as miss-interpretation. Contrarily we understand a very specific FT-catalyst-surface is being built during self-organization of the FT-regime.

Likewise, the "turnover concept" is not applicable in FT-synthesis, because relevant chemisorption measurements are not possible. For the FT-synthesis - with its several steps of reaction needed for only one CH_2 monomer to form – it remains undefined what one turn-over event could be and on which of the active sites it should happen.

The complex product composition (hundreds of compounds, gases, vapours, liquids, solids, an aqueous phase in addition to the organic phase, and compositional changes during formation of the Fischer-Tropsch regime) has been all the time a difficulty in Fischer Tropsch experimentation. But the order of this multiplicity embodies information about the elemental reactions and the nature of the active sites. It seems there is much to learn about catalysis from Fischer-Tropsch synthesis.

Modern tools of imaging on an atomic scale to be used in the future should provide complementary insight about the Fischer-Tropsch-catalyst surface architecture.

Summary

Fischer-Tropsch synthesis on iron and cobalt catalysts is compared on the basis of detailed product composition and structure of the catalyst as depending on time as in relation to self-organization of the FT-regime.

The comparison includes elemental reactions, as chain-growth, and –branching, alternative olefin- or paraffin product desorption and olefin reactions of hydrogenation, isomerization, re-adsorption on growth sites and CO-methanation.

Catalyst re-assembling with iron means reaction of iron with carbon from CO-dissociation to create the FT-active carbide surface (generally enriched with alkali as the essential iron promoter). With cobalt, restructuring means segregation of the metal surface and thus sites disproportionation for such of chain growth and such for CH_2- (monomer) formation.

The FT cobalt catalyst is of dynamic nature. All relative rates of elemental reactions are depending on time (run length), temperature and partial pressures (mainly CO and H_2). The structure of the sites appears to be thermodynamically controlled, as linked to CO-adsorption-induced surface segregation.

With the iron catalyst, the active sites appear to be static: During catalyst re-assembling the rate the FT-reaction increases, whereas the selectivity does not change. Among the parameters of reaction only the temperature is of significant influence, as controlling the average molecular weight of the products. Olefin secondary reactions and methanation tend to be merely present.

This comparison, aiming at a deeper understanding of the specific iron and cobalt catalyst behaviours in Fischer-Tropsch synthesis could also be useful for an advanced catalyst design for achieving best performance of the FT-conversion.

References

[1] H. Pichler, Twenty-five years of synthesis of gasoline by catalytic conversion of carbon monoxide and hydrogen in "Advances in Catalysis" Vol. 4, Eds. W. Frankenburg, E.Rideal, V. Komarewsky, Academic Press Inc., New York, (1952) 271
[2] H. Schulz, S. Nehren, Erdöl und Kohle-Erdgas-Petrochemie, 39 (1984) 45
[3] H. Schulz, W. Böhringer, C. Kohl, N. Rahman, A. Will, Entwicklung und Anwendung der Kapillar-GC-Gesamtprobentechnik für Gas/Dampf-Vielstoffgemische, DGMK-Forschungsbericht 320, DGMK, Hamburg (1984)
[4] A. Geertsema, Dissertation (University of Karlsruhe) Karlsruhe (1976)
[5] E. van Steen, Dissertation (University of Karlsruhe) Karlsruhe (1993)

[6] H. Schulz, K. Beck, E. Erich, Stud.Surf. Sci. Cat. 36 (1988) 457

[7] H. Schulz, K. Beck, E. Erich, Fuel Proc. Techn. 18 (1988) 293

[8] H. Schulz, K. Beck, E. Erich, Fischer Tropsch CO-hydrogenation, a non trivial surface polymerization: selectivity of branching, Proc. "9th Int. Congr. on Catalysis" Calgary, 1988, Vol. 2, Eds.: M. Phillips, M. Ternan, The Chemical Institute of Canada, Ottawa, (1988) 829

[9] H. Schulz, M. Claeys, Appl.Catal. A: Gen. 186 (199) 91

[10] H. Schulz, M. Claeys, Appl.Catal. A: Gen. 186 (199) 71

[11] T. Riedel. H. Schulz, G. Schaub, K. Jun, J. Hwang, K. Lee, Topics in Catalysis 26 (2003) 41

[12] J.Wilson, G. de Groot, J.Phys. Chem. 99 (1995) 7860

[13] A.Steynberg, R.Espinoza, B. Jager, A. Vosloo, Appl. Catal.,A: Gen. 186 (1999) 41

[14] H. Schulz, Flugstaub-Synthese, in "Chemierohstoffe aus Kohle", Ed J. Falbe, Georg-Thieme-Verlag, Stuttgart (1977), 272

[15] H. Schulz, Th. Riedel, G. Schaub, Topics in Catalysis 32 (2005) 117

[16] H. Schulz, H. Achtsnit, Revista Portuguesa de Quimica 19 (1977) 317

[17] C.Kellner, A. Bell, J. Catal. 70 (1981) 418

[18] H. Schulz, H. Gökcebay, Fischer-Tropsch CO-hydrogenation as a means for linear olefins production, in "Catalysis of Organic Reactions", Ed. J. Kosak, M. Dekker, New York (1984) 153

[19] H. Schulz, Topics in Catalysis 26 (2003) 73

[20] Zh. Nie, Dissertation (University of Karlsruhe), Karlsruhe (1996)

[21] H. Schulz, Zh. Nie, Stud.Surf. Sci. Cat. 136 (2001) 159

[22] H. Schulz, Zh. Nie, F. Ousmanov, Catalysis Today 71 (2002) 351

[23] B.W. Wojciechowski, Catal. Rev. Sci. Eng. 30 (1988) 629

[24] H. Schulz, E.van Steen, M. Claeys, Stud. Surf. Sci. Catal. 81 (1994) 455

[25] M. Claeys, Dissertation (University of Karlsruhe), Karlsruhe (1997)

[26] F. Fischer, K. Meyer, Brennstoff-Chemie 12 (1931) 225

[27] F. Fischer, H. Koch, Brennstoff-Chemie 13 (1932) 61

[28] M. Dry, The Fischer -Tropsch Synthesis, in "Catalysis",J. Anderson and M. Boudart (Eds.), Springer Verlag, Berlin , Vol. 1 (1981) 159

[29] E.Mabaso, E. van Steen, M. Claeys, Lecture at the 19th NACSM, Philadelphia May 22-27, 2005, Paper O-150

[30] G, Bezemer, A. van Dillen, K. de Jong, Lecture at the 19th NACSM, Philadelphia, May 22-27. 2005, Paper O-141

Fischer-Tropsch Synthesis, Catalysts and Catalysis
B.H. Davis and M.L. Occelli (Editors)

Fischer-Tropsch Synthesis: Compositional Modulation Study using an Iron Catalyst

Matthys Janse van Vuuren[1] and Burtron H. Davis[2]

[1]*Sasol Technology R&D, Sasolburg, South Africa,* [2]*Center for Applied Energy Research, University of Kentucky, 2540 Research Park Dr., Lexington, KY 40511*

1. Introduction

The use of compositional modulation of catalytic reactors is an established procedure [1]. In this approach, sudden and cyclic changes in the feed composition may lead to improvements in the catalyst productivity and/or selectivity.

The compositional modulation technique has been applied to the Fischer-Tropsch synthesis (FTS) reaction [2-5]. It was found that the cyclic feeding of CO/H_2 had an influence on the selectivity of the FTS products. Among the conclusions was that for an iron catalyst the selectivity for methane increased under periodic operation compared to the steady state operation [5]. In the study [5] it was found that the propane/propene ratio increased under periodic operation and the largest changes were with periods between one and ten minutes. Due to the limitations of the analytical technique utilized, they could not separate ethane and ethene so that the selectivity basis was for the C_3 hydrocarbons. In this study the analytical procedure permitted analysis of products only to the C_8-compounds.

For the reactor used in these studies [2-5] only low conversions could be utilized to obtain differential conversion levels ($\leq 2\%$ based on CO). This limitation on conversion was imposed because of the exotherm of the FTS reaction. The FTS reaction usually takes days to stabalize to make a correct assessment of the steady-state activity and selectivity of the catalyst. This stabalization period was not practiced in the work in references 2-5.

To determine the real effect of periodic operation on a FTS catalyst, it is important to do the assessment at conditions that can be compared to a normal steady-state assessment of the catalyst. Ideally, these conditions could include operation in a slurry reactor where heat removal is optimum (isothermal operation) for a minimum of 100 hours before a forced periodic operation is undertaken. The cyclic feeding of CO/H$_2$ apparently has not been attempted with a continuously stirred tank reactor (CSTR) due to its large dead volume, and therefore the dilution of the cyclic effect. In order to test the cyclic behavior for FTS in a CSTR, a system was assembled to determine the frequency of cycles that can be monitored with a large dead volume reactor. Results from the operation of this reactor are reported.

2. Experimental

When the change of gases in a CSTR is studied, it is important to know the concentration of the gas mixture in the reactor. This can be calculated for an ideal system when the gas flow, CSTR volume, pressure, temperature and gas flow are known. Unfortunately, a CSTR partially filled with molten wax is not an ideal system. It was therefore decided to practically determine the rate of gas concentration change in the actual FTS CSTR. A schematic of the reactor setup to accomplish this is shown in figure 1.

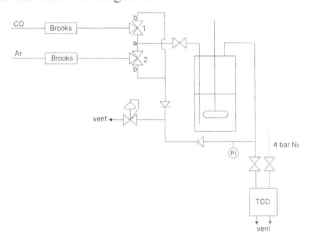

Figure 1. Schematic diagram of the CSTR as used in the gas change experiments.

For this determination, the Brooks mass flow meters for CO and Ar are set with similar flows. The pressure between the mass flow meters and the back pressure regulator is controlled by the back pressure regulator. With the 3-way valve (1) set to position (b) and 3-way valve (2) set to position (a), Ar flows through the reactor and CO flows directly to the back pressure regulator. A simultaneous change of 3-way valves (1) and (2) causes the gas to the reactor to be changed from Ar to CO without a change in the overall flow, and without a change in the total pressure of the system. As the thermal conductivity cell (TCD) of the g.c. is sensitive to a change in pressure, this setup works well. The TCD was operated continuously with N$_2$ as a reference gas so that any chanage in the gas composition from the reactor gave a change in the signal of the TCD. This change in signal against time as the valves were switched gave the cyclic response of the reactor system. To determine the sensitivity of the TCD under cyclic operation, the reactor was bypassed to minimize the dead volume (schematic shown in figure 2). With this configuration, a cyclic response of about 6 sec. could be achieved. The change in the gas composition could also be observed 14 sec. after a change in gas was made.

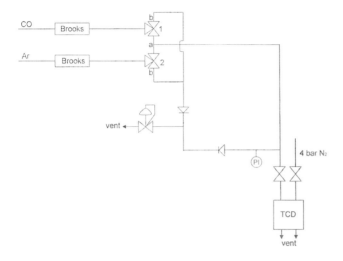

Figure 2. Setup when reactor was by-passed to determine response of cyclic operation.

When the reactor was included in the flow path, the replacement of one gas with the other was monitored with the TCD. It was observed that when H_2 and Ar were used, the replacement of Ar with H_2 was more rapid than the replacement of H_2 by Ar (figure 3, left). When the gas was changed from H_2 to CO the difference was noticeably smaller but still evident (figure 3, right). The response of the TCD vs. gas composition for CO/Ar was determined and is shown in figure 4 (left). It is obvious that the TCD signal is not linear to the mole fraction of Ar. A correction of the signal is needed to obtain the correct gas composition from the TCD signal. Likewise, the response of the TCD vs. gas composition for CO/H_2 was not linear (figure 4, right).

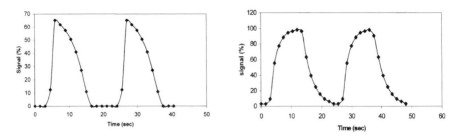

Figure 3. (left) Cyclic replacement of H_2 and Ar. (right) Cyclic replacement of CO and Ar.

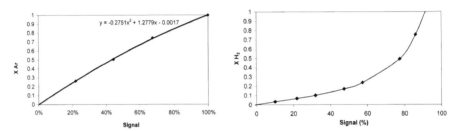

Figure 4. (left) Plot of mol fraction of Ar vs. signal of TDC. (right) Plot of mol fraction of H_2 vs. signal of TCD.

The frequency with which the gases could be changed and still obtain a replacement of the gas in the reactor was determined. For this experiment, the reactor was filled first with 680.9 g of Polywax 3000 (polyethylene narrow fraction with average M.W. of 3000) or 730.9 g. The latter amount of Polywax filled the reactor to a level such that 34 mm of gas space was available above it. With the periodic change of CO and Ar, the graph shown in Figure 5 was obtained. When the kinetic fit was determined to verify that the replacement of the CO by Ar, and vice versa, was according to a first order rate expression, the following graph was obtained (figure 6). Therefore, the replacement of the gas in the dead volume follows first order kinetics. When the frequency was changed with 730 g of Polywax 3000 in the reactor, the cycles shown in figure 7 were obtained. The data make it clear that when the cycle time increases, one gas can completely replace the other gas while with shorter cycle times, one gas can only partially replace the other.

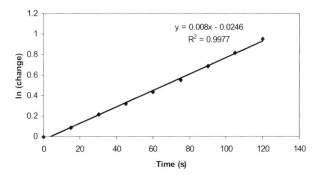

Figure 5. Fit for first order replacement of gas.

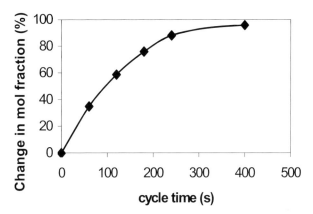

Figure 6. Change in the mol fraction of Ar vs. period of the cycle.

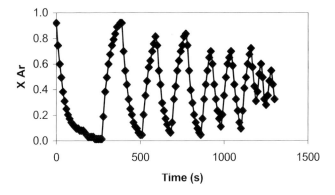

Figure 7. Change in the mol fraction of Ar vs. time when the frequency of the period flow is changed.

2.1. Study of FTS under forced periodic operation

2.1.1. Reactor Setup

A schematic of the reactor setup that was used for the FTS reactions is shown in Figure 8. It is important to obtain a cyclic flow of CO and H_2 by controlling the Brooks mass flow meters. This was obtained as follows. The Brooks mass flow meters are electronically controlled with a 0-15V signal and this signal can be overridden to close the mass flow meter. This override can then either open the mass flow meter fully if +15V is provided or fully close it if -15V is provided. Thus, the mass flow meter could be periodically closed and opened to the set point, as shown in figure 9. (Information about the response time of the mass flow meter when it is opened or closed is available on the internet at http://www.uhv.co.kr/pdf/ds-5850e-97-4.pdf.) The response time for the mass flow meter was within two seconds with an overshoot of approximately 15%. For slow periodic operation, this system worked well. An electronic timer with a relay function was used to provide the -15V for the manual override of the mass flow meter.

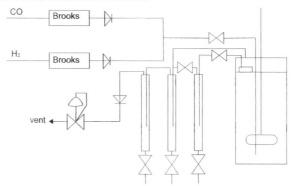

Figure 8. Schematic of the reactor setup for the FT reaction with periodic operation.

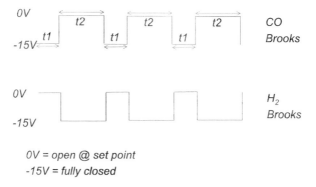

Figure 9. Electronic control for Brooks mass flow meters.

A potassium promoted precipitated iron catalyst was used (Fe:Si:K = 100:4.6:1.0). The reactor was loaded with 650 g (or other amount as indicated) solvent and 16 g of unreduced catalyst. The activation was accomplished using CO at 300°C at 175 psig and a flow of 2700 h^{-1}. After 24 hours the temperature was reduced to 270°C and the flow of H$_2$ and CO was increased. The ratio of H$_2$/CO was 0.7 and the total flow was 88 SL/h.

After activation, the run was operated under steady state conditions to ensure that the catalyst had the expected activity and the conversion was stable. When this was accomplished, the operation was changed to the cyclic mode. This was achieved as follows. For the gas feed mixture a flow of 88 SL/h was used and to maintain the same flow during the switching conditions, both the H$_2$ and CO flows were set at 88 SL/h. The flow was then switched between 100% CO and 100% H$_2$ by the control of the Brooks flow meter operation. Only the CO or H$_2$ was open at any one time during the cyclic operation. For instance, when the H$_2$ was open for 41.8% of the time and the CO was open for 58.2% of the time, the overall H$_2$/CO ratio was 0.7.

The liquid sample vessels were drained under reaction conditions. Before the wax sample was drained from the reactor, the stirrer was switched off to let the catalyst/wax mixture settle to a given level in reactor. The valve between the reactor and the hot (200°C) sample collection vessel was then slowly opened. After the sample was drained, the valve was closed and the sample was drained from the collection vessel by opening the valve at the bottom of the collection vessel.

The first run was made using a decene trimer solvent since this material was a liquid at room temperature, permitting easy handling of the sampling operations. However, when the reactor was opened following the cyclic operation, it was found that too much solvent had been removed, either because of the high vapor pressure of the solvent or withdrawing too much sample directly from the reactor liquid. To overcome this problem, a run was made with a much heavier startup solvent (PW 3000) that has a molecular weight of 3000 (average carbon number of 200). The vapor pressure of this solvent is sufficiently low that the volume of the solvent exiting the reactor in the vapor phase is close to zero.

The normal steady state flow operation was conducted for about 120 hours, then the cyclic operation, with several cycle times, was conducted and following this, the system was operated again at the initial steady state conditions (figure 10). As observed in the earlier run with the decene solvent, the FTS rate decreased with an increase in the period of the cycle (figure 11). The methane selectivity increased (figure 12) and the CO$_2$ rate decreased (figure 13). Thus, the CO efficiency (% CO converted to hydrocarbons) shows an increase with cycle period (figure 14).

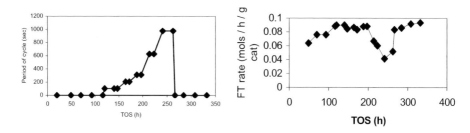

Figure 10. (left) Experimental track for cyclic operation with PW3000 solvent.

Figure 11. (right) Activity (FT rate) vs. time-on-stream.

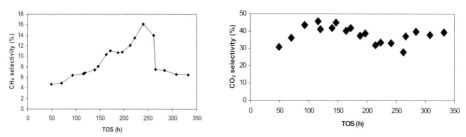

Figure 12. (left) Methane selectivity vs. TOS for MA T004.

Figure 13. (right) CO₂ selectivity for MA T004.

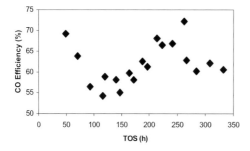

Figure 14. CO efficiency (CO to hydrocarbons/CO converted).

The olefin selectivity decreased with an increase in the period (figure 15). In this figure, the change in olefin selectivity is expressed as the change from the steady state conditions prior to the cyclic operations (ss in figure 15). When the periodic cycling started, there was a decrease in the percentage of olefins for the first period (419 sec). A further increase in the period resulted in a further decrease in the olefin selectivity (584 sec). A return to the steady state conditions showed that for the initial sample collected (ss2), only the shorter chain olefins returned to the steady state ratio. The next two steady state samples showed an increase toward the initial steady state conditions with respect to olefin content (ss 3 and ss4).

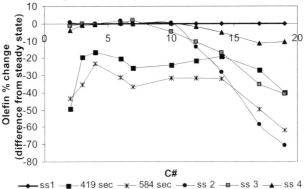

Figure 15. Effect of forced periodic operation on olefin percentage change.

The steady state Anderson-Schulz-Flory (ASF) plot shows a single alpha value for the C_1-C_{20} products; however for the sample collected for synthesis for the longest period time there is a distinct two-alpha ASF plot (figure 16). A plot of the lower alpha (for C_1-C_9 products) versus the time on stream clearly shows that this alpha value changes with periodic operation (figure 17). Furthermore, as the period length increases the alpha value for the C_1-C_9 products decreases. With a return to steady state operations for the last four sample collections, the alpha value returned to the value that it had prior to the periodic operation.

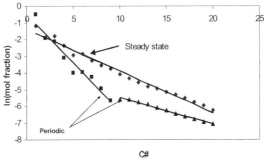

Figure 16. Change in alpha value from steady-state to period cycle.

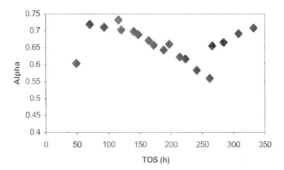

Figure 17. Change of alpha value for C_1-C_9 product.

3. Discussion

Under steady state conditions hydrogen and carbon monoxide are available in the reactor at all times. The ratio present in the reactor at a given time will, however, usually differ from that of the feed ratio and will depend upon the conversion level as well as the ratio of FTS to WGS to attain the given conversion level. Under periodic operation the ratios of H_2/CO are changing with time. CO replaces H_2 and then H_2 replaces CO, etc. There are therefore certain times when the partial pressure of H_2 is high and other times when it is low, and the same applies for CO. The longer the period, the longer the time where the partial pressure of only H_2 or CO will be high. In FTS a certain H_2/CO ratio is necessary for the reaction to prevail. This is why the activity decreases under increasing period.

The data in figure 15 make it clear that for the conditions utilized in this study, the periodic operation decreases the olefin fraction and increases that of the paraffins. The catalyst therefore appears to have a higher hydrogenation activity under periodic operation. This change is likely due to a change in the competitive adsorption of the CO and alkenes. When there is a low partial pressure of CO in the reactor, and a high pressure of H_2, the competitive adsorption of the alkene increases. The combination of higher surface concentration of alkene and hydrogen then leads to more alkene hydrogenation, and a lower alkene/alkane fraction. The competitive adsorption of the alkene is greatest for ethene and falls off with carbon number to approach a constant value. On the other hand, the vapor pressure decreases with increasing carbon number so that the higher carbon number alkenes will be retained in the reactor for longer relative times and will therefore undergo more secondary hydrogenation [6]. Thus, the hydrogenation of alkenes will be a complex situation with competitive adsorption dominating for low carbon number alkenes and holdup in the reactor for higher carbon number compounds [7]. In addition, the change in the period has an effect on the production of the heavier products as is apparent from the data in Tables 1-3. Due to the decrease in the wax and reactor wax production, the heavier FTS products stay in the reactor for a longer time, causing the alkenes to be hydrogenated to a greater extent. This is clearly visible for the C_{10} and heavier products in figure 18. An alpha plot is shown for total products from figure 16 and for the alkene, alkane and 1-alkene fractions. Except for ethene, the alkene fraction lies above the total product plot and then at a carbon number of about 15 the value begins to fall on or below the total ASF plot. The alkane fraction, apart from methane and ethane, falls below the ASF plot and only approaches it at about carbon number 15. The 1-alkene fraction is above the ASF plot initially but by at about carbon number 8 the value falls on the ASF line and then falls significantly below it with increasing carbon number.

Table 1. Sample Conditions for Run MAT004 and Liquid Product Production Rate

Sample	Date	Prod gas (slph)	Period (s)	TOS (hours)	Oil (g/h)	Wax (g/h)	Water (g/h)	Rewax (g/h)
001	1/11/2003	61.7	0	49	0.309	0.113	1.258	0.021
002	1/12/2003	56.68	0	70	0.410	0.308	0.915	0.082
003	1/13/2003	55.75	0	93	0.438	0.225	0.731	0.055
004	1/14/2003	50.72	0	116	0.433	0.541	0.546	0.089
005	1/14/2003	50.54	102	121	0.084	0.060	0.089	0.039
006	1/15/2003	50.71	102	140	0.364	0.344	0.289	0.188
007	1/15/2003	52.577	102	148	0.107	0.113	0.112	0.031
008	1/16/2003	53.53	206	165	0.197	0.151	0.225	0.031
009	1/16/2003	54.16	206	173	0.084	0.059	0.095	0.013
010	1/17/2003	53.5	311	189	0.133	0.062	0.193	0.027
011	1/17/2003	53.8	311	197	0.066	0.045	0.101	0.028
012	1/18/2003	62.8	628	214	0.073	0.063	0.156	0.008
013	1/18/2003	65.26	628	224	0.033	0.033	0.084	0.002
014	1/19/2003	72.2	975	242	0.036	0.040	0.108	0.033
015	1/20/2003	70.5	975	262	0.022	0.033	0.096	0.010
016	1/20/2003	55.12	0	267	0.026	0.018	0.048	0.010
017	1/21/2003	53.25	0	284	0.116	0.075	0.154	0.010
018	1/22/2003	52.58	0	308	0.170	0.645	0.192	0.019
019	1/23/2003	50.95	0	332	0.170	0.109	0.177	0.000

Table 2. Conversion Data for MAT004

SampleID	TOS (h)	X Syngas (%)	X H$_2$ (%)	X CO (%)
001	49.30	38.35	43.01	35.09
002	70.00	47.50	50.66	45.29
003	93.08	52.74	54.79	51.30
004	116.00	62.34	62.78	62.03
005	120.50	60.00	62.53	58.23
006	140.00	60.57	63.01	58.86
007	147.83	59.57	61.52	58.21
008	164.75	56.60	57.86	55.69
009	172.75	55.79	56.92	54.97
010	188.75	55.27	57.01	54.03
011	197.00	56.38	57.94	55.26
012	214.25	38.71	40.07	37.73
013	223.75	35.66	36.90	34.77
014	241.83	25.09	26.78	23.88
015	262.00	27.41	27.20	27.55
016	266.58	51.92	54.10	50.39
017	284.25	55.25	56.51	54.36
018	308.50	57.03	58.06	56.31
019	332.50	59.07	59.72	58.61

Experimental Conditions: P = 175 psig; T = 270°C; feed rate = 88 λh^{-1} (STP); H$_2$/CO = 0.7. Conversion = x = (flow in – flow out) / (flow in).

Table 3. Rate and Selectivity Data for MAT004.

Sample ID	TOS (h)	H₂:CO Usage	CO Rate (mols/h/g metal)	H₂ Rate (mols/h/g metal)	CH₄ Rate (mols/h/g metal)	CH₄ Selectivity (%)	FT Rate (mols/h/g metal)	Hydrocarbon Rate (mols/h/g metal)	CO₂ Rate (mols/h/g metal)	CO₂ Selectivity (%)	Efficiency[a] (%)
001	49	0.859	0.092	0.079	0.003	4.7	0.064	0.912	0.028	30.8	69.2
002	70	0.784	0.119	0.093	0.004	4.9	0.076	1.087	0.043	36.2	63.8
003	93	0.748	0.135	0.101	0.005	6.3	0.076	1.089	0.059	43.5	56.5
004	116	0.709	0.163	0.115	0.006	6.7	0.088	1.265	0.074	45.7	54.3
005	120	0.752	0.153	0.115	0.006	6.9	0.090	1.289	0.063	41.1	58.9
006	140	0.750	0.154	0.116	0.007	7.4	0.090	1.287	0.065	41.8	58.2
007	147	0.733	0.153	0.112	0.007	8.1	0.084	1.210	0.069	45.0	55.0
008	164	0.745	0.145	0.108	0.009	10.3	0.087	1.240	0.058	40.2	59.8
009	172	0.742	0.143	0.106	0.009	11.0	0.083	1.190	0.060	41.9	58.1
010	188	0.756	0.140	0.106	0.009	10.7	0.088	1.259	0.053	37.4	62.6
011	197	0.752	0.144	0.108	0.010	10.8	0.088	1.261	0.056	38.7	61.3
012	214	0.761	0.098	0.075	0.008	12.0	0.067	0.957	0.031	31.9	68.1
013	223	0.760	0.090	0.069	0.008	13.5	0.060	0.862	0.030	33.5	66.5
014	241	0.803	0.062	0.050	0.007	16.2	0.041	0.595	0.021	33.2	66.8
015	262	0.707	0.072	0.051	0.007	13.9	0.052	0.741	0.020	27.8	72.2
016	266	0.752	0.132	0.099	0.006	7.5	0.083	1.190	0.049	37.2	62.8
017	284	0.728	0.143	0.104	0.006	7.3	0.086	1.231	0.057	39.8	60.2
018	308	0.722	0.148	0.107	0.006	6.6	0.092	1.315	0.056	37.9	62.1
019	332	0.714	0.154	0.110	0.006	6.5	0.093	1.335	0.061	39.4	60.6

a. Efficiency $= [[CO]_{in} - <[CO]_{out} + [CO_2]_{out}) / <[CO]_{in} - [CO]_{out}]100$

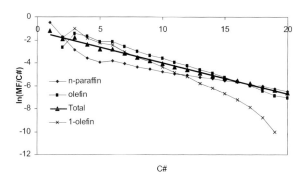

Figure 18. Alpha plot for n-paraffins, olefins and total product under steady-state conditions.

The phenomena where the longer FTS products took a longer time to react to a periodic change can be explained on the basis of their low partial pressure. In addition, the production of these higher carbon number products is lower than the lower carbon number products. Thus, the heavier the product, the longer it will stay in the reactor. The lighter products have a high vapor pressure and leave the CSTR almost as soon as they form; however, as the carbon number increases the vapor pressure decreases, resulting in an increase in the average residence time in the reactor. This effect is evident for the data in figure 15. For steady state FTS conditions, the straight line ASF plot is the expected result. Under periodic operation, two regions of linear ASF plot are obtained. One product grouping is produced during periodic operation and these are products lighter than are produced in the second grouping, the FTS products produced during steady state conditions. Thus, the products produced during the periodic operation exhibit a two-alpha plot, as is frequently reported in the literature. The alpha value for the lower carbon numbers (1 through 9) is lower than the steady state alpha value while the higher carbon number products have an alpha value that is higher than the steady state alpha value. The alpha value for the lower carbon number products is considered to be representative of the products produced during the periodic operation and the alpha value defined by the higher carbon number products during periodic operation are dominated by products produced during the steady state operation. With the lighter products in the fraction of C_{10} and above being removed in the gas phase more rapidly than the higher carbon number products, this results in an emphasis of the higher carbon number products. It is likely that these vapor pressure effects are responsible, or nearly so, for the observation of the two, or more, alpha plots that are commonly reported in the literature. This vapor pressure effect is augmented by the lower production rate of higher carbon number products during operation periods of low alpha values, as defined by C_1–C_9 products.

It is normal for the CO_2 selectivity to decrease when the conversion over an iron catalyst is decreased [8]. The change in selectivity that was observed in figure 13 is also observed under steady state when the conversion falls from 60% to 25% due to an increase in the space velocity. The efficiency (amount of CO converted to FTS products and not CO_2, fig 14) is also an indication that the change observbed with an increase in the period is normal for an iron catalyst. The periodic operation therefore only caused the conversion to decrease and a result of that decrease was that the selectivity for hydrocarbon products, relative to CO_2, increased.

With an iron catalyst, the higher the H_2/CO ratio the lower the alpha value. In general, the alpha value generated during cyclic operation was lower than obtained for the catalyst during steady state operation. For the portion of the cycle where CO is replacing hydrogen, the H_2/CO ratio will increase but eventually products will cease to form because the hydrogen is consumed and CO alone cannot produce hydrocarbons. For the other cycle, the H_2/CO ratio will increase and this will result in increasingly lighter products being formed.

The overall effect of the cyclic operation is therefore the production of lighter products than was formed during steady state operation.

In summary, periodic operation has an influence on the selectivity of the products from FTS. First, there is a decrease in the alpha value of FTS with increasing period. Secondly, the alkane/alkene ratio increases with an increase in the period. There did not appear to be a change in CO_2 selectivity that could not be attributed to a change in CO converesion. Finally, the fraction of the hydrocarbon product that was methane increased with increasing length of the period, as previously reported. In general, the periodic operation with the iron catalyst does not appear to provide effects in selectivity or activity that are desirable.

4. Acknowledgment

This work was supported by U.S. DOE contract number DE-FC26-98FT40308 and the Commonwealth of Kentucky.

• References

1. P. L. Silveston, Compositional Modulation of Catalytic Reactors, Gordon and Breach Sci. Pub., 1998.
2. P.L. Silveston, R.R. Hudgins, A.A. Adesina, G.S. Ross, J.L. Feimer, Chem. Eng. Science, 41 (1986) 923.
3. J.L. Feimer, P.L. Silveston, R.R. Hudgins, Can. J. Chem. Eng. 63 (1985) 86.
4. J.L. Feimer, P.L. Silveston, R.R. Hudgins, Can. J. Chem. Eng., 63 (1985) 481.
5. A.A. Adesina, R.R. Hudgins, P.L. Silveston, Catal. Today, 25 (1995) 127.
6. B. Shi and B.H. Davis, Catal. Today, 106 (2005) 129.
7. L-M. Tau, H.A. Dabbagh and B.H. Davis, Energy & Fuels, 4 (1990) 94.
8. W. Ngantsoue-Hoc, Y. Zhang, R.J. O'Brien, M. Luo and B.H. Davis, Appl. Catal. A: Gen., 236 (2002) 77.

Fischer-Tropsch Synthesis, Catalysts and Catalysis
B.H. Davis and M.L. Occelli (Editors)

Fischer-Tropsch synthesis: influence of support on the impact of co-fed water for cobalt-based catalysts

Gary Jacobs, Tapan K. Das, Jinlin Li, Mingsheng Luo, Patricia M. Patterson, and Burtron H. Davis

Center for Applied Energy Research, 2540 Research Park Drive, Lexington, KY 40511 USA

Co catalysts were prepared with variable cobalt oxide-support interactions through judicious selection of the cobalt loading, the type of support utilized, and the promoter employed, if any, along with its loading. For a comparable Co loading range, while a positive effect of water was found for catalysts identified to have supports that only weakly interacted with the cobalt clusters, an adverse impact of water was recorded when cobalt was supported on more strongly interacting supports, such as TiO_2 and especially, Al_2O_3. However, alumina supported cobalt catalysts were found to have much higher active site densities in the cobalt loading range explored, due to a smaller average crystallite size. More robust Co/Al_2O_3 catalysts, less sensitive to the negative effect of water, were obtained at higher Co loadings, where the average cluster size was > 10 nm.

1. Introduction

Supported cobalt catalysts are important for the slurry phase Fischer-Tropsch synthesis of hydrocarbons, which can be subsequently processed to produce an ultraclean, virtually sulfur-free, diesel [1]. Owing to their low selectivity for the water-gas shift reaction, cobalt catalysts are well-suited for the conversion of synthesis gas mixtures with a high H_2/CO ratio, such as natural gas-derived syngas. Iron-based catalysts, on the other hand, exhibit higher intrinsic water-gas shift selectivity and are more suitable for converting syngas with a lower H_2/CO ratio, such as that derived from coal gasification. There has been much research into the impact of water on the activity, selectivity, and stability of cobalt catalysts. Water has been reported to influence Co-based catalysts in a number of different ways, including most importantly, CO conversion, the selectivity of light products (especially methane), the selectivity of heavier products, the selectivity to 1-olefins versus paraffins, the selectivity of carbon dioxide, and the catalyst rate of deactivation. These, in turn, are reported to vary with H_2O partial pressure. It is important to note that H_2O is, of course, a primary product from the Fischer-Tropsch synthesis reaction:

$$CO + 2H_2 \rightarrow -[CH_2]_n- + H_2O$$

In this work, however, the focus is primarily on the impact of externally added H_2O co-fed with the reactant feed to the activity, selectivity, and stability of cobalt-based catalysts.

Although most researchers agree that the active sites for Fischer-Tropsch synthesis are surface Co^0 atoms, there appears to be less agreement as to the nature of the effect of water on cobalt-based catalysts. To demonstrate the disparity, a generally beneficial effect of H_2O has been reported, perhaps kinetic, for unsupported and silica-supported cobalt catalysts up to a certain threshold level [2-7], above which the effect is accompanied by a more rapid deactivation rate [6-8]. The beneficial effect includes an increase in CO conversion, decreased CH_4 and other light product selectivity (while keeping CO_2 at a low selectivity level), and increased selectivity to C_5+ (or $C_{10}+$) selectivity. However, the positive effect on CO conversion seems to be present only when the cobalt clusters are small enough to fit inside the pores, suggesting possibly a role of H_2O adsorption in preventing pore filling by heavier hydrocarbons, thereby alleviating intraparticle transport restrictions [9]. There is typically reported either no effect or a small, negative, reversible effect of added H_2O at low levels (e.g., at and below about 20 molar %), but an increase in the deactivation rate at higher levels of added H_2O for cobalt alumina catalysts, eventually leading to some irreversible and sometimes catastrophic deactivation at levels higher than about 25 molar % [7,11-15]. In fact, overcoming the deleterious effects of water associated with cobalt alumina catalysts was an important development that led to their commercial application [1,16]. It appears that more robust catalysts, less sensitive to permanent deactivation effects, can be made by increasing the average cobalt crystallite size through increasing the cobalt loading [1,15,16]. For titania-supported cobalt catalysts, both positive [2,3,4,7] and negative [17] effects on CO conversion for externally added H_2O have been reported, though it is important to note that the cobalt dispersions differed markedly between the studies reporting a positive effect (e.g., 3.8% [4] and 2.3% [7]) and those reporting a negative effect (e.g., 12.2%) for Co/titania. The results suggest the possibility that metal-support interactions may have played a role. All the cases referenced for the titania-supported cobalt catalysts reported some decrease in the methane selectivity.

It is deemed instructive, therefore, to examine more closely selected case studies in some detail from the literature, including both unsupported cobalt catalyst and a number of different supported cobalt-based catalysts. The supports chosen for study include silica, alumina, and titania, the most common supports reported in the literature.

Metal promoters are often used to enhance the reducibility of cobalt oxides supported on oxides whereby significant interactions with the support oxide occur. For example, ruthenium [17,18,19,20], platinum [13,14,20], and rhenium [2-4,7,10-12,18,20,21] are common choices. As an introduction, however, the contribution of the promoter to the effect of water will not be considered in the selected case studies. It is important to note that the manner in which the water effect has been examined by different groups varies. The case studies selected were all carried out using a fixed bed reactor system. Careful attention to essential details regarding the manner in which the data was reported, as well as the differences among conditions and configurations, is warranted. A brief assessment is then made after each case study is presented, and tables are used to highlight the essential findings from each of the four case studies.

1.1. Bulk cobalt catalyst Case Study

Two examples of the impact of added H_2O on bulk cobalt catalysts were reported in 1993 [2]. One of them is detailed below.

1.1.1. Preparation

The catalyst was prepared by adding an aqueous solution of ammonium carbonate to an aqueous solution of cobalt nitrate, resulting in a precipitation reaction. The precipitate was filtered, washed with deionized water, and dried at 120°C, with subsequent calcination in air at 500°C for 5 hours.

1.1.2. Characterization

XRD indicated that the calcined cobalt catalyst was essentially Co_3O_4. The surface area was 7.7 m^2/g.

1.1.3. Reaction testing

Reactor testing setup: The catalyst was tested in a down-flow 3/8" OD stainless steel tubular fixed bed reactor. The charge consisted of 6.8 g of the Co_3O_4 catalyst diluted with 18 g of quartz powder. Reduction of Co_3O_4 should then result in a 5 g yield of cobalt black. To even out the temperature profile, in addition to the dilution of the catalyst with quartz, an aluminum block jacket was fitted tightly around the reactor.

Catalyst activation: The catalyst was reduced overnight with 200 ccm H_2 at 450°C at close to atmospheric pressure.

Reaction conditions: P_{tot} = 20 atm, T = 200°C, space velocity = 3600 scm^3/g_{cat} hr on the basis of H_2 and CO only, feed composition = 63.1%H_2, 33.0%CO, 3.9%N_2, H_2/CO = 1.9.

Water addition methodology: H_2O was added in four separate runs, and data were, in each case, taken after 70 hours on stream. H_2O was added on a per 100 moles of syngas basis in increments of 0, 12.5, 27.7, and 50.0 and the total reactor pressure was changed in increments 20.7, 21.6, 24.9, and 28.6 atm. The syngas (i.e., H_2 + CO) partial pressure was decreased somewhat in increments 20.0, 18.7, 18.9, and 18.6 atm. The H_2O partial pressure increased in increments 0, 2.3, 5.2, and 9.3 atm.

Activity: The initial CO conversion of 12.1% was reported to increase favorably up to 27.7 molar % H_2O addition in the increments 12.1, 27.3, and 28.7 %. However, at 50.0% molar % added H_2O, the CO conversion declined to 18.8%.

Selectivity: The mole % carbon dioxide selectivity was found to be very low and was reported to vary only slightly with increasing molar % added H_2O in increments 0.27, 0.21, and 0.34. Data were not reported for CO_2 at the 50 molar % H_2O addition level. The mole %, methane selectivity was reported to decrease significantly in increments of 10.5, 7.1, 5.8, and 4.0, suggesting a favorable shift in the product distribution toward heavier products.

In weight percent, methane selectivity was reported to decrease in the increments 11.5, 8.1, 6.7, and 4.6%. The remaining data reported are also in weight percent. Light product selectivity in the C_2 - C_4 range likewise was reported to decrease in a beneficial manner in increments of 15.6, 8.7, 7.8, but then increased to 8.6. Products in the C_5 - C_9 range decreased in increments 17.4, 9.5, 9.0, and then increased to 9.7. Finally, C_{10}+ product selectivity was found to increase very favorably in increments 55.5, 73.7, 76.5, and 77.1.

Stability: Stability data as a function of time on stream were not reported.

Assessment: The data are intriguing and suggests that there is a limited, but significant advantage in co-feeding water up to about 28 molar % added H_2O for an unsupported cobalt catalyst. In summary, there is a positive increase in CO conversion, a favorable decrease in the selectivity of light products in the range of C_1 to C_9, and a desirable increase in the selectivity of C_{10}+ products. Since the CO_2 selectivity was reported to remain low, the data suggest that there is not a direct tradeoff between decreased methane selectivity and higher CO_2

selectivity. Rather, the data are consistent with a true shift in the product distribution to heavier products. A summary of the run is tabulated in Table 1.

Table 1. Bulk Cobalt Catalyst Case Study – Run summary of the effect of H_2O for an unsupported cobalt catalyst [2]. T = 200°C, space velocity = 3600 scm^3/g_{cat} hr on the basis of H_2 and CO only, feed composition = 63.1% H_2, 33.0% CO, 3.9% N_2, H_2/CO = 1.9.

Run #	#1	#2	#3	#4
Moles of H_2O added per 100 moles of syngas	0	12.5	27.7	50.0
Total pressure (atm)	20.7	21.6	24.9	28.6
Syngas partial pressure (atm)	20.0	18.7	18.9	18.6
H_2O partial pressure (atm)	0	2.3	5.2	9.3
Syngas molar % in feed	Decreasing With H_2O Addition	Decreasing With H_2O addition	Decreasing With H_2O addition	Decreasing With H_2O addition
CO conv. %	12.1	27.3	28.7	18.8
Moles CH_4 produced per 100 moles of CO converted	10.5	7.1	5.8	4.0
Moles CO_2 produced per 100 moles of CO converted	0.27	0.21	0.34	Not Reported
Wt.% CH_4 in product	11.5	8.1	6.7	4.6
Wt.% C_2-C_4 in product	15.6	8.7	7.8	8.6
Wt.% C_5-C_9 In product	17.4	9.5	9.0	9.7
Wt.% C_{10}+ In product	55.5	73.7	76.5	77.1

It is interesting to note that in another test, Kim [2] reported that the H_2O addition from 0 to 21 molar % not only favorably impacted the methane, light product, and C_{10}+ product selectivities, but also increased the 1-olefin to paraffin ratio for at least the C_2 through C_4 products, including an increase from 0.13 to 0.36 for C_2, from 2.47 to 3.68 for C_3, and from 1.87 to 2.52 for C_4. This trend in favor of 1-olefin production is also desirable.

1.2. Co/silica catalyst Case Study

Two studies of the impact of added H_2O on a cobalt catalyst supported on silica were reported by Krishnamoorthy et al. [5], one carried out at a higher total pressure of 19.7 atm, and one carried out at a lower total pressure of 4.9 atm. Some results for the study at the higher pressure are outlined below.

1.2.1. Preparation
A 12.7%Co/SiO$_2$ catalyst was prepared by incipient wetness impregnation of silica (PQ Corporation, CS-2133; treated in dry air at 400°C for 3 hours). The impregnated support was dried at 60°C. The catalyst was reduced in H$_2$ directly, without calcination, at 350°C for 1 hour. The catalyst was passivated in 1%O$_2$/He at room temperature.

1.2.2. Characterization
Though not reported in [5], our measurements on PQ silica indicated that the BET surface area is close to 365 m^2/g, with a pore volume of 2.44 cm^3/g and an average pore diameter of 27.8 nm. The metal dispersion, based on H$_2$ chemisorption measurements at 100°C after reduction at 325°C, was reported to be 5.8%. However, it is not clear if this is the true dispersion of the metal, or an uncorrected dispersion, as it is necessary for cobalt catalysts to determine the extent of cobalt reduction.

It appears that the authors assumed 100% reduction of cobalt, which may lead to an overestimation of the cluster size and consequently, an underestimation of the true metal dispersion. Assuming that 100% of the cobalt was reduced, a real dispersion of 5.8% would result in an average cobalt cluster size, assuming a spherical morphology, of approximately 18 nm. Since a lower extent of reduction would only increase the true dispersion of the metal, we believe it is safe to presume that the cobalt clusters were, on average, small enough to fit within the pores of the silica support studied.

1.2.3. Reaction testing
Reactor testing setup: The catalyst was tested in a plug flow reactor. The charge consisted of 1.75 g of Co/SiO$_2$ catalyst diluted with 2.8 g of silica, both particle sizes in the range of 100 to 180 μm in diameter.

Catalyst activation: The catalyst was reduced for 1 hour with 350 ccm H$_2$ at 325°C, presumably at atmospheric pressure.

Reaction conditions: P_{tot} = 19.7 atm, T = 200°C, space velocity = 3100 scm³/g$_{cat}$ hr, feed composition = 62%H$_2$, 31%CO, 7%N$_2$. H$_2$/CO = 2.0.

Water addition methodology: H$_2$O was added in a manner such that the average (over the length of the bed) molar % of H$_2$O, including that produced by the syngas conversion, in the reactor changed in increments 1.5, 12.0, 23.0, and 43.5. Since the total pressure was maintained constant with H$_2$O addition, then increasing the addition of H$_2$O would result in a decrease of the syngas reactant partial pressures. The average H$_2$O partial pressure was reported to increase in increments 0.3, 2.4, 4.5, and 8.6 atm.

Activity: CO conversion was reported to increase with increasing levels of average molar % H$_2$O up to 43.5 average molar % H$_2$O in the increments 8.4, 11.5, 17.6, and 20.9 %.

Selectivity: Carbon dioxide selectivity was not reported. In mole %, methane selectivity was reported to decrease favorably in increments 7.8, 4.9, 3.7, and 2.8, while at the same time, C$_5$+ selectivity was found to increase in a beneficial manner in increments 86.0, 90.1, 91.6, and 92.2%. 1-Olefin to paraffin ratios were also found to be enhanced by co-fed H$_2$O. For example, for C$_5$, the O/P ratio increased in increments of 2.08, 2.58, 2.76, and 2.89, while for C$_8$, the ratio increased in increments of 0.72, 1.57, 1.85, and 2.13.

Stability: No stability data as a function of time on stream was reported.

Assessment: From the standpoint of kinetics, the data suggests that there is an advantage of adding water, even up to levels as high as 43.5 average molar % H$_2$O for Co/SiO$_2$ catalyst. There is an increase in CO conversion, a decrease in the selectivity of methane, and an increase in the selectivity of C$_5$+ products observed. Although CO$_2$ selectivity was not presented, the tradeoff between decreasing methane selectivity and increasing C$_5$+ selectivity suggests that carbon was not lost to undesired carbon dioxide formation. A summary of the testing is provided in Table 2.

1.3. Co/titania Catalyst Case Study

Studies of the impact of added H$_2$O on a cobalt catalyst supported on titania were reported by Kim [2] including one carried out at a higher space velocity of 3540 cm³/g$_{cat}$ hr, and one carried out at a lower space velocity of 1180 cm³/g$_{cat}$ hr for unpromoted Co/TiO$_2$. Some results for the study at the higher space velocity are outlined below. Data for the impact of H$_2$O on Re promoted Co/TiO$_2$ catalysts are also included in the patent.

Table 2. Co/silica Catalyst Case Study – Run summary of the effect of H_2O for a silica-supported 12.7% cobalt catalyst [5]. $P_{tot} = 19.7$ atm, T = 200°C, space velocity = 3096 scm^3/g$_{cat}$ hr (changing with H_2O addition?), feed composition = 62% H_2, 31% CO, 7% N_2, H_2/CO = 2.0.

Sample #	#1	#2	#3	#4
Avg. molar % of H_2O present Across reactor	1.5	12.0	23.0	43.5
Total pressure (atm)	19.7	19.7	19.7	19.7
Syngas partial Pressure (atm)	~18.4	Decreasing With H_2O addition	Decreasing With H_2O Addition	Decreasing With H_2O addition
Avg. H_2O Partial pressure (atm)	0.3	2.4	4.5	8.6
Syngas molar % in feed	Decreasing With H_2O Addition	Decreasing With H_2O Addition	Decreasing With H_2O Addition	Decreasing With H_2O Addition
CO conv. (%)	8.4	11.5	17.6	20.9
Moles CH$_4$ Produced per 100 moles of CO converted	7.8	4.9	3.7	2.8
Moles CO$_2$ produced Per 100 moles of CO converted	Not Reported	Not Reproted	Not Reported	Not Reported
Moles of carbon in C$_5$+ produced per 100 moles of CO converted	22.67	32.56	50.55	60.43
1-olefin/paraffin ratio: C$_5$	2.08	2.58	2.76	2.89
1-olefin/paraffin ratio: C$_8$	0.72	1.57	1.85	2.13

1.3.1. Preparation

The titania support was Degussa P25, with a predominately rutile phase. The catalyst was prepared using cobalt nitrate as the precursor, with subsequent drying in air and calcination.

1.3.2. Characterization
Catalyst characterization data were not provided.

1.3.3. Reaction testing
Reactor testing setup: The system used was similar to the one utilized for the unsupported Co catalyst tests. The catalyst was diluted with rutile titania powder, with a ratio of 5 parts catalyst to 6 parts titania.

Catalyst activation: The catalyst was reduced in H_2, but details are not provided.

Reaction conditions: $P_{tot} = 20.7$ atm, T = 200°C, space velocity = 3540 scm^3/g_{cat} hr on the basis of H_2 and CO only, feed composition = 63.9%H_2, 32.1%CO, 4.0%N_2, H_2/CO = 2.0.

How was the water added: H_2O was added in the molar % (on a per 100 moles of syngas basis) increments 0 and 10.1 in such a manner that the syngas (i.e., H_2 + CO) composition decreased with H_2O addition from 96% of the feed to 86.3%.

Activity: CO conversion was reported to increase from 8.5% to 21.9% with adding H_2O.

Selectivity: In mole %, carbon dioxide selectivity was found to be similar before and after H_2O addition, at approximately 0.22 and 0.21%, respectively. In mole %, methane selectivity was reported to decrease sharply from 9.6% without added H_2O to 4.1% after H_2O addition. The O/P ratio was found to improve as well. For C_2, the O/P increased from 0.20 to 0.78. For C_3, the change was from 2.22 to 3.92 and for C_4, the change was from 1.24 to 2.38. In weight %, methane selectivity was reported to decrease from 10.9% to 4.6%, C_2 - C_4 decreased favorably from 9.8 to 5.0%, C_5 - C_9 decreased from 15.9 to 9.3%, and $C_{10}+$ increased impressively from 63.4 to 81.1%.

Stability: Stability data as a function of time on stream were not reported.

Assessment: Essentially identical trends were obtained for the titania supported cobalt catalyst as with the silica-supported and unsupported cobalt-based catalysts. The low levels of CO_2 reported clearly suggest that methane was not converted unfavorably to carbon dioxide. Rather, a shift in the product distribution toward heavier products is suggested. The results of the case study are summarized in Table 3.

Table 3. Co/titania Catalyst Case Study – Run summary of the effect of H_2O for a titania-supported 12% cobalt catalyst [2]. P_{tot} = 20.7 atm, T = 200°C, space velocity = 3540 scm^3/g_{cat} hr on the basis of H_2 and CO only, feed composition = 63.9% H_2, 32.1% CO, 4.0% N_2, H_2/CO = 2.0.

Run #	#1	#2
Moles of H_2O added per 100 moles of syngas	0	12.0
Total pressure (atm)	20.7	20.7
Syngas partial pressure (atm)	19.9	17.9
H_2O partial pressure (atm)	0	2.1
Syngas molar % in feed	96	86.3
CO conversion (%)	8.5	21.9
Moles CH_4 produced per 100 moles of CO converted	9.6	4.1
Moles CO_2 produced per 100 moles of CO converted	0.22	0.21
Wt.% CH_4 in product	10.9	4.6
Wt.% C_2-C_4 in product	9.8	5.0
Wt.% C_5-C_9 in product	15.9	9.3
Wt.% C_{10}+ in product	63.4	81.1

1.4. Co/alumina Catalyst Case Study

A number of studies of the impact of H_2O on alumina-supported cobalt catalysts have been carried out by Holmen et al. [7,10-12]. Of those works, one assessment of the impact of water on unpromoted Co/Al_2O_3 has been selected from Storsaeter et al. [7].

1.4.1. Preparation

The alumina chosen was Puralox SCCa-5/200 from Condea pre-calcined at 500°C in air for 10 hours. The support was impregnated with an aqueous solution of cobalt nitrate to the point of incipient wetness. The catalyst was dried at 120°C for 3 hours and calcined at 300°C for 16 hours. The loading was 12% of cobalt.

1.4.2. Characterization

The catalyst was well-characterized by a combination of methods. By TPR, the cobalt oxide species were found to interact strongly with the alumina support. A TPR carried out after reduction in hydrogen at 350°C for 16 hours still showed that a considerable fraction of Co oxide species remained unreduced. The extent of reduction, quantified by tallying the number of pulses of O_2 that were consumed by the catalyst, was found to be 53%. The dispersion was reported by both H_2 chemisorption and XRD to be 6.3 and 8.8%, respectively. The latter method was carried out by line broadening analysis of a peak for Co_3O_4, and then accounting for contraction of the cluster in converting to the metal, by utilizing the correction factor 3/4. Therefore, the average crystallite size in nm was found to be in the range of 10.9 to 15.2 nm. Since the average pore size reported was 6.7 nm, it would appear that the average cluster would be too large to fit in the pore. However, if the diameter is taken, from the other data reported for the support to be radius = $2V/A$ = $2*(0.34\ cm^3/g)/(161\ m^2/g)$, then the average pore diameter should be double that, and closer to 8.4 nm. Furthermore, if the average cobalt cluster size is corrected from the chemisorption data to included the extent of reduction, the true cluster size may in fact be smaller than the 15.2 nm reported.

Uncorrected % dispersion =	(# of Co^0 surface atoms)/(# of Co atoms in sample)
Corrected % dispersion =	(# of Co^0 surface atoms)/ (# of Co^0 atoms in the sample)
Corrected % dispersion =	(# of Co^0 surface atoms)/[(# of Co atoms in sample)(% reduction)]
Corrected % dispersion =	(6.3%)/(0.53) = 11.9%
Cluster size estimate =	8.0 nm.

Therefore, based on this correction, the conclusion is that the cluster size may be small enough to reside inside the pores of the alumina.

1.5. Reaction testing

Reactor testing: The catalyst was tested in a stainless steel fixed bed reactor with an inner diameter of 10 mm. The charge consisted of 1-2 g of Co/Al_2O_3 catalyst diluted to minimize temperature gradients with SiC (75 - 150 µm) in a weight ratio of between 0.2 - 0.5. An aluminum jacket was also utilized outside the reactor.

Catalyst activation: The catalyst was reduced in flowing hydrogen at atmospheric pressure at 350°C for 16 hours.

Reaction conditions: P_{tot} = 19.7 atm, T = 210°C, space velocity = 2982 cm^3/g_{cat} hr, feed composition = 63.9%H_2, 32.1%CO, 4.0%N_2, H_2/CO = 2.1.

How was the water added: H_2O was first degassed with helium, fed into a vaporizer kept at 375°C, and mixed with the reactants prior to the reactor inlet. 20 Molar % H_2O was added to the feed, although an inert balancing gas was not used. Therefore, the syngas reactant partial pressures should decrease with the addition of external H_2O.

Activity: CO conversion was reported to decrease from 42.6% to 39.5% with adding 20 molar % H_2O to the feed.

Selectivity: Carbon dioxide selectivity was not reported. In mole %, methane selectivity was reported to decrease favorably from 9.7% without added H_2O to 6.7% after H_2O addition, while concurrently, the C_5+ selectivity was reported to increase from 80.2% to 83.0%. In separate data, the 1-olefin to paraffin ratio for C_3 was reported to increase with increasing H_2O addition (Conditions: No H_2O at 44.8% conversion, 20% H_2O at 34.5% conversion, and 33%H_2O at 25.8% conversion) in increments of 2.3, 2.3, and 3.8, respectively.

Stability: Interestingly, while the deactivation rate was not influenced by increasing the conversion through varying the space velocity (which should thereby increase the partial pressure of H_2O), addition of external H_2O increased the deactivation rate, especially at higher partial pressures (e.g., 33 molar %) of added H_2O. Switching off the H_2O only led to a partial recovery, indicating some irreversible deactivation occurred.

Assessment: The same decrease in CH_4 selectivity was observed as in the other cases, accompanied by an enhancement toward C_5+ products in the product distribution. However, CO conversion was found to decrease by addition of H_2O. Moreover, addition of H_2O led to an acceleration in the catalyst deactivation rate, especially at higher partial pressures of added H_2O. The same phenomenon was not observed, however, solely by increasing the H_2O partial pressure via adjusting the space velocity downward to increase CO conversion. Although CO_2 selectivity was not reported, the decrease in light product selectivity is suggested to be due to a shift toward heavier product formation, and not to increased CO_2 production. The tests are summarized in Table 4.

In considering the above four case studies, some questions arise regarding choice of reactor and method of H_2O addition. Fixed bed reactor testing can suffer from holdup of heavy wax products in the catalyst and silica bed or other diluent. One must therefore wonder if switching from a stream without H_2O to a stream containing H_2O may result in the flushing out of wax product held up within the bed. The use of a slurry phase reaction system can avoid such unknowns, as the wax produced is easily solubilized in start-up wax at initial conditions, or in the reactor wax after considerable time on stream. To demonstrate the limitations of relying on fixed bed reactor test data, we recently carried out fixed bed reactor tests whereby we switched from a feed containing syngas + inert gas to one containing syngas + supercritical solvent, while maintaining constant reactant feed partial pressures, total pressure, and space velocity. Switching to the feed containing a tuned supercritical solvent mixture resulted in considerable wax extraction from the catalyst bed. That is, the wax product not only contained wax produced from the reaction carried out under supercritical conditions, but it also contained extracted wax produced earlier in the run, when the reaction was carried out in the absence of the supercritical solvent. One must therefore consider if switching to a feed containing H_2O may result in extraction of wax from the catalyst pores, and a temporary increase, therefore, in $C_{10}+$ selectivity.

Table 4. Co/alumina Catalyst Case Study – Run summary of the effect of H_2O for an alumina-supported 12% Co-catalyst [7]. Reaction conditions: $P_{tot} = 19.7$ atm, $T = 210^oC$, space velocity = 2982 cm^3/g_{cat} hr, feed composition = 63.9% H_2, 32.1% CO, 4.0% N_2, $H_2/CO = 2.1$.

Sample #	#1	#2
Moles of H_2O added per 100 moles of syngas	0	25.3
Total pressure (atm)	19.7	19.7
Syngas partial pressure (atm)	19.2	15.4
H_2O partial pressure (atm)	0	3.9
Syngas molar % in feed	96	86.3
CO conversion (%)	42.6	39.5
Moles CH_4 produced per 100 moles of CO converted	9.7	6.7
Moles CO_2 produced per 100 moles of CO converted	Not reported	Not reported
Moles of carbon in C_5+ produced per 100 moles of CO converted	80.2	83.0
1-olefin to paraffin ratio, C_3	2.3	3.3

Another disadvantage to the above methodology is that an inert balancing gas was not utilized to keep the reactant partial pressures, compositions, and total gas space velocity constant during addition of H_2O. That is to say, an equivalent molar % of inert gas was not replaced by the same amount of H_2O during the H_2O addition step. In the first case, for the unsupported cobalt catalyst, the partial pressures of the reactants were kept relatively constant, but the total pressure of the system was increased. In all other cases, the total pressure was maintained constant, but again, no balancing gas was utilized. Therefore, in this situation, the syngas reactant partial pressures must change during water addition, making it difficult to assess the real kinetic impact of water. A better method would have been to replace a known partial pressure of inert gas with the same partial pressure of steam such that the total gas space velocity, the syngas reactant feed partial pressures, and the overall total pressure would remain constant.

Certainly, the above studies are very interesting, highly suggestive, and were chosen as a basis of comparison for our studies. We attempt to characterize the water effect over a number of these materials in such a manner as to overcome some of the obstacles previously highlighted, and to place the conclusions drawn from the previous studies on a firmer footing. For example, fixed bed reactors are typically avoided for Fischer-Tropsch synthesis, as there are problems of wax product holdup in the catalyst/dilution agent beds, leading to poor product solubilization and extraction. Therefore, a slurry phase type configuration is usually preferred (e.g., CSTR, slurry bubble column reactor, etc.), where product solubilization is better carried out (e.g., either in the start-up solvent, typically a narrow cut of a high molecular weight wax, or at greater time on stream, the reactor wax itself). For kinetic analyses, another problem of using fixed bed reactors is that the partial pressures of reactants and products will vary along the length of the bed, meaning that assumptions must be made regarding the calculation of average partial pressures. Use of a CSTR alleviates this problem, as there is generally a more uniform composition within the reactor, such that the exiting product stream will represent the average of the contents of the reactor, leading to more meaningful conversion and selectivity results. Also, the Fischer-Tropsch synthesis reaction is exothermic to the point that it is important to consider that localized temperature fluctuations (e.g., hot or cool spots) in a catalyst bed can adversely impact reaction data. For example, in the previous studies, since an inert balancing gas was not used, H_2O was fed in such a manner that the total gas velocity increased, and this alone will impact the temperature gradients along the length of a fixed bed reactor. Use of a CSTR will lead to a more uniform temperature distribution, and therefore, more meaningful results. CSTRs have the additional benefit that one does not have to consider channeling effects, often present in a packed bed situation.

To allow us to better assess the impact of water, instead of adding external water to the reactant feed stream directly, a situation that causes the partial pressures of the syngas reactants to be altered, or the total pressure of the reactor to increase, we utilized a method that leads to constant reactant partial pressures, constant total gas space velocity, and constant total pressure before, during and after H_2O addition. That is, we replaced an equivalent amount of co-fed inert gas with the same amount of steam in an incremental manner.

Before selecting catalyst materials for use in our study, particular attention was paid to characterization results. Regarding the use of Co/Al_2O_3 catalysts, from the standpoint of activity, it is often reported that the metal oxide-support interaction is a major problem. Certainly, it is true that after a standard reduction at 350°C, only a fraction of the cobalt is reduced [7,20,21]. However, the number of active sites depends not only on the degree of reduction, but also on the size of the cobalt crystallites. In comparing 15%Co/SiO$_2$ with 15%Co/Al$_2$O$_3$ catalysts, which were prepared by standard impregnation and calcination, while cobalt silica offered much higher degree of reduction, the number of active sites was found to be lower in comparison with cobalt alumina, due to the larger average diameter of the Co crystallites [20]. When alternate preparation procedures are used, smaller Co crystallites are formed on SiO$_2$, so interactions are observed in TPR [6,22].

For a given support, the metal oxide-support interaction essentially provides a basis for design by varying cobalt loading [20]. Co crystallites should not be so small that they are too difficult to reduce. On the other hand, the interaction should not be completely overcome, as the crystallites will be too large to provide adequate surface [20]. An intermediary cobalt crystallite size (10-20 nm range) for Co/Al$_2$O$_3$ provides adequate reducibility as well as surface area. Our better Co/Al$_2$O$_3$ catalysts were those with a loading of 25% and an average crystallite size of 11 nm; the degree of Co reduction was approximately 42% after 10 hour hydrogen treatment at 350°C [20].

The use of promoters (Re, Pt, Ru) was explored to facilitate the reduction of supported Co catalysts for which there are strong Co oxide interactions (Co/TiO$_2$ and Co/Al$_2$O$_3$). The aim was to achieve more active cobalt catalysts with less Co loading than typically used. In this study, we examine the influence of co-fed water on four selected Co-based catalysts, including 12.4%Co/SiO$_2$ [6], 0.2%Ru-10%Co/TiO$_2$ [17], 0.5%Pt-15%Co/Al$_2$O$_3$ [13,14], and 25%Co/Al$_2$O$_3$ [15]. Particular attention is also paid to the influence of H$_2$O on the changing catalyst structure during the course of a reaction test. In separate tests, catalyst samples were retrieved from the reactor such that the catalyst was collected under inert gas in the wax matrix. These samples were scanned at the synchrotron utilizing X-ray absorption spectroscopic methods, including XANES and EXAFS [14,15].

2. Experimental

2.1. Catalyst Preparation

A 12.4%Co/SiO$_2$ catalyst [6] was prepared by two step incipient wetness impregnation (IWI) of a SiO$_2$ support (PQ Co. CS-2133, BET SA 353 m^2/g, pore volume 2.36 cm^3/g, average pore diameter 25.8 nm). The sample was dried under vacuum. Calcination was not carried out between loading steps. Half of the preparation was calcined at 350°C in air. The other half was dried but not calcined, and the direct reduction of the nitrate procedure was used.

A 10.0%Co/TiO$_2$ [17] catalyst was prepared by IWI of a TiO$_2$ support (Degussa P-25 TiO$_2$, BET SA 45 m^2/g, pore volume 0.369 cm^3/g, average pore diameter 31 nm). The sample was dried and calcined at 300°C. Half was impregnated with a solution of ruthenium nitrosylnitrate, dried, and calcined at 300°C to give a 0.2%Ru loading.

A 15.0%Co/Al$_2$O$_3$ [13] catalyst was prepared by a three step IWI of an Al$_2$O$_3$ support (Condea Vista B γ-alumina, BET surface area 200 m^2/g, pore volume 0.51 cm^3/g, average pore diameter 9.4 nm). The sample was dried under vacuum and although the sample was not calcined between cobalt nitrate loading steps, the final sample was calcined at 400°C. Half was promoted with Pt by IWI of the tetraammineplatinum (II) nitrate salt and calcined at 350°C.

A 25.0%Co/Al$_2$O$_3$ catalyst [15] was prepared by a slurry impregnation method, and cobalt nitrate was used as precursor. In this method [1], the ratio of the volume of solution used to the weight of alumina was 1:1, such that approximately 2.5 times the pore volume of solution served as the loading solution. Two impregnation steps were used, with interval drying, prior to calcination at 400°C.

2.2. Catalyst Characterization

2.2.1. Temperature programmed reduction

Temperature programmed reduction (TPR) profiles were recorded using a Zeton Altamira AMI-200 unit which employs a thermal conductivity detector. Samples were first purged at 350°C in flowing Ar to remove traces of water. TPR was performed using a 10%H$_2$/Ar mixture and referenced to Ar at a flow of 30 cm^3/min. The sample was heated from 50 to 800°C using a heating ramp of 10°C/min [20].

2.2.2. Hydrogen chemisorption/pulse reoxidation

Hydrogen chemisorption/pulse reoxidation was conducted using temperature programmed desorption (TPD) [23], also measured with the Zeton Altamira unit. The sample weight was 0.220 g. Catalysts were activated in a flow of 10 cm^3/min of H$_2$ mixed with 20 cm^3/min of Ar at 623 K for 10 h and then cooled under flowing H$_2$ to 100°C. The sample was held at 100°C under flowing Ar to remove and/or prevent adsorption of weakly bound species prior to increasing the temperature slowly to 350°C. The catalyst was held under flowing Ar to desorb any remaining chemisorbed hydrogen until the TCD signal returned to baseline. The TPD spectrum was integrated and the number of moles of desorbed hydrogen determined by comparing its area to the areas of calibrated hydrogen pulses. Dispersion calculations were based on the assumption of a 1:1 H:Co stoichiometric ratio and a spherical crystallite morphology. In estimating cobalt crystallite size from hydrogen chemisorption measurements [20,21], it is necessary to estimate the fraction of cobalt existing in the metallic phase. Therefore, after hydrogen TPD, the sample was reoxidized at 350°C using O_2 pulses. The percent reduction was calculated assuming the metal reoxidized to Co_3O_4 [20,21].

2.2.3. X-ray absorption spectroscopic methods

XAS measurements on references and catalyst samples were conducted in transmission mode at NSLS at BNL, using beamline X18b, which is equipped with a Si(111) channel cut monochromator. EXAFS spectra were recorded at close to the boiling temperature of N_2 in a cell. Data reduction of EXAFS spectra was carried out with WinXAS and BAN. Standard data reduction were carried out. Fitting of the Co-Co was carried out by using a Hanning window and taking the reverse Fourier transform of r-space data. After converting to Chi(k), the fitting of the spectra was carried out in k-space using FEFFIT. The L_{III} edge of Pt was also examined, in order to determine if Pt was in direct contact with Co for 0.5%Pt-15%Co/Al_2O_3 after standard hydrogen reduction treatment [24].

2.3. Reaction testing using continuously stirred tank reactors

For a typical reactor run, the calcined catalyst (ca. 15-20 g) was reduced ex-situ in a fixed bed reactor with H_2 at 60 NL/min. The reactor temperature was increased from room temperature to 100°C at a ramp rate of 2°C/min. The temperature was then increased to 350°C at a rate of 1°C/min and kept at 350°C for at least 10 h. The catalyst was then transferred under inert atmosphere to a CSTR to mix with Polywax 3000. The catalyst was then reduced in-situ in the CSTR using H_2 at atmospheric pressure. The in-situ reduction was carried out at 280°C with H_2 at 30 NL/min for 24 hours. After activation, the reactor temperature was decreased to the reaction condition and synthesis gas ($2H_2$/CO) was introduced to increase the reactor pressure to the working reaction condition. Conversion of CO and H_2 and the formation of products were measured following a period of 24 hours at each condition.

3. Results

3.1. Impact of co-fed H_2O on activity and stability

Inert balancing gas enabled us to maintain constant partial pressures of reactants as well as constant space velocity such that the water effect could be monitored in an unambiguous manner. To establish the relative strength of the metal-support interaction, unpromoted calcined catalysts were prepared in a similar range of Co loading. TPR (Figure 1) and hydrogen chemisorption/pulse reoxidation measurements (Table 5) were carried out on unpromoted, calcined Co/SiO_2, Co/TiO_2, and Co/Al_2O_3 catalysts, and the relative ranking of the interaction was: $Co/Al_2O_3 > Co/TiO_2 > Co/SiO_2$. As expected, due to the interaction, for a similar loading of cobalt the relative average cobalt crystallite diameter was: $Co/Al_2O_3 < Co/TiO_2 < Co/SiO_2$. The Co/TiO_2 and Co/Al_2O_3 catalysts were promoted with Ru and Pt, respectively, and tested for the impact of water. Since the cobalt crystallites on Co/SiO_2 were too large to provide adequate activity for study, a special procedure was used, circumventing calcination, whereby direct reduction of the cobalt nitrate precursor was performed, a technique which improved (decreased) the average cobalt crystallite size [22]. This catalyst was also tested for the impact of water. As shown in this study, the relative strength of the metal-oxide support interaction among supports may account for the differences observed for the water effect.

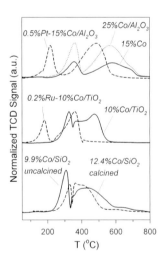

Figure 1. TPR profiles of unpromoted and promoted Co catalysts.

Table 5. H_2 chemisorption (TPD) and pulse reoxidation [6,13-15,17]

Catalyst	Tr (K)	Mmol H_2 Desorbed Per g cat	Uncorr %Disp	O_2 uptake (µmol/g)	% Red	Corr %Disp	Corr Diam (nm)
12.4% Co/SiO$_2$	623	18.2	1.7	902	64	1.7	38.2
Co/SiO$_2$ nitrate Reduct route	623	25.6	3.0	433	39	3.0	13.2
10.0% Co/TiO$_2$	573	42.8	5.1	593	52	9.7	10.6
0.2% Ru- 10% Co/TiO$_2$	573	66.6	7.8	722	64	12.2	8.5
15% Co/Al$_2$O$_3$	623	66.9	5.3	509	30	17.5	5.9
0.5% Pt-15% Co/Al$_2$O$_3$	623	140.6	11.0	1024	71	18.4	5.6
25% Co/Al$_2$O$_3$	623	77.7	77.7	1174	42	8.7	11.8

The results for the 12.4%Co/SiO$_2$ catalyst, the 0.2%Ru-10%TiO$_2$ catalyst, the 0.5%Pt-15%Co/Al$_2$O$_3$ catalyst, and the 25%Co/Al$_2$O$_3$ catalyst are summarized in Tables 6 - 9 for comparative purposes among the different catalyst systems, and for comparison with the four case studies previously summarized.

3.2. SiO$_2$ supported Co catalyst

As shown in Figure 2 and Table 6, for the support exhibiting the weakest interaction with Co oxide species during reduction, Co/SiO$_2$, a measurable increase in CO conversion was observed when 5 molar % water was co-fed, from 22.5% to 28.1% [6]. A similar increase was observed from 8-20% by volume H$_2$O addition as well. At 25% added water, however, although an increase was observed initially, this initial enhancement was followed by accelerated deactivation of the catalyst, which did not recover after the water was switched off (i.e., replaced by the inert argon balancing gas). By examining the filled circles in Figure 2, one can estimate the impact of H$_2$O on the aging rate of the catalyst. In the initial period (0-120 hours), prior to H$_2$O addition, the deactivation rate was approximately 0.26% per day. Up to 20 vol % added H$_2$O (150 - 450 hours), the catalyst was quite resilient, and the deactivation rate only increased to 0.37% per day. However, at 25% H$_2$O addition (475 - 510 hours), and even after the water was switched off (530 - 650 hours), the catalyst rapidly deactivated at a rate of 1.5% per day.

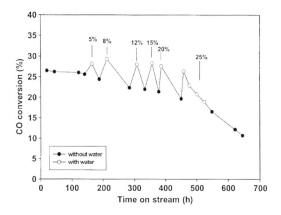

Figure 2. Effect of water on CO conversion for Co/SiO$_2$ (T = 483K, P = 20 atm (2.02 MPa), P$_{H_2}$ + P$_{CO}$ = 2.0, SV = 8 SL/g$_{cat}$/h, % is vol.% of total feed.

Table 6. Run summary of the effect of H$_2$O for a silica-supported 12.4% Co-catalyst [6]. Reaction conditions: P$_{tot}$ = 20 atm, T = 210°C, space velocity = 8000 scm^3/g$_{cat}$ hr, feed composition = 46.7% H$_2$, 23.3% CO, 30.0% Ar, H$_2$/CO = 2.0.

Sample#	#1	#2	#3	#4	#5	#6	#7
Moles of H$_2$O added per 100 moles of syngas	0	7.1	11.4	17.1	21.4	28.6	35.7
Total pressure (atm)	20	20	20	20	20	20	20
Feed syngas partial pressure (atm)	14	14	14	14	14	14	14
Feed H$_2$O partial pressure (atm)	0	1.0	1.6	2.4	3.0	4.0	5.0
Syngas molar % in feed	70	70	70	70	70	70	70
H$_2$O molar% H$_2$O in feed	0	5	8	12	15	20	25
CO Conversion (%)	22.5	28.1	29.2	28.1	28.9	28.6	27.4
Reactor H$_2$O partial pressure (atm)	1.0	2.3	3.0	4.0	4.8	5.8	6.6
Moles CH$_4$ produced per 100 moles of CO converted	5.6	4.8	4.3	4.3	4.1	4.0	3.8
Moles CO$_2$ produced per 100 moles of CO converted	<0.01	<0.01	0.01	0.09	0.11	0.19	0.24
Moles CH$_4$+CO$_2$ produced per 100 moles of CO converted	5.6	4.8	4.31	4.39	4.21	4.19	4.04
Moles of carbon in C$_5$+ produced per 100 moles of CO converted	89.1	90.0	91.6	91.8	92.0	91.9	91.5
1-olefin to paraffin ratio, C$_3$	0.70	0.73	0.74	0.75	0.75	0.77	0.76
1-olefin to paraffin ratio, C$_4$	0.61	0.62	0.66	0.65	0.66	0.70	0.69
Deactivation rate (%/day)	0.26	0.37	0.37	0.37	0.37	0.37	1.50

3.3. TiO₂ supported Co catalyst

In contrast to Co/SiO_2, the Ru-Co/TiO_2 catalyst (Figure 3 and Table 7) exhibited a slightly negative effect on CO conversion when water was co-fed with the syngas [17]. Hardly detectable at higher space velocities (e.g., 8000 and 4000 $scm^3/g_{cat}hr$), the negative impact was more pronounced at lower space velocities (e.g., 2000 and 1000 $scm^3/g_{cat}hr$), where the conversion and consequently, the H_2O partial pressure, were higher. At the lowest space velocity condition, there was a steeper decline in CO conversion with added H_2O, perhaps due to a higher H_2O partial pressure arising from the combination of intrinsic H_2O and external H_2O (about 14.5 molar % in the feed). Aging rates at space velocities of 8000 and 2000 $scm^3/g_{cat}hr$ were similar (0.38 and 0.47%, respectively), if one considers the filled circles (Figure 3) corresponding to the CO conversion without H_2O addition. However, at space velocity 1000 $scm^3/g_{cat}hr$, the catalyst did not recover after H_2O addition, and the deactivation rate was much more rapid (about 4.2% per day).

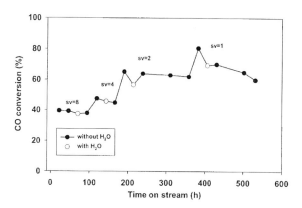

Figure 3. Effect of water on CO conversion for 0.2% Ru-10% Co/TiO₂ (T = 503K, P = 20 atm (2.02 MPa), P_{H_2} + P_{CO} = 17 atm (1.72 MPa), H₂/CO = 2.0, SV varied, % vol H₂O added = 15%.

Table 7. Run summary of the effect of H_2O for a titania-supported 10% Co-catalyst [17]. Reaction conditions: P_{tot} = 20 atm, T = 230°C, space velocity = varies, feed composition = 57.1% H_2, 28.6% CO, 14.3% Ar, H_2/CO = 2.0.

Sample#	#1	#2	#3	#4	#5	#6	#7	#8
Space velocity ($scm^3/g_{cat}hr$)	8000	8000	4000	4000	2000	2000	1000	1000
Moles of H_2O added per 100 moles of syngas	0	16.7	0	16.7	0	16.7	0	16.7
Total pressure (atm)	20	20	20	20	20	20	20	20
Feed syngas partial pressure (atm)	17.1	17.1	17.1	17.1	17.1	17.1	17.1	17.1
Feed H_2O partial pressure (atm)	0	2.9	0	2.9	0	2.9	0	2.9
Syngas molar % in feed	85.7	85.7	85.7	85.7	85.7	85.7	85.7	85.7
H_2O molar% in feed	0	14.5	0	14.5	0	14.5	0	14.5
CO Conversion (%)	39.2	37.5	47.7	46.3	65.3	56.5	80.6	69.4
Moles CH_4 produced per 100 moles of CO converted	7.60	6.50	7.60	6.44	6.69	6.42	6.33	6.26
Moles CO_2 produced per 100 moles of CO converted	0.17	0.49	0.34	0.64	0.94	1.58	2.27	3.50
Moles CH_4+CO_2 produced per 100 moles of CO converted	7.7	6.99	7.94	7.08	7.63	8.00	8.60	9.76
Deactivation rate (%/day)	0.38	0.38	Not Rptd.	Not Rptd.	0.47	0.47	4..2	4.2

3.4. Al_2O_3 supported Co catalyst

The Co/Al_2O_3 catalysts exhibited a negative effect on CO conversion during water co-feeding. The water effect was reversible for the 0.5%Pt-15%Co/Al_2O_3 catalyst up to about 25% added water [13]. Above that threshold, an important irreversible and catastrophic component to the decline in CO conversion occurred (Figure 4 and Table 8). The 25%Co/Al_2O_3 catalyst was less susceptible to water over the entire range up to 20% H_2O [15]. Although a slightly irreversible contribution to the impact was observed above 20% H_2O

addition (Figure 5 and Table 9), no catastrophic irreversible decline occurred, even after 30% H_2O addition, indicating the catalyst was more robust.

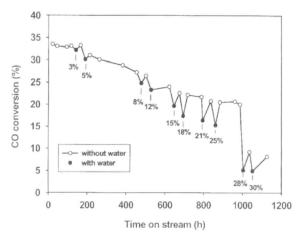

Figure 4. Effect of water on CO conversion for 0.5% Pt-15%Co/Al_2O_3 (T = 483K, P = 28.9 atm (2.9 MPa), P_{H_2} + P_{CO} = 20.0 atm (2.02 MPa), H_2/CO = 2.0, SV = 8 SL/g_{cat}/h, vol% is % of total.

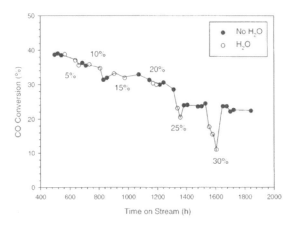

Figure 5. Effect of water on CO conversion for 25% Co/Al_2O_3 (T = 493K, P = 28.9 atm (2.9 MPa), P_{H_2} + P_{CO} = 20.0 atm (2.02 MPa), H_2/CO = 2.0, SV = 3.5 SL/g_{cat}/h, vol% is % of total.

Table 8. Run summary of the effect of H_2O for a 0.5 Pt promoted alumina-supported 15% Co-catalyst [13,14]. Reaction conditions: P_{tot} = 28.9 atm, T = 210°C, space velocity = 8000 $scm^3/g_{cat}hr$, feed composition = 46.7% H_2, 23.3% CO, 30.0% Ar, H_2/CO = 2.0.

Sample#	#1	#2	#3	#4	#5	#6	#7	#8	#9	#10
Moles of H_2O added per 100 moles of syngas	0	4.3	7.1	11.4	17.7	21.4	30	35.7	40	42.9
Total pressure (atm)	28.9	28.9	28.9	28.9	28.9	28.9	28.9	28.9	28.9	28.9
Feed syngas partial pressure (atm)	20.2	20.2	20.2	20.2	20.2	20.2	20.2	20.2	20.2	20.2
Feed H_2O partial pressure (atm)	0	0.87	1.5	2.3	3.5	4.3	6.1	7.2	8.1	8.7
Syngas molar % in feed	70	70	70	70	70	70	70	70	70	70
H_2O molar% in feed	0	3	5	8	12	15	21	25	28	30
CO Conv. (%)	33.1	32.2	30.1	24.8	23.3	19.6	16.4	15.3	5.1	4.5
Reactor H_2O Partial pressure (atm)	2.14	3.25	3.64	3.88	4.77	5.77	7.36	8.37	8.37	8.39
Moles CO_2 produced per 100 moles of CO converted	0.26	0.66	0.80	0.84	1.06	1.46	2.74	3.80	11.5	12.4
Deactivation rate (%/day)	0.31	0.31	0.31	0.31	0.31	0.31	0.31	0.31	*	*

*Catastrophic deactivation observed (i.e., > 75% loss of activity).

Table 9. Rate summary of the effect of H_2O for alumina-supported 25% Co-catalyst [15]. Reaction conditions: P_{tot} = 19.7 atm, T = 220°C, space velocity = 5000 $scm^{3/g}_{cat}$ hr, feed composition = 43.3% H_2, 21.7% CO, 3% Ar, H_2/CO = 2.0,

Sample #	#1	#2	#3	#4	#5	#6
Time on stream (h)	334	431	455	503	599	617
Moles of H_2O added per 100 moles of syngas	0	39	39	39	0	0
Total pressure (atm)	19.7	19.7	19.7	19.7	19.7	19.7
Feed syngas partial pressure (atm)	12.8	12.8	12.8	12.8	12.8	12.8
Feed H_2O partial pressure (atm)	0	4.9	4.9	4.9	0	0
Syngas molar % in feed	65	65	65	65	65	65
H_2O molar % in feed	0	25	25	25	0	0
CO conversion (%)	46.6	38.9	35.2	26.2	34.0	33.9
Reactor H_2O partial Pressure (atm)	2.40	7.72	7.38	6.64	1.65	1.65
Moles CH_4 produced per 100 moles of CO converted	7.43	5.37	5.40	5.14	7.01	7.06
Moles CO_2 produced per 100 moles of CO converted	0.49	1.44	2.44	2.64	0.41	0.20
Moles $CH_4 + CO_2$ produced Per 100 moles of CO converted	7.92	6.81	7.84	7.78	7.42	7.26
Deactivation rate (%/day)	0.166	Some irreversible deactivation (~20%) occurred.				

The irreversible impact was explored further using EXAFS and XANES spectroscopy of used catalyst samples removed from the reactor in the wax product under inert atmosphere. For the 0.5%Pt-15%Co/Al_2O_3 catalyst, evidence for the formation of an irreducible Co-aluminate like species was obtained (Figure 6 and Figure 7) for the high H_2O condition [14], as a feature in the XANES derivative spectra at 7717 eV matched that of cobalt aluminate reference (Figure 6). The EXAFS (Figure 7) showed a loss in Co-Co coordination, accompanied by a growth of peaks for Co-O and Co-Co of what is likely a cobalt aluminate surface complex. However, for the 25%Co catalyst, while a decrease in Co-Co coordination was observed during H_2O co-feeding,

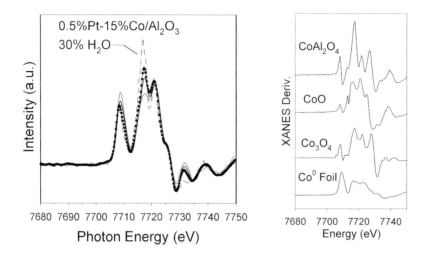

Figure 6. XANES derivative spectra for 0.5% Pt-15% Co/Al$_2$O$_3$ suggests formation of cobalt-aluminate like species with 30% H$_2$O.

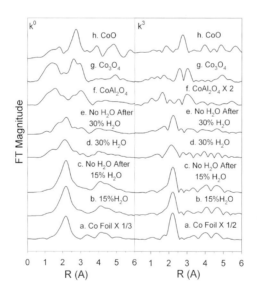

Figure 7.   For 0.5% Pt-15% Co/Al$_2$O$_3$ Co-Co coordination decreases at higher H$_2$O concentration, and does not recover when H$_2$O is switched off, indicating irreversible deactivation. Evidence for Co-O bond formation during water additon at 30 molar % H$_2$O is clearly observed with the k^0-weighted data.

the Co-Co coordination recovered when H_2O was switched off (Figure 8) [15], in contrast to the Pt promoted catalyst. XANES (Figure 9) suggested CoO formed during the high water condition, but re-reduced after water was turned off. For the $0.5\%Pt-15\%Co/Al_2O_3$ catalyst, direct evidence from EXAFS was obtained to show that the Pt promoter was in contact with Co, and not other Pt atoms [24]. The first peak of Pt-Pt coordination in the Pt^o foil is at ~2.7 Å, while the first peak for the Pt promoted catalyst is loaded at ~2.1 Å, correlating with Pt-Co metal coordination. The 2nd, 3rd and 4th coordination shells are likewise contracted, indicating the presence of Co in those shells, as well. EXAFS spectra were recorded after Pt was reduced, as indicated by the low white line intensity in XANES (Figure 10).

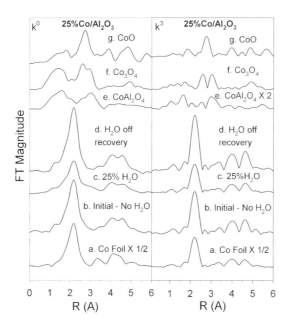

Figure 8. For 25% Co/Al_2O_3, the Co-Co coordination recovers after H_2O addition. However, the slight increase in Co-Co metal coordination during recovery indicates that cluster growth took place by the oxidation-reduction cycle.

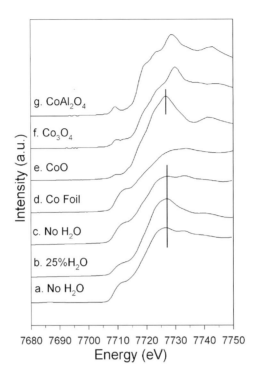

Figure 9. Normalized XANES spectra for 25% Co/Al$_2$O$_3$ suggests formation of CoO with 25% H$_2$O, followed by re-reduction when H$_2$O is switched off.

The aging rate for the 0.5%Pt promoted 15%Co/Al$_2$O$_3$ catalyst was assessed over the 800 hours prior to the catastrophic decline in activity at 28% H$_2$O addition. Below that level, the aging rate was quite constant (even after considering the reversible recovery periods following the brief periods of H$_2$O addition) at approximately 0.3% per day. The 25%Co/Al$_2$O$_3$ catalyst, by contrast, was more robust, with an average decline of about 0.17% per day up to about 20 vol % H$_2$O addition. To compare the two catalysts regarding their sensitivities to irreversible deactivation above 20 vol % H$_2$O addition, the 25%Co/Al$_2$O$_3$ catalyst lost about 20% of its activity after recovering from 30% H$_2$O addition relative to the period of recovery after 20 % H$_2$O addition (i.e., prior to the region where some irreversible deactivation is observed). By comparison, the 0.5%Pt-15%Co/Al$_2$O$_3$ catalyst lost 76% of its activity. Clearly, the 25%Co/Al$_2$O$_3$ is more robust.

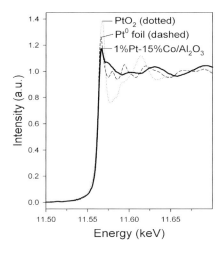
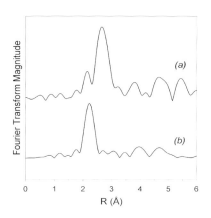

Figure 10. (left) XANES at Pt L_3 edge shows that Pt is in the reduced state. (right) EXAFS spectrum **a** shows Pt-Pt coordination in a Pt foil reference, while spectrum **b** shows that after reduction, Pt in the 1% Pt-15% Co/Al$_2$O$_3$ catalyst is coordinated to Co.

3.5. Impact of co-fed H$_2$O on selectivity

3.5.1. SiO$_2$ supported Co catalyst

In mole %, carbon dioxide selectivity was very low and was found to increase only slightly with increasing molar % added H$_2$O from < 0.01 with no added H$_2$O to approximately 0.23 with 25 molar % H$_2$O addition (Table 6). In mole %, methane selectivity decreased in small increments from about 5.6% without H$_2$O addition to approximately 3.8% with 25 molar % H$_2$O addition. Finally, C$_5$+ product selectivity was found to increase slightly after 5 molar % H$_2$O addition from 89.1% without H$_2$O to 90%. At 8 molar % H$_2$O addition, the C$_5$+ selectivity further increased to about 91.6%, and leveled off at close to 92% above that condition. A detectable increase in the 1-olefin to paraffin ratios in C$_3$ and C$_4$ products was also observed. For C$_3$, the O/P ratio increased from 0.70 without added water to about 0.76 at 25 molar % H$_2$O addition. For C$_4$, the change observed was from 0.61 to about 0.69. Although there is a slight increase in CO$_2$ selectivity with a decrease in CH$_4$ selectivity, if one considers their sum, it appears to be that there may not be a direct tradeoff between the two products.

3.5.2. TiO$_2$ supported Co catalyst

In mole %, carbon dioxide selectivity was found to increase with addition of about 14.5 molar % H$_2$O to the feed (Table 7). The effect was more pronounced at lower space velocity, where the conversion was higher, and the partial pressure of H$_2$O was therefore higher. Above space velocity 2000 cm^3/g$_{cat}$hr, the CO$_2$ selectivity remained below 1%. However, at space velocity 2000 cm^3/g$_{cat}$hr, the CO$_2$ selectivity increased from 0.94 to 1.58 molar % with external H$_2$O addition, and at 1000 cm^3/g$_{cat}$hr, from 2.27% without H$_2$O to 3.50% with H$_2$O. In mole %, methane selectivity was impacted only slightly at all of the space velocities studied. At space velocity 8000 cm^3/g$_{cat}$hr, the methane selectivity decreased slightly from 7.6% without added H$_2$O to about 6.5% with external H$_2$O addition. As with silica, it appears that there is not a direct tradeoff between decreased methane selectivity and increased CO$_2$ production during H$_2$O addition. Considering the sum of CO + CH$_4$ at each space velocity, at space velocity 8000 scm^3/g$_{cat}$hr, the CH$_4$ decrease is greater than the increase in CO$_2$, while at 1000 scm^3/g$_{cat}$hr, the opposite trend is observed. At space velocity 1000 cm^3/g$_{cat}$hr, the effect was even less pronounced, changing from 6.33% without H$_2$O to 6.26% with H$_2$O addition.

3.5.3. Al$_2$O$_3$ supported Co catalysts

0.5%Pt-15%Co/Al$_2$O$_3$: In mole %, carbon dioxide selectivity was found to increase with addition of H$_2$O to the feed (Table 8). The effect was more pronounced after reaching the point of catastrophic deactivation, at 28 molar % H$_2$O addition. For example, with no H$_2$O addition, the CO$_2$ selectivity was approximately 0.26%, increasing to 3.8% with 25 molar % H$_2$O addition. By contrast, after co-feeding 28 molar % H$_2$O, the CO$_2$ selectivity jumped to 11.5%.

25%Co/Al$_2$O$_3$: Carbon dioxide selectivity was also found to increase for the more heavily loaded Co/Al$_2$O$_3$ catalyst, increasing from 0.49% to 1.4% (and increasing to 2.64% with time on stream after that) after addition of 25 molar % H$_2$O. The CO$_2$ selectivity decreased to its previous levels after H$_2$O was switched off. The methane selectivity was found to reversibly decrease somewhat with H$_2$O addition from 7.43% without added H$_2$O to approximately 5.4% with externally added H$_2$O. Methane increased again after H$_2$O was switched off to about 7.1%, while CO$_2$ selectivity receded to previous low levels (~0.41%). Interestingly, before and after H$_2$O addition the sum of the CO$_2$ and CH$_4$ selectivities is very close. It is not clear whether or not there is a direct inverse relationship between the two products in this case.

4. Discussion

4.1. SiO₂ supported Co catalyst

In considering our results with those of Kim [2] (Bulk Cobalt Catalyst Case Study) and Krishnamoorthy et al. [5] (Co/silica Catalyst Case Study) for unsupported and Co/SiO_2 (weakly interacting support), the trends are quite consistent, although the degree of the promoting effects previously reported in [2,5] were not as pronounced in our CSTR studies [6]. An increase in CO conversion is observed, methane selectivity is decreased, and shift in the product distribution to heavier product is observed by considering either the $C_{10}+$ [2] or C_5+ selectivity [5,6]. The water-gas shift activity, reflected in the CO_2 selectivity, was low in both the study by Kim [2] and our study [6]. In their SSITKA (steady state isotopic transient kinetic analysis) studies of an unsupported cobalt catalyst, Bertole et al. [4] described the positive effect of H_2O in terms of an increase in the reactivity of adsorbed CO on the catalytic surface (i.e., cobalt surface), leading to increased surface concentrations of the monomeric carbon precursors to the formation of hydrocarbon intermediates. At the same time, they report that CO surface coverage was not influenced significantly, indicating that H_2O does not compete effectively with CO for the cobalt surface sites [4]. The results of Krishnamoorthy et al. [5] appear to be consistent with this interpretation. In in-situ infrared spectroscopy studies using Co/silica catalyst, water was not found to significantly influence either the number or the chemical characteristics of the CO binding sites on the cobalt catalytic surface available during the Fischer-Tropsch synthesis reaction. However, another possibility that must be considered is that transport restrictions (for example, liquid filled pores) to CO diffusion to the cobalt catalyst surface may be present, which, if removed, would also lead to enhanced activity, decreased CH_4 selectivity, and a higher C_5+ selectivity. Interestingly, in a recent study [9], we found that when cobalt clusters were large enough that they must be outside the pores of the silica, the positive water effect was not observed. We only observed the phenomena when the cobalt clusters were small enough to fit inside the pores. This discussion demonstrates the importance of an accurate average cobalt cluster size measurement.

Regarding the increased deactivation rates and irreversible nature of the deactivation [26,27] at higher vol % added H_2O levels, the cause may be due to cobalt surface reoxidation effects [25,28], oxidation involving the formation of irreducible cobalt support complexes (e.g., cobalt silicates) [29], or pore collapse or breakdown leading to encapsulation of cobalt clusters [9], or a combination of phenomena. As pointed out by Krishnamoorthy et al. [5] and van Berge et al. [25,28], bulk oxidation of cobalt metal to either CoO or Co_3O_4 is not thermodynamically allowed under Fischer-Tropsch reaction conditions. However, metal-oxygen bonds at surfaces are stronger than those in bulk oxides, such that oxidation may be possible. Another consideration is that the metal support interaction will influence the thermodynamics, such that small metal clusters may be oxidized under conditions in which oxidation of the bulk metal is not possible. Following this reasoning and regarding the TPR profiles, it would appear that TiO_2 and moreso Al_2O_3, would be more susceptible to oxidative deactivation processes.

4.2. TiO₂ supported Co catalyst

Moving to titania, a support which interacts considerably with supported metal clusters [30], we [17] did not obtain the same positive effects as observed by other groups [2,7]. Instead of a significant increase in CO conversion as observed by Kim [2] (Co/titania Catalyst Case Study), who noted a sharp 13.4% increase, or Storsaeter et al. [7], who reported an 8% increase, we obtained virtually no effect at high space velocities (i.e., low conversions), and even a slight negative effect at lower space velocities (i.e., higher conversion levels). Bertole et al. [4] suggested an increase in the CO activation (i.e., dissociation) rate could be attributed to an interaction between CO and TiO_X fragments decorating the cobalt catalytically active surface. We did observe a slight decrease in the methane selectivity at higher space velocities, but not close to the extent observed by others [2,7].

The discrepancy in CO conversion may be attributed to the fact that our cobalt dispersion was much higher (we reported 12.2% [17] relative to the 2.3% reported in [7]), indicating that the cobalt clusters in our study were much smaller than in the other studies, and even below 10 nm. This is why we utilized a small amount of ruthenium promoter (i.e., 0.2%), to promote the reduction of the cobalt species interacting with the support from 52 to 64%, which increased the active cobalt surface by a factor of approximately 1.6 (see Table 5). It is possible that the decline in activity was due to oxidation effects, resulting in a decreased CO conversion, and an increase in CO_2 selectivity. Interestingly, oxidized cobalt species [31,32] have been reported to be active for the water gas shift reaction. Recently, Jongsomjit et al. [33], utilizing a combination of TPR and Raman spectroscopy, reported the formation of a cobalt titanate-like compound during reduction of 20%Co/TiO_2, which did appear to increase with H_2O partial pressure. It is anticipated that our lower loading 10% Co catalyst would be more susceptible to these effects, due to an enhanced surface interaction with the support.

4.3. Al₂O₃ supported Co catalyst

By our TPR (Figure 1) and hydrogen chemisorption/pulse reoxidation measurements (Table 5), alumina is the support with the strongest interactions with cobalt oxide species, making it difficult to reduce them even at 15% cobalt loading levels. This is why, in one example, we tried 0.5% Pt promoter, to enhance their reduction (from 30% to 71%), and thereby increasing the amount of cobalt metal surface available by a factor of about two. Since the resulting cobalt clusters were well below 10 nm on average (~ 6 nm), it is not surprising that the catalyst performed differently than the other supports in the presence of H_2O. Consistent with the results of [7] (Co/alumina Catalyst Case Study), we observed a decline in CO conversion with the addition of H_2O, which was for the most part reversible up to 25 molar % added H_2O (Figure 4 and Table 8). It is not clear at this time if the reversible decrease is due to surface oxidation of the cobalt clusters, or due to an adsorption inhibiting effect on the rate by H_2O for the Co/Al₂O₃ catalyst system. It is strongly suggested by the results of XANES and EXAFS, however, that irreducible cobalt support complexes were formed after the catastrophic decline in CO conversion following 28 molar % and higher H_2O addition. The remarkable increase in CO_2 selectivity from 3.8 to 11.5% between 25 and 28 molar % H_2O addition levels provides further evidence, based on our earlier arguments, that an oxidized form of cobalt has formed, exhibiting enhanced water-gas shift activity (Table 8). Formation of irreducible cobalt support complexes has been reported by others during TPR studies in which H_2O was co-fed [19,34,35].

Increasing the loading to 25% cobalt improved the robustness of the catalyst to catastrophic deactivation, and even at 30 molar % H_2O addition (Figure 5 and Table 9), the catalyst largely recovered its activity. In XANES, we found evidence that CoO was formed during 25 molar % H_2O addition, likely confined to the surface, and re-reduced to the metal when water addition was terminated. At that stage, EXAFS showed that some increase in cobalt-cobalt coordination in the metal occurred, indicating some cluster growth, following the oxidation-reduction cycle [15]. The irreversible deactivation following exposure to 30 molar % water addition was about 20%, compared to the catastrophic loss in activity of > 75% observed for the 0.5%Pt-15%Co/Al₂O₃ catalyst.

A number of studies have strongly suggested an effect of H_2O is the oxidation of cobalt clusters for cobalt alumina catalysts, including the use of gravimetric techniques, TPD, pulse adsorption, and XPS [10-12]. For the 25%Co/Al₂O₃ catalyst, the reversible effect of H_2O may be due to a surface reoxidation process. The increase in CO_2 selectivity (Table 9) suggests increased WGS activity, as discussed in previous cases, potentially caused by a structural change from Co^0 to an oxidized form. It is also interesting to note that the increase in CO_2 selectivity almost matches the corresponding drop in CH_4 selectivity, as their sum is almost constant, which could suggest conversion of metallic cobalt to an oxidized form (e.g., CoO, Co_3O_4, or a cobalt support compound). However, it should be noted that the increase in CO_2 selectivity observed for 25%Co/Al_2O_3 was much lower than in the case of the 0.5%Pt-15%Co/Al_2O_3 catalyst, where the catastrophic deactivation occurred.

Interestingly, new membrane-type reactor configurations are currently being considered in order to remove H_2O in-situ and thereby alleviate somewhat the effects of H_2O on activity, as well as decelerate the deactivation rate for these catalysts [36,37].

5. Conclusions

Co catalysts in a comparable loading range were prepared with variable cobalt oxide-support interactions. Up to 20 molar % co-fed H_2O, a positive effect of co-fed water was found for cobalt supported on a weakly interacting support, silica, which included an increase in CO conversion, a decrease in CH_4 selectivity, and a corresponding improvement in C_5+ production. At the same time, the CO_2 selectivity remained low. At 25% H_2O addition, the deactivation rate of the catalyst accelerated by a factor of four.

In contrast, an adverse impact of water on CO conversion was recorded when catalysts were prepared on more strongly interacting supports (e.g., TiO_2 and especially Al_2O_3). However, Al_2O_3, despite the stronger interaction with cobalt, displayed higher active site densities, due to the smaller crystallite size, and the negative effect of water was less pronounced at higher Co loadings. For a 15%Co/Al_2O_3 catalyst prepared with 0.5% of Pt promoter, used to assist in promoting the reduction of smaller cobalt species interacting more strongly with the support, the catalyst was more susceptible to the deactivating effect of H_2O. While a reversible loss in CO conversion was observed up to 25% co-fed H_2O, the catalyst catastrophically lost activity (more than 75%) at 28% H_2O addition. EXAFS and XANES results suggested the formation of irreducible cobalt support compounds was responsible for the deactivation. Consistent with this interpretation, a significant increase in water-gas shift activity suggested conversion of Co^0 to an oxidized form of cobalt. The unpromoted 25%Co/Al_2O_3 catalyst offered a higher extent of reduction than an unpromoted 15%Co catalyst, due to a larger cluster size which eased the surface interaction with Al_2O_3. A largely reversible effect of H_2O was observed, even up to levels as high as 30% H_2O. Even so, a 20% irreversible loss of activity was noted at 25% H_2O addition. EXAFS and XANES results suggested formation of CoO, likely confined to the surface layers, which re-reduced when H_2O addition was switched off. After switching off the H_2O, following recovery, increased Co-Co coordination was observed by EXAFS suggesting that the oxidation-reduction cycle led to a slight growth in the cluster size, which would explain the irreversible loss in activity. The 25% Co/Al_2O_3 catalyst was still much more robust than the 0.5%Pt-15%Co/Al_2O_3 catalyst. A cluster size effect is likely the reason for the decreased sensitivity (11.8 nm for 25%Co/Al_2O_3 versus 5.6 nm for 0.5%Pt-15%Co/Al_2O_3). At lower levels of added H_2O (< 25% H_2O), the deactivation rate of the 25%Co/Al_2O_3 catalyst was about half that of the 0.5%Pt-15%Co/Al_2O_3 catalyst.

It is clear that metal-support effects play a critical role in determining how water will affect the performance parameters of supported cobalt catalysts when co-fed H_2O is added.

6. Acknowledgments

This work was supported by U.S. DOE contract no. DE-FC26-98FT40308 and the Commonwealth of KY. We are especially thankful to Professor Mark Dry for helpful discussions.

- **References**

(1) van Berge, P.J., Barradas, S., van Loosdrecht, J., Visagie, J.L., Erdgas Kohle 2001, 117, 138.
(2) Kim, C.J., U.S. Patent No. 5,227,407 (1993) to Exxon Res. Eng. Co.
(3) Kim, C.J., Eur. Appl. Patent No. 89,304,092 (1989) to Exxon Res. Eng. Co.
(4) Bertole, C.J., Mims, C.A., and Kiss, G., J. Catal. 2002, 210, 84.
(5) Krishnamoorthy, S., Tu, M., Ojeda, M.P., Pinna, D., and Iglesia, E., J. Catal. 2002, 211, 433.
(6) Li, J., Jacobs, G., Das, T.K., Zhang, Y., and Davis, B.H., Appl. Catal. A: General 2002, 236, 67.
(7) Storsæter, S., Bort, Ø., Blekkan, E.A., and Holmen, A., J. Catal. 2005, 231, 405.
(8) Huber, G.W., Guymon, C.G., Conrad, T.L., Stephenson, B.C., and Bartholomew, C.H., Stud. Surf. Sci. Catal. 2001, 139, 423.
(9) Dalai, A.K., Das, T.K., Chaudhari, K.V., Jacobs, G., and Davis, B.H., Appl. Catal. A: General 2005, 289, 135.
(10) Hilmen, A.M., Lindvåg, O.A., Bergene, E., Schanke, D., Eri, S., Holmen, A., Stud. Surf. Sci. Catal. 2001, 136, 295.
(11) Hilmen, A.M., Schanke, D., Hanssen, K.F., Holmen, A., Appl. Catal. A: General, 1999, 186, 169.
(12) Schanke, D., Hilmen, A.M., Bergene, E., Kinnari, K., Rytter, E., Ådnanes, E., and Holmen, A., Catal. Lett. 1995, 34, 269.
(13) Li, J., Zhan, X., Zhang, Y., Jacobs, G., Das, T., Davis, B.H., Appl. Catal. A: General 2002, 228, 203.
(14) Jacobs, G., Das, T.K., Patterson, P.M., Li, J., Sanchez, L., and Davis, B.H., Appl. Catal. A: General 2003, 247, 335.
(15) Jacobs, G., Patterson, P.M., Das, T.K., Luo, M., and Davis, B.H., Appl. Catal. A 2004, 270, 65.
(16) Espinoza, R.L., Visagie, J.L., van Berge, P.J., and Bolder, F.H., U.S. Patent No. 5,733,839 (1998).
(17) Li, J., Jacobs, G., Das, T.K., and Davis, B.H., Appl. Catal. A: General 2002, 233, 255.
(18) Bertole, C.J., Mims, C.A., and Kiss, G., J. Catal. 2004, 221, 191.
(19) Jongsomjit, B., Panpranot, J., and Goodwin, Jr., J.G., J. Catal. 2001, 204, 98.
(20) Jacobs, G., Das, T., Zhang, Y., Li, J., Racoillet, G., Davis, B., Appl. Catal., 2002, 233, 263.
(21) Vada, S., Hoff, A., Adnanes, E., Schanke, D., Holmen, A., Catal. Lett. 1996, 38, 143.
(22) Davis, B.H. and Iglesia, E., DOE Quarterly Report #8, July-September 2000.
(23) Jones, R.D. and Bartholomew, C.H., Appl. Catal., 1988, 39, 77.
(24) Jacobs, G., Chaney, J.A., Patterson, P.M., Das, T., Maillot, J., Davis, B., J. Synch. Rad., 2004, 11, 414.

(25)   Van Berge, P.J., van de Loosdrecht, J., Barradas, S., van der Kraan, A.M., Catal. Today, 2000, 58, 321.

(26)   Reuel, R.C., Bartholomew, C.H., J. Catal. 1984, 85, 63.

(27)   Reuel, R.C., Bartholomew, C.H., J. Catal. 1984, 85, 78.

(28)   Van Berge, P.J., van de Loosdrecht, J., Barradas, S., van der Kraan, A.M., Symposium on syngas.conversion to fuels and chemicals, Div. Petrol. Chem., 217th National Meeting of the ACS, Anaheim, CA, March 21-25, 1999, p. 84.

(29)   Kogelbauer, A., Weber, J.C., and Goodwin, J.G., Jr., Catal. Lett. 1995, 34, 259.

(30)   Haller, G.L. and Resasco, D.E., Adv. Catal. 1989, 36, 173.

(31)   Newsome, D.S., Catal. Rev. Sci. Eng. 1980, 21, 27.

(32)   Fu, L., Bartholomew, C.H., J. Catal. 1985, 92, 36.

(33)   Jongsomjit, B., Sakdamnuson, C., Goodwin, J.G., Jr., Praserthdam, P., Catal. Lett. 2004, 94, 209.

(34)   Sirijaruphan, A., Horvath, A., Goodwin, J.G., Jr., and Oukaci, R., Catal. Lett. 2003, 91, 89.

(35)   Zhang, Y., Wei, D., Hammanche, S. and Goodwin, J.G., Jr., J. Catal. 1999, 188, 281.

(36)   Khassin, A.A., Sipatrov, A.G., Yurieva, T.M., Chermashentseva, G.K., Rudina, N.A., Parmon, V.N., Catal. Today 2005, 105, 362.

(37)   Khassin, A.A., Sipatrov, A.G., Chermashetseva, G.K., Yurieva, T.M., Parmon, V.N. Topics in Catal. 2005, 32, 39.

Fischer-Tropsch Synthesis, Catalysts and Catalysis
B.H. Davis and M.L. Occelli (Editors)

255

Identification of cobalt species during temperature programmed reduction of Fischer-Tropsch catalysts

Øyvind Borg,[a] Magnus Rønning,[a] Sølvi Storsæter,[a,1] Wouter van Beek,[b] Anders Holmen[a,*]

[a] *Department of Chemical Engineering, Norwegian University of Science and Technology, NO-7491 Trondheim, Norway.*
[b] *The Swiss-Norwegian Beam Lines at ESRF, SNBL/ESRF, BP 220, F-38043 Grenoble, Cédex, France.*
[*] *Corresponding author. Telephone: +47 73 59 41 51; Fax: +47 73 59 50 47.*
E-mail address: anders.holmen@chemeng.ntnu.no (Anders Holmen).
[1] *Permanent address: Statoil R&D, Research Centre, Postuttak, NO-7005 Trondheim, Norway.*

Abstract

An *in situ* X-ray absorption spectroscopic (XAS) study of the reduction behaviour of a series of rhenium-promoted supported cobalt Fischer-Tropsch catalysts has been performed at the cobalt K absorption edge. The catalysts were prepared using incipient wetness impregnation to give 12 or 20 wt.% cobalt and 0.5 wt.% rhenium. The catalysts were reduced at 673 K for 3 h in 5% hydrogen in helium while simultaneously recording XANES spectra. Time-resolved XANES experiments show that Co_3O_4 is reduced in two steps to cobalt metal with CoO as the intermediate species on all the supports (SiO_2, TiO_2, α-Al_2O_3, γ-Al_2O_3). The reduction temperature for the first step coincides well with the temperature for the transformation of Co_3O_4 to CoO observed in standard temperature programmed reduction. At the end of the reduction process, cobalt is present as CoO and cobalt metal. The degree of reduction, as measured by XAS and oxygen titration, is support dependent and follows the order SiO_2 > TiO_2 > α-Al_2O_3 > γ-Al_2O_3.

Key words: Fischer-Tropsch synthesis, XAS, TPR, Oxygen titration, Cobalt, Rhenium, Silica, Titania

1. Introduction

Supported cobalt is considered to be the most favourable catalytic material for the synthesis of long-chain hydrocarbons from natural gas-based synthesis gas because of its high activity, high selectivity for linear paraffins, low water-gas shift activity, and low price compared to noble metals. Common supports for Fischer-Tropsch catalysts include alumina, silica, and titania.

Metallic cobalt is the active material for Fischer-Tropsch synthesis. Catalyst preparation involves impregnation of a metal precursor on the support, drying, calcination, and finally, in order to transform the inactive oxide into the metallic state, the catalyst is reduced *in situ* prior to use.

In some cases, the cobalt precursor tends to interact with the support. This interaction impedes the generation of active cobalt sites during reduction. Normally, it leaves a fraction of the cobalt chemically inactive. According to Jacobs *et al.* [1], the strength of the interaction for the three most common supports follows the order γ-alumina > titania > silica. The presence of a promoter such as rhenium facilitates reduction of cobalt species interacting with the support [2-4]. However, cobalt is usually not completely reduced after the normal reduction procedures. The effect of rhenium for Fischer-Tropsch synthesis selectivity was recently described in detail by Storsæter *et al.* [5]. It was concluded that presence of rhenium shifts the product distribution to heavier compounds, quantified by the C_{5+} selectivity.

Temperature programmed reduction (TPR) is a valuable method for gathering information about the reduction process. The reduction properties of cobalt oxide deposited on different support materials have been extensively investigated. According to most reports [6-10], unsupported, as well as supported Co_3O_4, are reduced to cobalt metal in two-steps;

$$Co_3O_4 + H_2 \rightarrow 3CoO + H_2O \tag{1}$$

$$3CoO + 3H_2 \rightarrow 3Co + 3H_2O \tag{2}$$

X-ray absorption spectroscopy can offer information about the nature and quantity of different cobalt species present during reduction. For supported cobalt, these species usually include Co_3O_4, CoO, and metallic cobalt. In addition, a part of the cobalt may interact with the support, making complete reduction difficult. XAS can conveniently be performed *in situ* providing the possibility to follow the dynamics of the process. Accordingly, it is not limited to pre- and post analysis of the catalyst.

The extent of reduction depends on the reduction conditions, the promoter, and on the choice of support. Although the reducibility of different cobalt Fischer-Tropsch catalysts has been studied thoroughly, there are often large differences in the experimental conditions, making direct comparisons difficult. Parameters such as the rate of heating, composition of the reducing agent, gas flow rate and particle size can all have a significant influence on the reduction properties. To minimise the influence of experimental conditions, all catalysts were subjected to exactly the same reactor set-up and gaseous environment during reduction.

Several techniques can be used to measure the extent of reduction. TPR, oxygen titration and *in situ* X-ray absorption spectroscopy are techniques that are able to give information on the fraction of reduced cobalt. However, the measurements are fundamentally different and, indeed, they often give different results. This work deals with a comparison of results obtained from these methods in order to verify the reliability of the techniques. Five supports have been examined; SiO_2, TiO_2, α-Al_2O_3, and two γ-Al_2O_3 supports.

2. Experimental

2.1. Catalyst preparation

A series of five supported catalysts containing 12 or 20 wt.% cobalt and 0.5 wt.% rhenium were prepared by one-step incipient wetness co-impregnation of the different supports with aqueous solutions of $Co(NO_3)_2 \cdot 6H_2O$ (Acros Organics, 99%) and $HReO_4$ (Alfa Aesar, 75-80%).

The following supports have been included: SiO_2 (PQ corp. CS-2133), TiO_2 (Degussa P25), α-Al_2O_3 (Puralox), γ-Al_2O_3[§] (Puralox) with mean pore diameter 12.3 nm, and γ-Al_2O_3 (Puralox SCCa-5/200) with mean pore diameter 7.1 nm. Before impregnation, the supports were calcined in flowing air at different temperatures as shown in Table 1.

After impregnation, the catalysts were dried in air at 393 K for 3 h before calcination in air at 573 K for 16 h. The temperature was increased by 2 K/min from room temperature to 573 K. Sieving the oxidised catalyst precursors to particle sizes 53-90 μm completed the preparation process.

Table 1. BET surface area, mean pore diameter, and pore volume for the supports and catalysts after calcination. For the nitrogen sorption data, the experimental error ($\pm 2\sigma$) is ± 5 m^2/g for the surface areas, ± 0.2 nm for the mean pore diameters, and ± 0.02 cm^3/g for the pore volumes.

Sample	Surface area (m^2/g)[a]	Mean pore diameter (nm)	Pore volume (cm^3/g)	Calcination temperature (K)	Calcination time (h)
SiO$_2$	333	25.2[a]	2.28[a]	773	10
12CoRe/SiO$_2$	302	12.0[a]	1.10[a]	573	16
TiO$_2$	9.3	844[b]	1.41[b]	973	10
12CoRe/TiO$_2$	12	790[b]	0.90[b]	573	16
α-Al$_2$O$_3$	17	150[b]	0.38[b]	1403	10
20CoRe/α-Al$_2$O$_3$	23	150[b]	0.27[b]	573	16
γ-Al$_2$O$_3$[§]	186	12.3[a]	0.73[a]	773	10
20CoRe/γ-Al$_2$O$_3$[§]	148	11.6[a]	0.50[a]	573	16
γ-Al$_2$O$_3$	196	7.1[a]	0.49[a]	773	10
12CoRe/γ-Al$_2$O$_3$	155	6.8[a]	0.36[a]	573	16

[a] Determined by nitrogen adsorption/desorption.
[b] Determined by mercury intrusion.

Fischer-Tropsch synthesis activity and selectivity data for the different catalysts have been presented elsewhere, for the catalysts containing 12 wt.% cobalt by Storsæter et al. [5], and for the alumina supported catalysts containing 20 wt.% cobalt by Borg et al. [11]. A summary of the results are presented in Table 2.

2.2. Porosity

Nitrogen adsorption-desorption isotherms for all catalysts were measured with a Micromeritics TriStar 3000 instrument and the data were collected at liquid nitrogen temperature, 77 K. The samples were outgassed at 573 K overnight prior to measurement. The BET surface areas [12] are given in Table 1. For samples 12CoRe/SiO$_2$, 20CoRe/γ-Al$_2$O$_3$[§], and 12CoRe/γ-Al$_2$O$_3$, the total pore volume and average pore size were calculated applying the Barrett-Joyner-Halenda (BJH) method [13].

For catalysts 12CoRe/TiO$_2$ and 20CoRe/α-Al$_2$O$_3$, the pore size measurements were performed using a Carlo Erba Porosimeter 2000 by mercury intrusion. Each sample was evacuated and dried at 423 K prior to analysis. A cylindrical pore model was assumed.

Table 2. Activities and selectivites measured in a fixed-bed reactor at Fischer-Tropsch conditions (T = 473 K, P = 20 bar, H_2/CO = 2.1).

Catalyst	CO conversion (%)	Hydrocarbon formation rate (g/(g cat · h))	Selectivity (%)	
			C_{5+}	CH_4
12CoRe/SiO$_2$[a]	40	0.33	83.4	8.7
12CoRe/TiO$_2$[a]	43	0.30	84.8	8.9
20CoRe/α-Al$_2$O$_3$[b]	41	0.54	85.1	8.8
20CoRe/γ-Al$_2$O$_3$[§b]	49	0.87	81.6	9.3
12CoRe/γ-Al$_2$O$_3$[a]	43	0.42	80.8	8.8

[a] From Storsæter et al. [5].
[b] From Borg et al. [11].

2.3. X-ray diffraction

X-ray diffraction patterns were recorded at room temperature by a Siemens D5005 X-ray diffractometer using CuKα radiation (λ = 1.54 Å). The scans were recorded in the 2θ range between 10 and 90° using a step size of 0.04°. The samples were crushed prior to measurement. The average, cobalt oxide crystallite thickness was calculated from the Scherrer equation [14] using the (311) Co_3O_4 peak located at 2θ = 36.9°. The average Co_3O_4 particle size was calculated multiplying the crystallite thickness by a factor of 4/3 which implies the existence of spherical Co_3O_4 particles [15]. Lanthanum hexaboride was used as reference material to determine the instrumental line broadening.

2.4. Temperature programmed reduction

TPR experiments were performed in a U-shaped tubular quartz reactor [16]. The samples were exposed to a reducing gas mixture consisting of 7 percent hydrogen in argon while the temperature was increased from ambient to 1203 K at a heating rate of 5 K/min.

A cold trap containing a mixture of 2-propanol and dry ice was used to eliminate water and other condensable compounds from the product gas mixtures. The consumption of hydrogen during reduction (H:Co ratio) was measured by analysing the effluent gas with a thermal conductivity detector. Calibration was done by reduction of Ag_2O powder.

2.5. XAS measurements

Transmission XAS data were collected at the Swiss-Norwegian Beamline (SNBL) at the European Synchrotron Radiation Facility (ESRF), France. Spectra were obtained at the cobalt K edge (7 709 eV) using a channel-cut Si(111) monochromator. Higher order harmonics were rejected by means of a chromium-coated mirror angled at 3.5 mrad with respect to the beam to give a cut-off energy of approximately 14 keV. The beam currents ranged from 130 - 200 mA at 6.0 GeV. The maximum resolution ($\Delta E/E$) of the Si(111) bandpass is $1.4 \cdot 10^{-4}$ using a beam of size $0.6 \cdot 7.2$ mm. Ion chamber detectors with their gases at ambient temperature and pressure were used for measuring the intensities of the incident (I_0) and transmitted (I_t) X-rays.

The amounts of material in the samples were calculated from element mass fractions and the absorption coefficients of the constituent elements [17] just above the absorption edge to give an absorber optical thickness close to 2.0 absorption lengths. The samples were loaded into a Lytle *in situ* reactor-cell [18] and reduced in a mixture of H_2 (5%) in He (purity: 99.995%: flow rate 30 ml/min) by heating at a rate of 5 K/min from ambient temperature to 673 K. The samples were kept at 673 K for 3 h. Two EXAFS scans were recorded and summed for each sample before and after reduction. XANES scans were taken continuously during reduction.

The energy calibration was checked by measuring the spectrum of a cobalt foil (thickness 0.0125 mm) with the energy of the first inflection point being defined as the edge energy.

2.6. XANES data analysis

The XAS data analysis program WinXAS v. 3.1 [19] was used for examining the time-resolved Co K absorption edge XANES data. The identification of the number of phases present during *in situ* reduction was done by a principal component analysis (PCA) of the experimental spectra [20]. Reference spectra were then used in a linear combination fitting procedure to determine the quantity of each phase present. The algorithm uses a least squares procedure to refine the sum of a given number of reference spectra to an experimental spectrum.

2.7. EXAFS data analysis

The XAS data were converted to k-space, summed and background subtracted to yield the EXAFS function $\chi(k)$ using WinXAS v. 3.1 [19]. Model fitting was carried out with *EXCURV98* using curved-wave theory and *ab initio* phase shifts [21, 22]. A cobalt metal foil (0.0125 mm), CoO, Co_3O_4, and $CoAl_2O_4$ were used as model compounds to check the validity of the *ab initio* phase shifts and establish the general amplitude reduction factor. Cobalt K-edge EXAFS data were fitted in the range $\Delta k = 3.0 - 14.0$ Å^{-1} using a Fourier filtering window $\Delta R = 1.0 - 3.2$ Å.

The extent of reduction from EXAFS was obtained by looking at the fractional coordination number of the nearest Co-O coordination shell. The bulk cobalt oxides have Co-O coordination of 6 (N_{ox}). The Co-O coordination number for the reduced catalysts (N_{red}) will be lower, depending on the extent of reduction. For a completely reduced catalyst, N_{red} will be zero. The degree of reduction is thus given by the expression $1 - N_{red}/N_{ox}$.

3. Results and discussion

3.1. Catalyst characterisation

3.1.1. Porosity
The BET surface area, pore volume, and mean pore diameter of the supports and the calcined samples are given in Table 1. Impregnation/calcination reduces the measured surface area except for the low surface area supports TiO_2 and α-Al_2O_3 where it increases. Table 1 also shows that for the SiO_2 sample, the mean pore diameter and the pore volume are substantially reduced upon impregnation/calcination.

3.1.2. X-ray diffraction
X-ray diffraction patterns of all catalysts confirm presence of Co_3O_4 [23]. Apart from the peaks indicative of Co_3O_4 and the various supports, the supported catalysts did not show any other peaks. Accordingly, Co_3O_4 is the dominant crystalline cobalt species after calcination. The average cobalt oxide particle sizes are included in Table 3.

3.1.3. Temperature programmed reduction
TPR profiles are presented in Figure 1, and the hydrogen consumption is given in Table 3. Integration was done using the trapezoidal rule [24]. Consumption of hydrogen for the possible reduction of rhenium species has not been taken into account.

Table 3. The degree of reduction as determined by oxygen titration and XAS measurements. The H:Co ratio is obtained from TPR measurements and the Co_3O_4 particle size is calculated from X-ray diffraction data. The experimental error ($\pm 2\sigma$) is $\pm 2 \cdot 10^{-2}$ mol H/mol Co for the H:Co ratio, $\pm$ 1% for the degree of reduction calculated from oxygen titration, and ± 1 nm for the Co_3O_4 particle size.

Catalyst	H:Co ratio[a] (mol H/mol Co)	Degree of reduction (%)			Co_3O_4 particle size (nm)
		Oxygen titration	XANES	EXAFS	
12CoRe/SiO$_2$	2.6	67	85	79	20
12CoRe/TiO$_2$	2.8	69	75	75	51
20CoRe/α-Al$_2$O$_3$	2.4	63	74	71	25
20CoRe/γ-Al$_2$O$_3$[§]	2.5	60	71	70	14
12CoRe/γ-Al$_2$O$_3$	2.6	59	63	65	15

[a] H:Co ratio for complete reduction: 2.67 mol H/mol Co.

It is usually observed that bulk Co_3O_4 exhibits two reduction peaks [4, 6]. The hydrogen consumption ratio between the two peaks is 1:3, consistent with the stoichiometry of Equations (1) and (2).

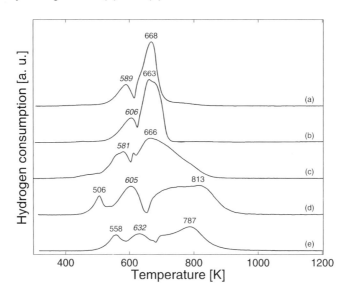

Figure 1. Temperature programmed reduction of 12CoRe/SiO$_2$ (a), 12CoRe/TiO$_2$ (b), 20CoRe/α-Al$_2$O$_3$ (c), 20CoRe/γ-Al$_2$O$_3$[§] (d), 12CoRe/γ-Al$_2$O$_3$ (e) after calcination. The temperatures in italic font represent the temperatures for maximum reduction rate for transformation of Co_3O_4 to CoO.

The TPR profiles of catalysts 12CoRe/SiO$_2$ and 12CoRe/TiO$_2$ show very similar behaviour (Figures 1a and b) and two distinct, but slightly overlapping reduction peaks are apparent. The two peaks can probably be assigned to the two-step reduction of Co$_3$O$_4$ to metallic cobalt with CoO as the intermediate species. The TPR profile of catalyst 20CoRe/α-Al$_2$O$_3$ given in Figure 1c also shows two large peaks resembling the two-step reduction of Co$_3$O$_4$, but in this case, the broad high temperature peak indicates a stronger metal-support interaction. The small, but sharp peak at about 625 K is also a characteristic feature. For samples 12CoRe/SiO$_2$, 12CoRe/TiO$_2$, and 20CoRe/α-Al$_2$O$_3$, the hydrogen consumption ratio between the two peaks is 0.4, 0.3 and 0.3, respectively, showing that the experimental values are close to the theoretical value of 1/3. However, overlapping of the two peaks and the fact that the second reduction step is not complete make accurate calculations difficult.

TPR profiles of samples 20CoRe/γ-Al$_2$O$_3^\S$ and 12CoRe/γ-Al$_2$O$_3$ deviate from the profiles of the SiO$_2$ and TiO$_2$ supported catalysts as shown in Figures 1d and 1e. However, the first reduction step, namely reduction of Co$_3$O$_4$ to CoO, takes place at similar temperatures for all the catalysts included in the study. As shown in Figure 1, a local maximum reduction rate (indicated with italic font) is observed in the interval 581 to 632 K for the five samples. This is in agreement with Castner et al. [6] who showed that for a series of SiO$_2$ supported catalysts, the initial reduction step was independent of particle size, pore size and surface area, and occurred in the same temperature interval. The hydrogen consumption ratio for both 20CoRe/γ-Al$_2$O$_3^\S$ and 12CoRe/γ-Al$_2$O$_3$ is 0.3, indicating that the reduction occurs as described by Equations (1) and (2) also for these catalysts.

Castner et al. [6] found that the second step of the reduction scheme (Equation 2) occurred at different temperatures depending on the interactions of small particles with the support stabilising the oxide phase. Stabilisation of the oxide phase due to metal-support interactions are most likely the cause for the broad feature starting at about 655 K in Figure 1d and at about 680 K in Figure 1e. Thus, it can be attributed to reduction of cobalt oxide phases interacting with the support [1, 5].

The peaks appearing at 506 and 558 K in Figures 1d and 1e are most likely arising from the reductive decomposition of residual nitrate after calcination [5, 25-27]. The general consensus is that during calcination in air at temperatures from 473 to 673 K, the supported cobalt nitrate is converted to supported Co$_3$O$_4$ [25-31]. This temperature range corresponds well with the peak temperatures observed in Figures 1d and 1e. The catalysts 12CoRe/SiO$_2$, 12CoRe/TiO$_2$, and 20CoRe/α-Al$_2$O$_3$, on the other hand, do not show any presence of left-over nitrate after calcination (Figures 1a - 1c). These catalysts contain larger pores that allow for easier removal of water and NO_x during calcination of the supports impregnated with cobalt nitrate hexahydrate [31]. The nitrate species are not stabilised on the catalyst surface, and hence easier to decompose.

3.1.4. Extent of reduction

Oxygen titration has previously been used to measure the extent of reduction for supported cobalt catalysts [2]. In Table 3, the extent of reduction obtained by oxygen titration is compared to XANES and EXAFS values. Table 3 shows that the degree of reduction obtained by oxygen titration is between 59 and 69 percent, the lowest values are obtained for the γ-Al_2O_3 supported samples and the highest values for cobalt supported on TiO_2 and SiO_2. The degree of reduction ranges from 63 to 85 percent using linear combination of XANES profiles, while EXAFS analysis yields values from 65 to 79 percent. The consistency between values obtained from XANES profiles and EXAFS analysis is good and within experimental errors.

The extent of reduction obtained by the XAS techniques follows the order $12CoRe/SiO_2$ > $12CoRe/TiO_2$ > $20CoRe/\alpha$-Al_2O_3 > $20CoRe/\gamma$-Al_2O_3§ > $12CoRe/\gamma$-Al_2O_3. For cobalt supported on SiO_2, TiO_2 and γ-Al_2O_3, the same order with respect to percentage reduction was observed by Jacobs et al. [1]. Jacobs et al. [1] reduced the TiO_2, SiO_2 and Al_2O_3 based catalysts for 10 h at 573, 623 and 623 K, respectively, and performed the oxygen titration at the reduction temperature. The degree of reduction for $10Co/TiO_2$ was 52 percent, while $15Co/SiO_2$ gave a percentage reduction of 64 percent. Finally, oxygen titration of $15Co/Al_2O_3$ resulted in an extent of reduction of only 29 percent.

The values from oxygen titration in this study are systematically lower than the values from the XAS techniques, but roughly the same order between the supports is observed. The only difference is that, within experimental errors, $12CoRe/SiO_2$ and $12CoRe/TiO_2$ have the same degree of reduction. To verify the assumptions underlying the oxygen titration technique, a separate TPR study was done as shown in Figure 2. The catalyst subjected to this investigation resembled the sample $20CoRe/\gamma$-Al_2O_3§, but a slightly different γ-Al_2O_3 support was used (labelled γ-Al_2O_3*; BET surface area 205 m^2/g; mean pore diameter 10.5 nm; pore volume: 0.67 cm^3/g). TPR profiles of $20CoRe/\gamma$-Al_2O_3* were recorded after calcination, after reduction and pulse oxidation and, finally, after reduction and recalcination. The H:Co ratios are given in Table 4. It is evident that reoxidation of the reduced catalyst with pulses of oxygen does not completely oxidise the sample. The H:Co ratio is significantly less than the ratio

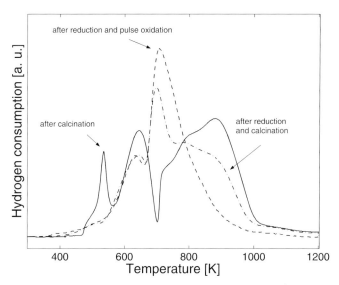

Figure 2. Temperature programmed reduction profile of 20CoRe/γ-Al₂O₃* after calcination (573 K, 16 h), after reduction (623 K, 16 h) and pulse oxidation (673 K), after reduction (623 K, 16 h) and calcination (573 K, 16 h).

for the calcined catalyst (2.2 mol H/mol Co vs. 2.6 mol H/mol Co), indicating that oxygen titration underestimates the extent of reduction. However, the TPR profile after reduction and pulse oxidation shows a 1:3 ratio (experimental value: 0.3) for the two separate peaks, indicating a two-step reduction of Co_3O_4.

A comparison of the TPR profile after calcination and the profile after reduction and pulse oxidation shows that the major difference is in the high-temperature peak. It is possible that the strength of interaction between the cobalt and the support is increased during reduction and hence stabilising a fraction of the metal towards oxidation. This theory is supported by Zhang *et al.* [32], Jongsomjit *et al.* [33] and Sirijaruphan *et al.* [34] who observed formation of a cobalt-support compound during reduction of an alumina supported catalyst. The formation was enhanced by small quantities of water vapour.

The TPR profile after reduction and subsequent recalcination closely resemble the profile of the calcined sample. Calcination, therefore, oxidises the reduced sample more completely than a pulse oxidation treatment. However, the H:Co ratio is still lower than the ratio for the original calcined sample (2.4 mol H/mol Co vs. 2.6 mol H/mol Co).

Table 4. Hydrogen consumption for 20CoRe/γ-Al$_2$O$_{3*}$ after calcination (573 K, 16 h), after reduction (623 K, 16 h) and pulse oxidation (673 K), and after reduction (623 K, 16 h) and calcination (573 K, 16 h). The experimental error ($\pm 2\sigma$) is $\pm 2 \cdot 10^{-2}$ mol H/mol Co for the H:Co ratio.

Catalyst state	H:Co ratio (mol H/mol Co)
After calcinations	2.6
After reduction and pulse oxidation	2.2
After reduction and calcination	2.4

Table 4 shows that the reduced catalyst is not completely oxidised when exposed to pulses of oxygen. About 83% (2.2/2.6) of the theoretical hydrogen consumption was observed during TPR of the reduced and pulse oxidised catalyst. In order to correctly compare the values obtained from oxygen titration with the XAS values, all values obtained by from oxygen titration can be normalised by dividing by 0.83. This normalisation procedure gives better agreement between the oxygen titration values and the XAS values. It shifts the previous values from 59 to 69% to 71 to 84%. Thus, the XAS techniques seem to be adequate methods for determining the degree of reduction of cobalt supported Fischer-Tropsch catalysts.

Khodakov et al. [35] also indicated that oxygen titration may underestimate the extent of reduction. They showed that in an inert atmosphere at temperatures higher than 623 K the supported CoO phase could be more stable than Co$_3$O$_4$. Oxygen titration conducted in helium at 673 K could therefore result in oxidation of cobalt metal phases to CoO instead of Co$_3$O$_4$ or to the mixture of CoO and Co$_3$O$_4$. The result would be that oxygen titration underestimates the extent of reduction, consistent with the conclusions drawn from Figure 2 and Table 4.

3.2. XAS

The results from the cobalt K-edge EXAFS analysis are listed in Table 5. The Fourier transform of the k^3 fitted cobalt K-edge EXAFS spectrum of 20CoRe/γ-Al$_2$O$_3$[§] is shown in Figure 3.

X-ray diffraction patterns and the cobalt EXAFS spectra of Co$_3$O$_4$ and the calcined catalysts were indistinguishable within experimental error, consistently showing that Co$_3$O$_4$ is the predominant cobalt species after calcination.

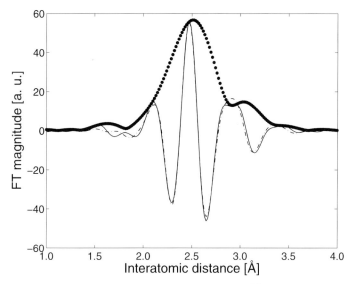

Figure 3. Fourier transform of the Fourier filtered (Δk = 1.0-3.2 Å^{-1}) data of the 20CoRe/γ-Al$_2$O$_3$[§] catalyst (- = FT experiment, -- = FT theory). The data are fitted using a Co-O and a Co-Co shell.

Time-resolved XANES experiments at the Co K absorption edge were recorded during *in situ* reduction. According to a principal component analysis of the spectra, three primary components were needed to adequately reconstruct the spectra of all catalysts. Co$_3$O$_4$, CoO, and a cobalt metal foil were chosen as model compounds. CoAl$_2$O$_4$ was initially included as a candidate model compound. However, its contribution to the reconstruction fit was insignificant.

Table 5. Data from the EXAFS analysis of the catalysts. The data are fitted in k space using k^3 weighting. Fitting range: Δk = 3.0 – 14.0 Å^{-1}, Fourier filtering range: ΔR = 1.0 – 3.2 Å.

Catalyst	N_{Co-O}[a]	R_{Co-O} (Å)	$2\sigma^2_{Co-O}$ (Å^2)	N_{Co-Co}	R_{Co-Co} (Å)	$2\sigma^2_{Co-Co}$ (Å^2)	Fit index
12CoRe/SiO$_2$	1.26	2.13	0.021	9.32	2.49	0.010	0.14
12CoRe/TiO$_2$	1.52	2.12	0.021	9.13	2.49	0.012	0.28
20CoRe/α-Al$_2$O$_3$	1.72	2.11	0.016	7.88	2.49	0.012	0.15
20CoRe/γ-Al$_2$O$_3$[§]	1.81	2.11	0.016	7.77	2.49	0.012	0.18
12CoRe/γ-Al$_2$O$_3$	2.11	2.10	0.021	6.97	2.49	0.014	0.19

[a] Coordination number.
[b] Interatomic distance.
[c] Fit index (*FI*). The Fit index is defined as: $FI_{EXAFS} = \sum_i \dfrac{1}{s_i}\{Exp(i) - Theory(i)\}^2$.

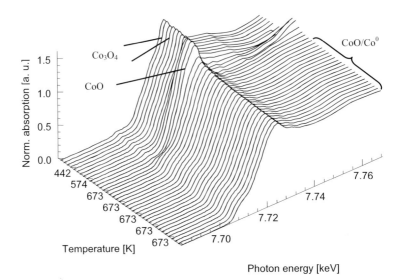

Figure 4. XANES profiles at different reduction stages for 12CoRe/γ-Al₂O₃. Scans were taken continuously with a time gap of five minutes.

This is consistent with earlier results stating that the cobalt rich alumina phase is not the $CoAl_2O_4$ spinel, but most likely cobalt atoms randomly distributed in vacant tetrahedral positions in the alumina lattice [36].

An example of the change in XANES profiles during reduction of 12CoRe/γ-Al₂O₃ is shown in Figure 4. The pre-edge feature typical for cobalt in partially tetrahedral coordination, which is the case for bulk Co_3O_4, gradually decreases. This change is consistent with the transformation to octahedrally coordinated cobalt as in CoO. Finally, the profiles closely resemble that of metallic cobalt showing a less intense white line and a characteristic pre-edge shoulder.

Figures 5 and 6 display changes in Co_3O_4, CoO and cobalt metal fractions with temperature for the catalyst 12CoRe/γ-Al₂O₃ and 20CoRe/α-Al₂O₃, respectively. The shape of the component fraction plots for the other catalysts was similar to the catalyst presented in Figures 5 and 6. The only significant differences for the various catalysts are the extent of reduction and the temperature in which the major transformation from Co_3O_4 to CoO occurs as shown in Tables 3 and 6.

For all catalysts the initial quantity of cobalt (II, III) oxide, Co_3O_4, was completely transformed into other cobalt species before reaching the final reduction temperature of 673 K. Furthermore, CoO is the dominant intermediate during reduction for all supported cobalt catalysts included in this investigation.

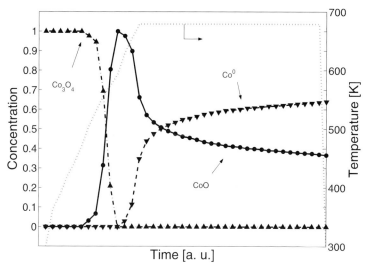

Figure 5. Temperature programmed reduction of 12CoRe/γ-Al₂O₃ using the *in situ* XAS cell. Temperature ramping is indicated in the figure (▲ = Co₃O₄, ● = CoO, ▼ = cobalt metal, ... = temperature).

The temperature for maximum CoO concentration coincides well with the temperature of the TPR peaks previously attributed to the reduction of Co₃O₄ to CoO as shown in Figure 1. The temperatures for transformation of Co₃O₄ to CoO obtained by the two techniques are given in Table 6. The XAS and TPR results are also in agreement for the further reduction of cobalt, namely the reduction of CoO to cobalt metal. According to TPR, the reduction of CoO starts immediately after the reduction of Co₃O₄ is completed. The XAS results show a sharp drop in the fractional amount of CoO accompanied by an immediate rise in the cobalt metal content (Figures 5 and 6).

Table 6. Temperature for transformation of Co₃O₄ to CoO. The experimental error (± 2σ) of the temperature from TPR is less than ± 5 K.

Catalyst	Temperature from XANES (K)	Temperature from TPR (K)
12CoRe/SiO₂	580	589
12CoRe/TiO₂	633	606
20CoRe/α-Al₂O₃	602	581
20CoRe/γ-Al₂O₃§	602	605
12CoRe/γ-Al₂O₃	630	632

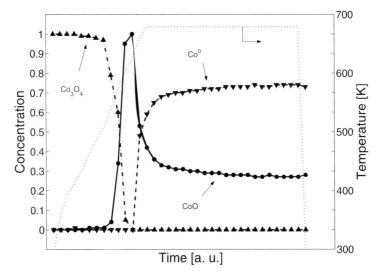

Figure 6. Temperature programmed reduction of 20CoRe/α-Al$_2$O$_3$ using the *in situ* XAS cell. Temperature ramping is indicated in the figure ($\blacktriangle$ = Co$_3$O$_4$, $\bullet$ = CoO, $\blacktriangledown$ = cobalt metal, ... = temperature).

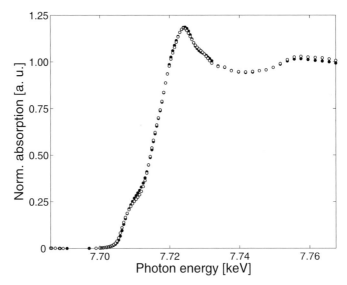

Figure 7. Experimental XANES spectra (-) and reconstructed XANES spectra (o) at the end of the reduction process for 12CoRe/γ-Al$_2$O$_3$.

In general, the linear combination of XANES and the principal component analysis give very good fit to the experimental data. Figure 7 shows a close agreement between the experimental spectrum and the reconstructed spectrum at the end of the reduction process for $12CoRe/\gamma-Al_2O_3$. However, due to the limited time-resolution of the XANES spectra (~ 5 minutes), the XANES profiles at the stages where the reduction evolves rapidly will contain more noise. This is because the sample composition has changed considerably during the course of the scan. As a result, some data points in Figures 5 and 6 have been interpolated between adjacent data points to give a smooth curve.

4. Conclusions

The reduction behaviour of a series cobalt catalysts containing 12 or 20 wt.% Co and 0.5 wt.% Re supported on TiO_2, SiO_2, $\gamma-Al_2O_3$, and $\alpha-Al_2O_3$ has been studied by TPR, O_2 titration and *in situ* XAS. The present study shows that the XANES TPR is capable of identifying the different cobalt species during the reduction and thus provides additional information to the TPR results. Linear combination of XANES profiles with those of selected reference compounds gives a relatively precise estimate of the extent of reduction of the cobalt particles. The reduction of cobalt deposited on various supports proceeds by a two-step reduction sequence going from Co_3O_4 via CoO to cobalt metal. The study indicates that O_2 titration underestimates the extent of reduction.

Acknowledgements

The authors thank the Norwegian Research Council for financial support and acknowledge the project team at the Swiss-Norwegian Beamlines at ESRF for experimental assistance (Experiment number CH-1632). Finally, Statoil is acknowledged for financial support and for supplying the alumina supports.

References

1. G. Jacobs, T. K. Das, Y. Zhang, J. Li, G. Racoillet, B. H. Davis, Appl. Catal. A 233 (2002) 263.
2. S. Vada, A. Hoff, E. Ådnanes, D. Schanke, A. Holmen, Top. Catal. 2 (1995) 155.
3. S. Eri, J. G. Goodwin, G. Marcelin, T. Riis, US Patent 4801573 (1989).
4. A. M. Hilmen, D. Schanke, A. Holmen, Catal. Lett. 38 (1996) 143.
5. S. Storsæter, Ø. Borg, E. A. Blekkan, A. Holmen, J. Catal. 231 (2005) 405.
6. D. G. Castner, P. R. Watson, I. Y. Chan, J. Phys. Chem. 94 (1990) 819.
7. B. A. Sexton, A. E. Hughes, T. W. Turney, J. Catal. 97 (1986) 390.
8. B. Viswanathan, R. Gopalakrishnan, J. Catal. 99 (1986) 342.
9. H. F. J. van't Blik, R. Prins, J. Catal. 97 (1986) 188.
10. G. Jacobs, J. A. Chaney, P. M. Patterson, T. K. Das, J. C. Maillot, B. H. Davis, J. Synchrotron Rad. 11 (2004) 414.
11. Ø. Borg, S. Storsæter, E. A. Blekkan, S. Eri, E. Rytter, A. Holmen, in preparation.

12. S. Brunauer, P. H. Emmett, E. Teller, J. Am. Chem. Soc. 60 (1938) 309.
13. E. P. Barrett, L. G. Joyner, P. P. Halenda, J. Am. Chem. Soc. 73 (1951) 373.
14. H. P. Klug, L. E. Alexander, X-ray Diffraction Procedures for polycrystalline and amorphous materials, John Wiley & Sons, New York, 1954.
15. J. L. Lemaitre, P. G. Menon, F. Delannay, in: F. Delannay (Ed.), Characterization of Heterogeneous Catalysts, Marcel Dekker, Inc., New York, 1984, p. 299.
16. E. A. Blekkan, A. Holmen, S. Vada, Acta Chem. Scand. 47 (1993) 275.
17. International Tables for X-ray Crystallography, Vol. 3 (Kynoch Press, Birmingham, 1962) 175.
18. F. W. Lytle, R. B. Greegor, E. C. Marques, D. R. Sandstrom, G. H. Via, J. H. Sinfelt, J. Catal. 95 (1985) 546.
19. T. Ressler, J. Synchrotron Rad. 5 (1998) 118.
20. T. Ressler, J. Wong, J. Roos, I. L. Smith, Environ. Sci. Technol. 34 (2000) 950.
21. EXCURV98: CCLRC Daresbury Laboratory computer program.
22. S. J. Gurman, N. Binsted, I. Ross, J. Phys. C. 17 (1984) 143.
23. DIFFRACplus EVA Release 2001, Bruker AXS.
24. E. Kreyszig, in Advanced Engineering Mathematics, John Wiley & Sons, Inc., New York, 8th edn., 1999, p. 870.
25. P. Arnoldy, J. A. Moulijn, J. Catal. 93 (1985) 38.
26. A. Hoff, E. A. Blekkan, A. Holmen, D. Schanke, Stud. Surf. Sci. Catal. 75 (1993) 2067.
27. E. van Steen, G. S Sewell, R. A. Makhothe, C. Micklethwaite, H. Manstein, M. de Lange, C. T. O'Connor, J. Catal. 162 (1996) 220.
28. G. A. El-Shobaky, G. A. Fagal, A. M. Dessouki, Egyptian J. Chem. 41 (1988) 317.
29. A. Lapidus, A. Krylova, V. Kazanskii, V. Borovkov, A. Zaitsev, J. Rathousky, A. Zukal, M. Jancalkova, Appl. Catal. 73 (1991) 65.
30. C. H. Mauldin, D. E. Varnado, Stud. Surf. Sci. Catal. 136 (2001) 417.
31. J. van de Loosdrecht, E. A. Barradas, E. A. Caricato, N. G. Ngwenya, P. S. Nkwanyana, M. A. S. Rawat, B. H. Sigwebela, P. J. van Berge, J. L. Visagie, Top. Catal. 26 (2003) 121.
32. Y. Zhang, D. Wei, S. Hammache, J. G. Goodwin, Jr., J. Catal. 188 (1999) 281.
33. B. Jongsomjit, J. Panpranot, J. G. Goodwin, Jr., J. Catal. 204 (2001) 98.
34. A. Sirijaruphan, A. Horváth, J. G. Goodwin Jr., R. Oukaci, Catal. Lett. 91 (2003) 89.
35. A. Y. Khodakov, Griboval-Constant, A., Bechara, R., V. L. Zholobenko, J. Catal. 206 (2002) 230.
36. A. Moen, D. G. Nicholson, M. Rønning, H. Emerich, J. Mater. Chem. 8 (1998) 2533.

Fischer-Tropsch Synthesis, Catalysts and Catalysis
B.H. Davis and M.L. Occelli (Editors)

Determination of the Partial Pressure of Water during the Activation of the BP GTL Fischer-Tropsch Catalyst and its Effects on Catalyst performance.

Joep J H M Font Freide[a], Barry Nay[a], Christopher Sharp[b]

[a] *BP Exploration Operating Company, Chertsey Road, Sunbury on Thames, Middlesex, TW16 7LN*
[b] *BP Exploration Operating Company, Saltend, Hedon, Hull, HU12 8DS*

1. Abstract

The effects of hydrogen concentration, temperature ramp rate, pressure and Gas Hourly Space Velocity (GHSV) were all shown to be of high importance in the reduction of the BP Gas to Liquids Fischer-Tropsch catalyst. A system was developed to accurately measure the water partial pressure and hydrogen concentration simultaneously at the exit of the catalyst bed, during reduction. The importance of the effect of the partial pressure of water measured during reduction, on final catalyst performance, was clearly demonstrated. Low water partial pressure levels (<0.1 bar) seem to be beneficial to performance, whilst higher levels are detrimental.

2. Introduction

With the ever-increasing drive towards commercialisation of Gas to Liquids (GTL) technology, attention has focused on the efficiency of the three stages that make up the conversion process, namely, carbon monoxide and hydrogen (or syngas) generation, syngas conversion into liquid hydrocarbons and product upgrading. All three of these processes must be highly efficient, but none more so than the conversion of syngas into liquid hydrocarbons, or the Fischer-

Tropsch reaction as it is better known. Although there are now several types of reactor in use, which vary in volume, it is still in the interest of the producer to maximise catalyst C_5^+ productivity no matter which reactor is used.

The Fischer-Tropsch reaction is highly exothermic and therefore advanced heat management designs can enhance heat transfer leading to improved productivity. Each particular reactor design has its own advantages and disadvantages with regards to heat transfer and catalyst productivity. However in an ideal world, where heat transfer and other process variables don't matter, it is the catalyst activity/selectivity that drives the economics. Since the discovery of the Fischer-Tropsch reaction in the early part of last century, scientists have constantly striven to improve the activity/selectivity of the catalyst systems employed. Without going into too much detail, cobalt is the metal of choice these days for FT, and can be supported on a variety of metal oxides including Al_2O_3, SiO_2, TiO_2 and ZnO. Availability of active metal sites is the key to high activity/selectivity catalysts and workers usually attempt to prepare catalysts with high metal dispersions, in order to maximise this (1). What needs to be understood is that the cobalt is usually deposited on the support as a substrate, which is then chemically treated to liberate cobalt metal. Only when the cobalt is deposited by a technique called metal vapour synthesis (MVS) is no further chemical treatment required. The most common form of substrate is usually a form of cobalt oxide, which is then reduced to cobalt metal. It would then follow that the better dispersed the cobalt oxide the better dispersed the cobalt metal will be, following reduction. However this is not necessarily true, since severe reduction conditions can often sinter the cobalt metal leading to a large reduction in metal surface area and a subsequent fall in catalytic activity. It therefore follows that activation, which is often just the last stage in what might have been a very complicated catalyst preparation is not just a simple reduction of a cobalt oxide to cobalt metal.

Over the years scientists have come to realise that although reduction with hydrogen is on the face of it simple, it is indeed anything but, and in many cases, is perhaps the most critical stage of ensuring a highly active/selective catalyst results. Many different catalyst activation procedures, aimed at maximising activity/selectivity, have been developed for each particular FT catalyst. Companies and academic groups alike have studied catalyst reduction in great detail, hoping to gain a deeper understanding that would enable them to increase activity/selectivity even further. Thousands of publications cover this subject, each claiming that certain specialist techniques employed can greatly influence catalyst performance. A large number of process variables can influence the degree of reduction, the key ones being listed below;

1. Hydrogen concentration
2. Reduction gas flow rate
3. Catalyst bed dimensions (short and fat or long and thin)
4. Fixed or fluid bed
5. Catalyst particle size
6. Temperature ramp rate
7. Dwell times at reduction temperature
8. Maximum temperature employed
9. Pressure

Any one of these variables can have a large influence on the degree of reduction or active metal dispersion. Taking the above variables into consideration it is deemed that only two factors actually influence the reduction process, however these factors, namely, temperature and amount of water produced are affected by changing any of the above. Obviously water partial pressures can be varied by changing any of the above variables, and keeping the level as low as possible at all times during the reduction is potentially the preferred option. It is noteworthy that catalyst damaged by high partial pressures of water during reduction cannot be recovered, whereas those damaged by high temperature or under reduction can usually be regenerated. Catalyst sintered by high temperature can be re-oxidised and reduced again, whilst under-reduced catalysts can be re-reduced under more forcing conditions. However catalysts subjected by high partial pressure of water during reduction are irreversibly damaged. It is postulated that the water, re-oxidises the cobalt into a form that is not easily reduced or maybe occludes the cobalt into the surface of the support or forms cobalt/support compound (2,3). This is particularly pertinent on large scale FT plants which exert higher pressure drops, which in turn leads to higher partial pressures of water. This is especially true in multitubular reactors where the long thin packed tubes would lead to high water concentrations especially at the bottom of the bed. Here the catalyst would be subject to water produced further up the bed as well as the water being generated at that particular site. Obviously fluid bed operations would be much less effected by water, however bed shape would be important.

The aim of the work described in this paper was an attempt to try to predict what adverse effects hydrogen partial pressures during activation would have on FT catalysts both in fixed bed and Continuously Stirred Tank Reactors. The opportunity was also taken to study the extent of catalyst reduction under different hydrogen concentrations and different temperature ramp rates.

3. Experimental

The initial challenge was to devise a method of measuring the water concentration in the gas stream, directly it leaves the activation reactor. Various devices were considered including moisture meters and Draeger tubes, however pressure, temperature and concentration levels make the use of these instruments impractical. The solution was to use an on-line ESS GeneSys Quadrupole mass spectrometer, with the sample capillary positioned directly within the exit end of the reactor. This instrument was calibrated for accurate measurement of very low levels of water and also hydrogen concentrations in the reactor exit gas. The positioning of the probe at the exit of the reactor, as near to the catalyst as possible was critical for the water measurements. An ONIX Prima δB magnetic sector on-line mass spectrometer further downstream was used to check the hydrogen concentration in the exit gas stream. Although this instrument could detect water, the concentration was always slightly lower than that measured directly at the exit of the reactor. This was attributed to slight losses of water on the pipe work and collection pots, between the bottom of the reactor and where the sample was taken. Water and hydrogen concentration measurements were taken from both fixed and fluid bed reactors during the course of this work. The catalysts used in all the tests were a 38 - 46 μ 20%Co/ZnO. For fixed bed tests the catalyst was diluted 1:4 with 110 μ aluminium oxide particles (ex Johnson Matthey.

Since water concentration was only measured as close as possible to the top or bottom (depending on whether fluid or fixed bed; i.e. downstream of the bed) of the catalyst bed, the assumption has to be made that this concentration of water cannot be exceeded anywhere in the bed. The best way of checking this was to measure both water and hydrogen consumption in the effluent gas and check that they always balanced throughout the reduction. Any discrepancy between the two would indicate that water was being left on the catalyst and that the partial pressure could have exceeded that measured at the reactor exit. It would not be possible to design experiments covering changes in every variable, so it was decided to assess the effect of hydrogen concentration on water formation using medium and slow temperature ramp rates. The other variables were fixed as follows:

Reduction gas flow rate	3000 h^{-1} GHSV
Reduction gas	hydrogen concentrations of 1%, 7%, 25% and 100, balance nitrogen
Catalyst bed dimensions	short and fat or long and thin
Fluid Bed	l = 2.0 cm, d = 2.4 cm
Fixed Bed	l = 24 cm, d = 0.96 cm
Catalyst particle size	90 micron
Temperature ramp rate	See below
Dwell time at reduction temperature	See below
Maximum temperature employed	See below

The medium and slow ramp rates employed were as follows:

Medium	**Slow**
2 hours at 50°C under reduction gas reduction gas	2 hours at 50°C under
Ramp 120°C h^{-1} to 180°C	ramp 120°C h^{-1} to 180°C
Dwell 0.5 h	Dwell 0.5 h
Ramp 60°C h^{-1} to 410°C	Ramp 10°C h^{-1} to 410°C
Dwell 6.0 hours	Dwell 6.0 hours

Following activation catalysts were tested for activity in the conversion of syngas into hydrocarbons, both in fixed bed and CSTR reactors. The reactors consisted of a 316 stainless steel tube contained within a heat shrunk copper/aluminium block, which was then wrapped length ways with electrical heating elements. In most tests the catalyst was diluted 1:4 with aluminium particles of a similar size. Following reduction the catalyst was assessed for syngas conversion potential with H$_2$:CO = 2:1, 30 bar, 1250 h^{-1} GHSV. Liquid and wax products were collected and weighed, whilst the exit gas was let down to atmospheric pressure prior to analysis with a Prima δb magnetic sector on-line mass spectrometer. The CO conversion % and selectivities % to methane, carbon dioxide, ethylene, ethane, propene, propane, butene and C4's were determined at a particular temperature. From this date the relative performance was determined.

The CSTR unit operated on exactly the same feed gas and used the same analysis system, however the reactor itself was more complicated. In this case the catalyst particles, were suspended in a heavy hydrocarbon oil (300 cm^3) contained within a rapidly stirred 800 cm^3 capacity autoclave. Syngas was bubbled into the autoclave through a bottom sparger and the products removed via a catalyst wax separator. A schematic of the stirred autoclave, inxcluding the catalyst separator is shown below in Figure 1.

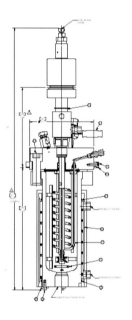

AUTOCLAVE ENGINEERS 800 cm AUTOCLAVE

The CSTR reactor was a 800 cm autoclave fitted with a Magnadrive powered gas induction stirrer. Gas was introduced just below the tip of the stirrer to give maximum mixing potential. The catalyst and the starting wax medium was added after external activation, via a 12 mm addition port in the lid. Heating/cooling was via a circulating oil filled jacket and also via the internal cooling coil. The temperature was measured by an internal PRT. Unreacted feed, gaseous and liquid products were removed from the catalyst using a specially designed separating device, fitted within the autoclave. A schematic of the separation device is shown below.

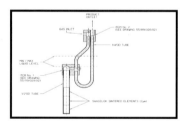

Figure 1: CSTR Test Equipment

4. Results and Discussion

4.1. Reduction Profile

It is assumed that the cobalt oxide reduces according to the following equation.

Co_3O_4 + $4H_2$ ---------------------------- $3Co$ + $4H_2O$ (equation 1)

In fact the reduction proceeds in two steps, reducing first to CoO and then through to Co. The twin peaks signifying a two-stage reduction are readily apparent in the reduction data. (Figures 2 & 3). Figures 2 & 3 show data from the Temperature Programmed Reduction (TPR) in a fluidised bed, of a 20% Co/ZnO catalyst. From the reduction data on the initial test it is apparent that the water production exactly matches hydrogen consumption data, confirming that the water measurement is correct and that none is retained in the bed. Hence the maximum water concentration measured is the highest partial pressure of water reached during the reduction.

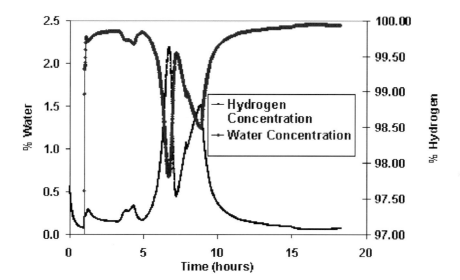

Figure 2: Hydrogen consumption and water production for Co/ZnO catalyst in a fluidised bed

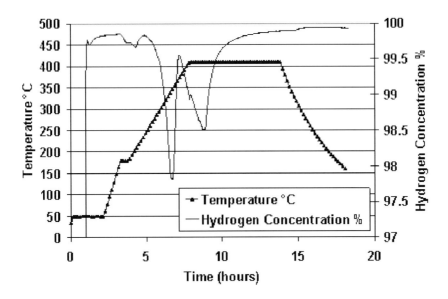

Figure 3: Bed temperature and hydrogen consumption for Co/ZnO catalyst in a fluidised bed

4.2. Results and Test Description

Measuring the partial pressure of water and hydrogen concentration at any point in the activation reactor is virtually impossible, due to the adverse conditions employed. However it was possible to measure the water partial pressure and hydrogen concentration very accurately at the reactor exit and as near to the catalyst bed as possible. It could be argued that water would be retained on parts the catalyst bed and that the concentration would not be the same throughout the bed. However water concentration is proportional to hydrogen consumption and it is very unlikely that hydrogen would be retained on the bed. The assumption was made that if the water concentration corresponds exactly to the hydrogen consumption as measured at the exit of the bed, then there must be no water retained on the bed, and that the maximum concentration of water measured must be the maximum concentration achieved at any point in the catalyst bed.

4.3. The Effect of Hydrogen Concentration

Assuming that the cobalt oxide is reduced according to equation (1) it is possible by integrating the hydrogen concentration data to calculate the amount of hydrogen consumed over time and hence calculate the extent of catalyst reduction. In view of the fact that water partial pressure may affect the degree of reduction and ultimately the catalyst performance, a series of experiments was undertaken to measure the degree of catalyst reduction whilst varying hydrogen concentration and temperature ramp rates. It was envisaged that varying the temperature ramp rate would have the largest effect, since this should have the most effect on water partial pressure. Two temperature ramp rates were studied (medium & slow) at hydrogen concentrations of 1%, 7%, 25% and 100%.

Medium	**Slow**
2 hours at 50°C under reduction gas	2 hours at 50°C under reduction gas
Ramp 120°C h^{-1} to 180°C	Ramp 120°C h^{-1} to 180°C
Dwell 0.5 h	Dwell 0.5 h
Ramp 60°C h^{-1} to 410°C	Ramp 10°C h^{-1} to 410°C
Dwell 6.0 hours	Dwell 6.0 hours

The water concentration for each of these ramp rates at the four hydrogen concentrations is shown in Table 1.

Table 1: Water partial pressure as a function of hydrogen concentration

H_2 Feed Conc.%	Temp. Ramp Rate °C h^{-1}	1st Max H_2O	Partial Pressure (bar)	2nd Max H_2O	Partial Pressure (bar)	% Reduction	Reaction Time (h)
1.0	60	0.9	0.009	0	0	30	2.8
1.0	10	0.27	0.0027	0.17	0.0017	41	20
7.0	60	1.7	0.017	0.35	0.0035	45	6.4
7.0	10	0.3	0.003	0.24	0.0024	50	21
25.0	60	1.3	0.013	0.60	0.006	65	10
25.0	10	0.23	0.0023	0.40	0.004	67	22
100.0	60	2.18	0.028	1.50	0.015	83	10
100.0	10	0.37	0.0037	0.33	0.0033	80	27

As was predicted, at all four hydrogen concentrations the slower ramp rate always gave rise to the lower partial pressure of water and larger extent of calculated reduction. However at 25% and 100% there is no significant change in the degree of reduction, whilst using both the medium and fast ramp rates, even though the water partial pressure is much larger when using the fast ramp rate. Indeed when using the low ramp rate the water partial pressure is below 0.01 bar at all hydrogen concentrations. At no time did the water partial pressure exceed or get even close to 0.05 bar (see Figure 4). It would appear that at lower hydrogen concentrations the ramp rate has a much greater effect on the degree of catalyst reduction, but at higher hydrogen concentrations the extent of calculated reduction is much higher.

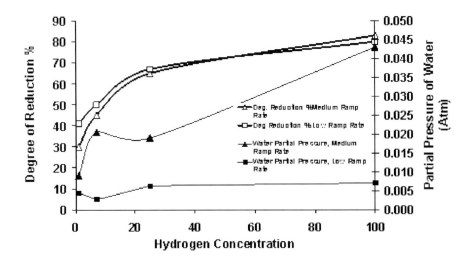

Figure 4: Hydrogen concentration effect on degree of reduction

4.4. The Effect of Water Partial Pressure

Since one of the aims of this work was to assess the effect of water partial pressure on performance, it was now necessary to compare the performance of a catalyst reduced using methods designed to maximise this effect. Two methods of increasing the partial pressure of water were attempted, namely carrying out the reduction at elevated pressure and heating under N_2 to 410°C and then

admitting the reduction gas   In a series of experiments not reported here, the temperature ramp rate was increased up to 10°C min^{-1} but in no instance did the water partial pressure exceed 0.1 bar. In order to achieve a higher water partial pressure, the hydrogen had to be introduced to a catalyst already maintained at 410°C under nitrogen. In this case the actual reduction period is very short (<15 min) with the partial pressure of water reaching 0.7 bar for a very short period.

The reduction to deliberately increase the partial pressure of water was undertaken at 6 bar, which based on the results at atmospheric pressure should have resulted in a partial pressure of water of 4.2 bar. However it was not possible to measure the partial pressure of water within the pressurised reactor, so it is assumed that it was attained. Measurement at the exit of the backpressure regulator showed the partial pressure of water and consumption of hydrogen to be as would have been expected at atmospheric pressure. The two catalyst samples subjected to these unusual reductions were then compared in syngas conversion tests with a sample activated with a medium ramp rate at atmospheric pressure. The activation conditions for the three tests are summarised as follows:

Test A: -   100% N_2 at 3000 h^{-1} GHSV
Ramp 60°C h^{-1} to 410°C, Dwell for 0.2 h
25%H_2/75%N_2 at 3000 h^{-1} GHSV
Dwell 6 h, then cool to Room Temperature

Test B: -   6 bar pressure
100% N_2 at 3000 h^{-1} GHSV
Ramp 60°C h^{-1} to 410°C, Dwell for 0.2 h
25%H_2/75%N_2 at 3000 h^{-1} GHSV
Dwell 6 h, then cool to Room Temperature

Test C: -   25%H_2/75%H_2 at 3000 h^{-1} GHSV
Ramp 60°C h^{-1} to 410°C
Dwell 6 h, cool to Room Temperature

A and B are first heated under N2 before being directly subjected to 25%/75% gas. A is at atmospheric, whilst B is at 6 bar.

Fixed bed tests were undertaken with H$_2$:CO = 2:1, 30 bar, 1250 h^{-1} GHSV. In order to best compare the catalytic performance, the temperature in each test was adjusted so as to achieve 62% carbon monoxide conversion. When the performance was steady over at least 36 h the data was recorded for comparative purposes. A brief summary of the performance of all three tests is shown in Table 2.

Table 2: FT performance in fixed bed as a function of activation conditions

Test No.	HOS	Temp. °C	CO Conv. %	C$_5$+ Sel. %
A	75	237	63	57
B	75	237	62	57
C	75	212	63	74

The data shows that the two samples activated in a manner so as to maximise the partial pressure of water in the catalyst bed (Samples A & B), exhibited similar performance characteristics (same operating temperature and C5+ hydrocarbom selectivity), whilst the sample activated in a manner such to minimise water partial pressure performed much better, operating at a lower temperature and achieving a higher C5+ hydrocarbon selectitvity. Interestingly the performances of samples A & B were almost identical, even though sample B had in theory been subject to six times the partial pressure of water than that experienced by sample A. This tends to suggest that there may be a theoretical maximum of water that is detrimental, and anything above that level does not enhance the deactivation rate. More experiments at various pressures would be needed to substantiate this claim. However it is clear that designing the activation to keep the partial pressure of water at a minimum at all times, is indeed beneficial to catalyst performance. At slow ramp rates at atmospheric pressure, the partial pressure of water is always <0.1 bar.

4.5. Effect of Gas Feed Rate (or GHSV)

The last parameter to be studied in respect of its effect on water partial pressure and ultimately on catalyst performance was gas feed rate or GHSV. It was thought that the faster the gas flow though the bed during reduction the more easily any water produced would be removed, hence keeping the partial pressure of water low throughout the catalyst bed. Two reductions using the same catalyst were undertaken in the fluid bed reactor at 2200 h^{-1} and 3500 h^{-1} GHSV respectively. Conditions employed for these reductions are shown below; whilst the water partial pressure profiles for the two temperature programmed reductions are shown in Figure 5 and 6.

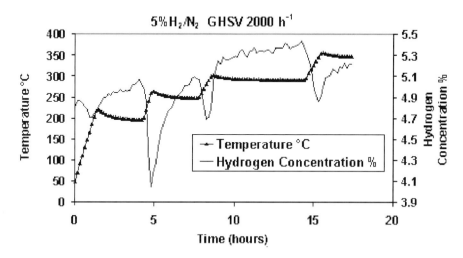

Figure 5: *Temperature Programmed Reduction Profiles at 2000 h^{-1}*

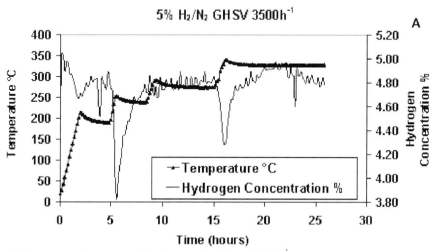

Figure 6: *Temperature Programmed Reduction Profiles at 2000 h^{-1}*

A	B
5% H$_2$/95%N$_2$ 2200 h^{-1} GHSV	5% H$_2$/95%N$_2$ 3500 h^{-1} GHSV
Ramp 120°C h^{-1} to 240°C, dwell 3h	Ramp 120°C h^{-1} to 240°C, dwell 3h
Ramp 120°C h^{-1} to 300°C, dwell 3h	Ramp 120°C h^{-1} to 300°C, dwell 3h
Ramp 60°C h^{-1} to 350°C, dwell 6h	Ramp 60°C h^{-1} to 350°C, dwell 6h
Ramp 60°C h^{-1} to 410°C dwell 6h	Ramp 60°C h^{-1} to 410°C dwell 6 h
Cool to Room Temp.	Cool to Room Temp

A gas mixture of 5% H$_2$/95%N$_2$ was used for this series of tests. It was anticipated that in test B where the reducing gas flow is faster, the water partial pressure would be lower and hence less detrimental to the catalyst performance. The performance test in a CSTR data is Table 3.

Table 3: FT performance in CSTR as a function of activation conditions

Test No.	HOS	Temp. °C	CO Conv. %	C$_5$+ Sel. %
A	80	261	30	75
B	80	227	40	88

As can be seen from the results the catalyst reduced at the higher GHSV gives a far superior performance, it operates at a lower temperature and has a higher selectivity to the desirable C5+ hydrocarbons. This result supports the hypothesis that catalysts reduced with lower water partial pressure are more active for syngas conversion. The data does suggest that it would be preferential to use higher gas flow rates for catalyst activations.

5. Conclusions

For Co/ZnO catalysts the degree of reduction is dependent on the concentration of hydrogen in the reducing gas, under the conditions studied. Concentrations below 25% are much less effective. For all the reductions undertaken at atmospheric pressure with gradual temperature ramping, over a range of hydrogen concentrations, both in fluid and fixed bed, the partial pressure of water never exceeded 0.05 bar. Extremely forceful conditions had to be employed (instant 100% hydrogen pulse at 410°C) to cause the water partial pressure to exceed 0.1 bar and these conditions would be hard to replicate in an industrial unit. Catalysts that were subjected to higher levels of water partial pressure, as generated by the consumed hydrogen, were much less active when tested for syngas conversion. Increasing the reduction gas flow rate resulted in a marked increase in catalyst activity, presumably due to lower water partial pressure over the catalyst bed. It would appear that there is possibly a maximum water partial pressure above which it has no further effect on catalyst performance.

References

1. Jacobs G., Das T. K., Zhang Y., Li J., Racoillet G. and Davis B.H. Appl.Catal. A, 233, 263, 2002
2. Schanke D, Hilmen A.M., Bergene E., Kinnari K., Rytter E., Adnanes E., and Holmen A., Catalysis Letters 34, 269, 1995
3. Huber G.W., Guymon C G., Conrad T L., Stephenson B C. and Bartholomew C H. Stud. Surf.Sci. Catal. 139, 423, 2001

in H_2O/H_2 mixtures at ratios much lower than expected for the oxidation of bulk cobalt. Oxidation as a deactivation mechanism for Co/Al_2O_3-based catalysts was supported by researchers at Sasol Technology and Delft University of Technology [4,5]. Davis, et al. [6-9], employing stirred tank slurry phase reactors, confirmed that water negatively affected the CO conversion for a Ru-promoted Co/TiO_2 and unpromoted and Pt promoted Co/Al_2O_3 catalysts. Furthermore, the latter workers reported that the impact of water on CO conversion was reversible up to a specific $P_{H_2O}/P_{CO} \leq 12$, suggesting a kinetic effect. This threshold ratio depended on the Co loading (and therefore, the average Co crystallite size), with more heavily loaded catalysts being more robust. That is, a wider range of P_{H_2O}/P_{CO} for the reversible effect of H_2O was observed. Yet few kinetic studies include the reversible impact of P_{H_2O} in the empirical rate expression.

Fischer-Tropsch synthesis is a complex reaction which produces a large number of products, and this in turn makes the development of kinetic expressions very challenging. A large number of kinetic data and rate expressions for FT synthesis have been reported in the literature [10-21]. Yang et al. [12], Pannell et al. [13], Wang [15] and Zennaro et al. [21] developed kinetic rate expressions for supported cobalt catalysts using data generated in fixed bed reactors by regression of a power-law equation of the general form:

$$R_{CO} = kP_{CO}^a P_{H_2}^b \tag{1}$$

They found the reaction order for H_2 (*b*) was positive but the reaction order for CO (*a*) was negative, suggesting inhibition by adsorbed CO. Sarup and Wojciechowski [16] compared six different possible mechanisms to FT synthesis data they obtained with a cobalt catalyst used in a Berty internal recycle reactor. They found the best fit was obtained with the expression:

$$r_{CO} = kP_{CO}^a P_{H_2}^b /(1 + \sum_i K_i P_{CO}^{c_i} P_{H_2}^{d_i})^2 \tag{2}$$

where k is the kinetic rate constant, *a* and *b* are the reaction orders, K_i is the adsorption constant for the *ith* adsorption term, and c_i and d_i describe the dependence of surface coverage of the *i*th adsorption term on its reactant partial pressure. From their experimental data, they reduced the six possible equations to two likely candidates.

Yates and Satterfield [18] determined the kinetics for a $Co/MgO/SiO_2$ catalyst operated in a CSTR. They represented their data by a Langmuir-Hinshelwood rate expression:

$$r_{CO} = kP_{CO}P_{H_2} /(1 + K_i P_{CO})^2 \qquad (3)$$

This equation had been reported previously by Sarup and Wojciechowski [16], where they assumed that hydrogenation of an adsorbed formyl species to form carbon and water was the rate determining elementary step in the mechanism.

In fact, the effect of water on CO conversion for supported cobalt catalysts is complex and has been investigated by several groups, with results ranging from negative, or deactivating [1-9] to negligible, (i.e., statistically able to neglect) [18,22,23] to even positive [3,24-29] deviations, depending on choice of support and dispersion of the metal, which is controlled in part by the loading. In our previous work it was found that the water effect was slightly negative for a 0.2 wt.% Ru-promoted 10 wt.% Co/TiO_2 [6] and that the negative effective was more pronounced for 25 wt.% Co/Al_2O_3 [9], and even moreso for 0.5 wt.% Pt-15 wt.% Co/Al_2O_3 [7,8] catalysts. The CO conversion did not change significantly at space velocities of 8 and 4 $NL/h/g_{cat}$ when water was added to the 10% Co/TiO_2 catalyst but when the space velocity was decreased to 2.0 and 1.0 $NL/h/g_{cat}$, the addition of water caused a decrease in the CO conversion. The activity of the catalyst recovered after the water addition was terminated at a space velocity of 2.0 $NL/h/g_{cat}$. Similarly, a negative reversible water effect was observed in case of 0.5 wt.% Pt-15 wt.% Co/Al_2O_3 catalyst in the range of 3-25 vol.% water added. The CO conversion decreased with increasing amounts of water and the activity of the catalyst recovered after water addition was terminated. An irreversible deactivation was observed, however, when the amount of added water was greater than 28 vol.%. For a comparable reaction test using 25 wt.% Co/Al_2O_3, the CO conversion was mainly reversible even at 30 vol.% H_2O addition, suggesting that a larger Co crystallite size may be more resistant to the irreversible deactivating effect of H_2O. One possibility for the temporary reversible decline in CO conversion with addition of water up to about 25 vol% H_2O (note: threshold dependent on catalyst formulation) may be due to a kinetic effect of water caused by adsorption inhibition of the reactants.

It is important to note that, in those cases where we observed the negative effect of water (i.e., Al_2O_3 and TiO_2 catalysts), strong interactions of cobalt with the support were identified by TPR, and the average cluster size for the reduced fraction was small (< 12 nm). Interestingly, in a previous study with a 12.4 wt.% Co/SiO_2 catalyst, the interactions between cobalt and the support were weak, and special precautions during pretreatment and activation (e.g., no calcination, slow temperature ramp) had to gbe carried out in order to maintain a small Co cluster size (~ 13.2 nm). In that case, we observed a positive effect of water on CO conversion when less than 25 vol.% water was added to the synthesis gas [27]. Ruling out poisons in the feed gas, possible causes for reversible and irreversible declines in CO conversion during Fischer-Tropsch synthesis include oxidation phenomena, which may or may not involve the formation of cobalt-support complexes [1-9,28,30-32], surface condensation, sintering [36] or coke formation. The oxidation of bulk phase metallic cobalt to CoO or Co_3O_4 should not be possible under Fischer-Tropsch conditions from thermodynamic considerations [4,5]. However, the oxidation of metallic cobalt to $CoAl_2O_4$ is allowed, but considered to be kinetically restricted under typical FT synthesis conditions [4,5]. It is important to consider the possibility that the support may influence the thermodynamics and/or kinetics of oxidation phenomena for small supported cobalt crystallites [4,5,9,28]. To illustrate this, in recent EXAFS and XANES studies on 0.5 wt.% Pt-15 wt.% CO/Al_2O_3 catalyst, it was suggested that the sudden irreversible loss of activity with greater than 25 vol.% added water was due, at least in part, to the formation of cobalt aluminate-like species [7,8]. The catalyst samples were withdrawn from the reactor during FT synthesis before, during and after H_2O addition for different levels of added water and cooled under inert gas so that the catalyst was embedded in the solid wax to prevent exposure to the atmosphere. No detectable changes in EXAFS and XANES spectra were noted, which all indicated the presence of cobalt crystallites, and not the transformation to an oxidized form, when less than 25 vol.% of water was added and thus the effect of water in this range was considered to most likely be kinetic. Previous workers have reported that cobalt catalyst oxidation occurs during Fischer-Tropsch synthesis [1-5], and that the extent of oxidation depends on the water partial pressure and the ratio of or P_{H_2O}/P_{H_2} or P_{H_2O}/P_{CO}. In some cases (e.g., 25% H_2O co-feeding with 25% Co/Al_2O_3 [9]) deactivation of the cobalt catalyst byoxidation was suggested to be, in large part, reversible and likely due to limited oxidation of cobalt, probably confined to surface layers, that was recovered by re-reduction when water addition was terminated. With Co/SiO_2 catalysts [27-29], on the other hand, water exhibited a positive effect on CO conversion at low levels of H_2O addition and, in this case, cobalt oxidation could not account for the impact of water. The complex interactions of H_2O with Co-based catalysts underscores the importance of sorting reversible (e.g., kinetic) and irreversible (e.g., structural) effects of water on a per catalyst basis.

The objective of this work was to obtain kinetic data over a well characterized, robust 25 wt.% Co/Al_2O_3 catalyst (i.e., close to commercial loadings) to assess quantitiatively the reversible effect of the water partial pressure on the rate of CO conversion during FT synthesis. Reactant orders were obtained by examining reactant partial pressure dependencies on the CO conversion rate. This was accomplished by maintaining the partial pressure of one reactant constant while varying the other through the use of Ar balancing gas. At each H_2/CO ratio (i.e., either at constant P_{CO} or P_{H_2}), CO conversion was measured for several space times. Regressed equations were obtained for CO conversion as a function of space time at each H_2/CO ratio (P_{CO} or P_{H_2} constant). The derivative of CO conversion versus space time was extrapolated to space time zero and substituted into the rate expression for r_{CO}. Reactant orders were then obtained by measuring the slope from a plot of r_{CO} vs. the reactant partial pressure (other reactant partial pressure held constant). Finally, the negative reversible effect of water was assessed in terms of an empirical parameter m multiplied by the ratio $P_{H_2O}/P_{reac\tan t}$ in the denominator of the rate expression, with final form:

$$r_{CO} = \frac{kP_{CO}^a P_{H_2}^b}{1 + m\frac{P_{H_2O}}{P_{H_2}}}$$

2. Experimental

2.1. Catalyst Preparation

The 25wt.%Co/Al$_2$O$_3$ catalyst was prepared using a slurry impregnation of Al$_2$O$_3$ (Condea Vista Sba-150 alumina, BET surface area 150 m^2/g) with a cobalt nitrate solution. In this method [33] the ratio of the volume of loading solution used to the weight of alumina is 1:1; this provides a solution volume that is approximately 2.5 times the pore volume of the catalyst. Two impregnation steps were used, loading 12.5 wt.% of Co for each step. Between the two steps the catalyst was dried under vacuum in a rotary evaporator, first at 333 K and then as the temperature was slowly increased to 373 K. After the second impregnation/drying step, the catalyst was calcined in air at 673 K for 4 hrs.

2.2. BET Surface Area and Temperature Programmed Reduction (TPR)

BET surface area measurements were conducted using a Micromeritics Tri-Star system to determine any loss of surface area after loading the metal. The catalyst sample was slowly heated to 433 K and held at this temperature for 4 hours under vacuum (~50 mTorr). Then the sample was transferred to the adsorption unit and the N_2 adsorption was measured at liquid N_2 temperature.

The temperature programmed reduction profiles of the calcined catalyst was recorded using a Zeton Altamira AMI-200 unit. The catalyst sample was first purged in a flow of argon at 573 K to remove the traces of water and then the temperature was reduced to 323 K. Temperature programmed reduction (TPR) was performed using a 10% H_2/Ar mixture referenced to Ar at a flow rate of 30 cc/min. The sample was heated from 323 K to 1073 K using a heating ramp of 10 K/min.

2.3. H_2 Chemisorption by TPD and percent reducibility by O_2 pulse titration

The amount of chemisorbed hydrogen was measured using a Zeton Altamira AMI-200 unit with a thermal conductivity detector (TCD). The calcined catalyst (~0.2 g) was activated using hydrogen at 623K for 10 h and then cooled under flowing hydrogen to 373K. The sample was held at 373 K under flowing argon to remove physisorbed and weakly bound hydrogen species prior to increasing the temperature slowly to the reduction temperature. At that temperature, the catalyst was held under flowing argon to desorb the remaining chemisorbed hydrogen until the TCD signal returned to the baseline. The TPD spectrum was integrated and the number of moles of desorbed hydrogen was determined by comparing to the areas of calibration pulses of hydrogen in argon. Prior to the experiments, the sample loop size was measured with pulses of N_2 in a helium flow and compared against a calibration curve produced from using gas tight syringe injections of N_2 into a helium flow.

After the TPD of H_2, the sample was reoxidized at the activation temperature by pulses of pure O_2 in helium carrier referenced to helium gas [34]. After oxidation of the cobalt metal clusters (when the entire O_2 pulse was observed by the TCD), the number of moles of O_2 consumed was determined, and the percentage reduction was calculated assuming that Co^0 reoxidized to Co_3O_4. To estimate the average cluster size, the percentage reduction was included in the dispersion calculation as follows, assuming a H:Co stoichiometric ratio of 1:1, as reported previously [34]. The uncorrected dispersion and uncorrected particle size were estimated assuming that the sample was completely reduced. To calculate the corrected dispersion and particle size the percent reduction of the sample was included in the calculation. For the particle size estimation, a spherical cluster morphology was assumed.

$$\%D_{uncorr.} = \text{(number of surface Co}^0 \text{ atoms} \times 100)/\text{(number of total Co atoms)}$$

$$\%D_{corr.} = \text{(number of surface Co}^0 \text{ atoms} \times 100)/[\text{(number of total Co atoms)(percentage reduction)}]$$

2.4. Water effect

The calcined catalyst (~20.0 g; weight was accurately known) was reduced ex-situ in a fixed bed reactor with a mixture of hydrogen and helium (1:2) at a flow rate of 70 SL/h at 623K and atmospheric pressure. The reactor temperature was increased from room temperature to 373K at the rate of 2K/min and held at 373K for 1h, then increased to 623K at a rate of 1K/min and kept at 623K for 10 h. The catalyst was transferred under protection of argon to a 1 L CSTR containing 300 g of melted Polywax 3000 (start-up solvent). The catalyst was then re-reduced in-situ in the CSTR in a flow of 30 SL/h hydrogen at atmospheric pressure. The reactor temperature was increased to 553K at a rate 1K/min and maintained at this activation condition for 24 h. Separate mass flow controllers were used to add H_2, CO and inert gas at the desired rate to a mixing vessel. Carbon monoxide passed through a lead oxide-alumina containing vessel to remove iron carbonyls. After the activation period, the reactor temperature was decreased to 453K, synthesis gas was introduced through a dip tube, and the reactor pressure was increased to 1.99 MPa (19.7 atm). The reactor temperature was then increased to 493K at a rate of 1K/min and the stirrer was operated at 750 rpm. The FT synthesis was started at a space velocity of 5.0 SL/h/g_{cat} and a H_2/CO ratio of 2.0, typical of a GTL process. The effect of water on CO conversion was determined at constant CO and H_2 partial pressure. To keep the partial pressure of hydrogen and carbon monoxide constant, 35 vol.% of argon was added to the feed to give a space velocity of 5.0 SL/h/g_{cat}. After 407 hrs on-stream, 5 vol.% water was added and a fraction of argon was replaced by water so that the sum of added water plus argon was 35 vol% of the total feed. A high pressure, precision ISCO syringe pump was used to add the water to the reactor. To define the impact of water addition on the activity and deactivation rate, the reaction conditions were switched back to the reference conditions (i.e., with 35 vol.% of argon). The added water and water produced by the FT synthesis were quantified to ensure that the water balance was accurate. Vapor products were continuously removed from the reactor and passed through two traps, one maintained at 373K and the other at 273K. The uncondensed vapor stream was reduced to atmospheric pressure through a back pressure regulator. The flow was measured using a bubble flow meter and the composition was quantified using an on-line GC (Refinery gas analyzer, HP Quad series Micro GC). The liquid products that accumulated in the reactor were removed every 24 h by passing through a 2 µm sintered metal filter, located at or below the liquid level in the CSTR, to a 573K trap. The liquid products from the 273K and 373K traps were combined, and the hydrocarbon and water fractions separated and analyzed by GC (HP 5890, Capillary column DB-5, 60m, i.d. 0.32 mm, for hydrocarbon; HP 5790 with TCD, for water). The reactor wax sample was analyzed by a high temperature GC (HP 5890, alumina clad, 25m, i.d.0.53 mm) to obtain a carbon number distribution for C_{20}-C_{80} products. At various intervals, the amount of water added was increased to include 10, 15, 20, 25 and 30 vol.%.

2.5. Kinetic testing

The kinetic study was performed at steady state conditions with the 25%Co/Al$_2$O$_3$ catalyst at 493K and 1.99 MPa (19.7 atm) total pressure. The catalyst loading and pretreatment procedure were the same as described above. The catalyst was allowed to achieve steady state during 500 hours time-on-stream, after which data for the kinetic experiments were collected. Two sets of experiments were carried out; one of these was at constant CO partial pressure (0.498 MPa) and variable H$_2$ partial pressure and the other at constant H$_2$ partial pressure (0.796 MPa) and variable CO partial pressure. The inlet gas compositions for constant CO partial pressure and constant H$_2$ partial pressure runs are presented in Tables 1 and 2, respectively. Four H$_2$/CO ratios (2.4, 2.0, 1.5 and 1.0) and six total space velocities (15.0, 12.5, 10.0, 8.0, 5.0 and 3.0 SL/h/g$_{cat}$) for each H$_2$/CO ratio were utilized. The kinetic experiment was started with constant CO partial pressure (Table 1) (0.498 MPa) with H$_2$/CO ratio 2.4 and a space velocity of 15.0 SL/h/g$_{cat}$ (25% CO, 60%H$_2$ and 15% Ar). The space velocities were then changed in decreasing order (12.5, 10.0, 8.0, 5.0 and 3.0 SL/h/g$_{cat}$). The deactivation was determined after returning to the reference conditions (space velocity =15 SL/h/g$_{cat}$, H$_2$/CO = 2.4, P$_{CO}$ = 4.93 atm) periodically, after completing the six space velocities for each H$_2$/CO ratio. A period of at least ~24 h was allowed between mass balances to ensure steady-state operation at each condition. The second set of runs (Table 2), at constant H$_2$ partial pressure (0.796 MPa), was started after finishing the first set of experiments at constant CO partial pressure (0.498 MPa). A repeat of the starting set of experiments (i.e., P$_{CO}$ = 0.498 MPa, H$_2$/CO = 2.4, space velocities =15.0, 12.5, 10.0, 8.0, 5.0 and 3.0 SL/h/g$_{cat}$) was carried out at the end of the kinetic study.

Table 1. Inlet Gas Composition for the Kinetic Studies (P_{CO} = 4.93 atm)

H_2/CO	1.0	1.5	2.0	2.4
CO%	25	25	25	25
H_2%	25	37.5	50	60
Ar%	50	37.5	25	15

Table 2. Inlet Gas Composition for the Kinetic Studies (P_{H_2} = 7.88 atm)

H_2/CO	1.0	1.5	2.0	2.4
y_{CO}, %	40	26.7	20	16.7
y_{H_2}, %	40	40	40	40
y_{Ar}, %	20	33.3	40	43.3

3. Results

3.1. Characterization

The results of the BET surface area measurements by adsorption of nitrogen at 77K are presented in Table 3. A loading of 25% metal is equivalent to 33.3% by weight Co_3O_4. If the Al_2O_3 is the only contributor to the area, then the area of the Co/Al_2O_3 catalyst should be 0.333 × 150 m²/g = 100 m²/g, slightly higher than the measured value of 89 m²/g. This indicates same pore blockage by cobalt oxide clusters. However, in several other comparisons of this and similar catalysts, the measured and calculated surface areas agreed within 5%, indicating little or more pore blockage.

Table 3. BET Surface Area, Dispersion and Particle Size of 25% Co/Al_2O_3 Catalyst

Catalyst	BET Surface Area (m²/g)	% Reduction	% Disp.	Average Particle Size (nm)
25% Co/Al_2O_3	89	42	8.7	11.8

The TPR profile of the 25%Co/Al_2O_3 catalyst is shown in Figure 1. The low temperature peak (500-650K) is typically assigned to the reduction of Co_3O_4 to CoO, although a fraction of the peak may include the reduction of the larger, bulk-like CoO species to Co° [34]. The second broad peak (700-100K) is attributed to the reduction of cobalt oxide species that interact with the support. In this TPR, the temperature was not ramped high enough to observe reduction of the bulk cobalt aluminate species, which has been shown to occur above 1073K for samples with up to 30% loading of cobalt [35]. The average particle size and the dispersion obtained from H_2 chemisorption were 11.8 nm and 8.7%, respectively.

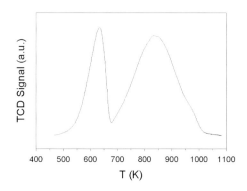

Figure 1. TPR profile of the 25 wt.% Co/Al$_2$O$_3$ catalyst.

3.2. Water effect

During a run at constant conditions, the conversion decreases more rapidly during the early period on-stream (ca. first 200 hrs) and then continues to steadily decline at a slower rate of 0.24% CO conversion/day (Figure 2). A similar run was made but additional water was co-fed at intervals after the second, slower activity period was attained; however, the feed for this run had 30% inert gas in addition to the H$_2$:CO = 2:1 syngas (Figure 3). Thus, water could be added by replacing a molar equivalent of the inert gas so that the total flow remained constant as well as the partial pressures of CO and H$_2$. As shown in Figure 3, the CO conversion was not altered as water was increased from 5 to 20% of the total feed by volume. The addition of 25 and 30 vol.% water caused the CO conversion to decrease and then recover to the expected value when water addition was terminated. During a repeat run, it was established that the decrease in CO conversion with the addition of even 30% water did not continue but rather reached a lower, constant value after about three days of water addition [9]. The CO conversion data obtained during periods of normal FTS (solid points, Figure 3) show a steady decline of 0.30% CO conversion/day; this value closely agrees with the value of 0.24% CO conversion/day shown in Figure 2.

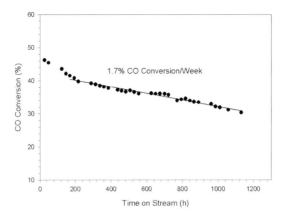

Figure 2. Representative deactivation curve for a 25 wt.% Co/Al_2O_3 catalyst when operated at 473 K, 3 SL/h/g_{cat}, 1.99 MPa (19.7 atm) and $H_2/CO = 2.0$.

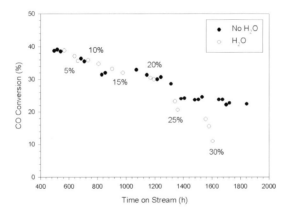

Figure 3. CO conversion at constant reaction conditions of temperature, pressure and total space velocity. Added water partial pressure was varied by replacing a fraction on the inert gas flow so that total water plus inert gas flow was constant.

3.3. Kinetic testing

In our water effect experiment, a primarily reversible decrease in CO conversion was observed when up to 30 vol% of water was added to the feed for this catalyst [9]. With the 0.5% Pt-15%Co/Al$_2$O$_3$ catalyst, a reversible water effect was obtained at a lower volume percent of water addition but irreversible deactivation occurred at > 25% vol. water addition [7]. One possibility for the effect of water is that the amount of catalytic active sites (i.e., surface cobalt metal atoms) available for the FT reaction changes with partial pressure of water, perhaps by a temporary oxidation process for cobalt [9]. Alternatively, competitive adsorption of water may decrease the surface concentration of CO and/or H$_2$ [9]. Thus, the following equation is proposed to described the reversible impact of water on the CO reaction rate:

$$r_{CO} = \frac{kP_{CO}^a P_{H_2}^b}{\left(1 + m\frac{P_{H_2O}}{P_{H_2}}\right)} \tag{4}$$

where k is a kinetic rate constant, a and b are the reaction orders of the reactants, and m is an empirical parameter for the impact of water vapor. In this equation, the partial pressures of H$_2$ (P_{H_2}), CO (P_{CO}) and water vapor (P_{H_2O}) are local values in the reactor.

The mass balances for similar runs, which include analyses of all condensed and noncondensed products, fall between 96 and 104%. The run at space velocity 15 SL/h/gcat with H$_2$/CO ratio 2.4 was repeated after completion of a set of runs at different space velocities (15.0, 12.5, 10.0, 8.0, 5.0 and 3.0 SL/h/gcat) at each H$_2$/CO ratio and the deactivation was defined (Figure 4). Over the longer time period for these measurements, the deactivation is not linear in time. The degree of deactivation depends on the space velocity; i.e., the deactivation rate is higher at lower space velocity. In most instances, a single point was taken at each space velocity for a given H$_2$/CO ratio. For the calculation of kinetic parameters, it is necessary to take the deactivation of the catalyst into account, and to correct the conversion to reflect a non-deactivated catalyst. To follow the decline in conversion due to aging effects, the CO conversion was checked periodically at a reference space velocity. The CO hydrogenation rates at varying space velocity conditions were adjusted to account for catalyst aging. This was carried out on an extent-of-reaction basis (i.e., total moles of CO reacted with time) by interpolating between each two reference conversion points measured at the same reference space velocity. Those reference points were measured prior to and after each set of space velocity conditions.

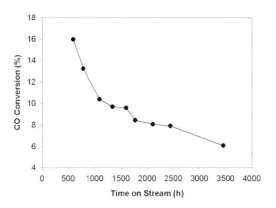

Figure 4. CO conversion for repeated conditions at P_{CO} = 4.93 atm (T = 493 K, P_{tot} = 1.99 MPa (19.7 atm), H_2/CO = 2.4 and space velocity = 15 SL/h/g_{cat}).

  As an example, consider a catalyst measured at a reference space velocity of 15 SL/h/g_{cat} at times-on-stream of 1 and 11 days where the total decline in activity was 4%. Between the two days, the space velocity was adjusted downward to 10, 8, 5, and 3 SL/h/g_{cat}. The number of CO moles converted at each space velocity was determined to be 100, 150, 200, and 250, respectively, giving a total of 700 moles converted over the 10 day period. The deactivation applied to the first condition would therefore be 100/700*0.04 = 0.0057. To adjust the conversion to account for aging, the value 0.0057 was added to the measured conversion to reflect a non-deactivated catalyst. For the second condition, the previous 0.0057 was added and an additional 150/700*0.04 was accounted for, and so on.

  This accounting procedure was verified by repeating a set of experiments conducted early in the run (at constant CO partial pressure, P_{CO} = 4.93 atm, H_2/CO = 2.4 and space velocities = 15.0, 12.5, 10.0, 8.0, 5.0 and 3.0 SL/h/g_{cat}) and at the end of the kinetic study with identical conditions. The deactivation-adjusted CO hydrogenation rates are very similar to the activity of the starting set of experiments (Table 4). The CO conversion and product distributions from the raw reaction data were adjusted for each individual sample to that of a fresh catalyst basis. Since reactor wax samples were not representative due to the short time at each reaction condition, the rate of C_{5+}

was calculated using the carbon balance and α value. The α value was calculated from the gas product analysis and the values ranged between 0.78-0.88, reasonable values for this cobalt catalyst. Under these reaction conditions, hydrocarbon products may be separated into two phases, liquid and vapor, in the reactor. It was assumed that H_2, CO, H_2O, and CO_2 are insoluble in the reactor liquid, a reasonable assumption considering the volume flow of each gas and their limited solubility in the liquid hydrocarbon phase. The fraction of hydrocarbons in the vapor phase was obtained under reaction conditions, assuming vapor/liquid compositions are applicable. Thus, the gas composition and, therefore, partial pressures of CO, H_2, and H_2O at the reactor outlet can be calculated from measured values.

Table 4. Deactivation Adjusted Activity Data for a Repeat Set of Experiments at the Beginning and End of the Kinetic Study at $P_{CO} = 0.498$ MPa (4.93 atm).

H_2/CO	Inlet Gas Composition	Space Velocity	X_{CO}	
			Beginning of the Run	End of the Run
	$y_{CO} = 25\%$	15.0	15.98	15.56
	$y_{H_2} = 60\%$	12.5	17.84	18.61
		10.0	20.64	20.41
2.4	$y_{Ar} = 15\%$	8.0	27.6	28.34
		5.0	39.21	38.34
		3.0	51.66	49.32

3.3.1. Determination of a and b

To determine the order of the FT reaction for H_2, constant CO inlet partial pressure was used at each H_2/CO ratio. The CO reaction rate is calculated using the relation:

$$r_{CO} \propto \frac{dX_{CO}}{d(1/GHSV)} \text{ or } r_{CO} \propto \frac{dX_{CO}}{d\tau} \tag{5}$$

where X_{CO} is the CO conversion and GHSV is the inlet total gas hourly space velocity ($SL/h/g_{cat}$).

In the first set of experiments, at constant inlet CO partial pressure, the CO conversion as a function of space time is shown in Figure 5. The CO conversion increases with an increase in H_2 partial pressure (i.e., an increase of

H$_2$/CO ratio of the inlet gas). The regressed equations were obtained using a polynomial function and extrapolated to the origin. The regressed equations determined from the data in Figure 5 for different H$_2$/CO ratios are summarized in Table 5. It is assumed that at very low CO conversion levels [*i.e.* $(dX_{CO}/d\tau)_{\tau=0}$], the partial pressure of water is zero so that the partial pressure of water did not affect the CO reaction rate (m = 0). Thus, equation (5) reduces to equation (6):

$$r_{CO} = kP_{CO}^a P_{H_2}^b \tag{6}$$

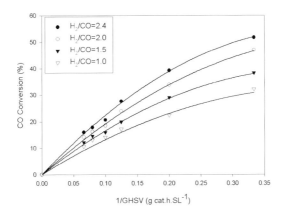

Figure 5. CO conversion as a function of space time at constant CO partial pressure 0.498 MPa (4.93 atm) over 25 wt.% Co/Al$_2$O$_3$ catalyst (T = 493 K, 1.99 MPa (19.7 atm) pressure).

Table 5. Regression of X_{CO} vs. τ (1/GHSV, h.g$_{cat}$/SL) at 1.99 MPa (19.7 atm) and 493 K

H$_2$/CO	P_{CO} = 4.93 atm	P_{H_2} = 7.88 atm
2.4	$X_{CO} = -284.82\tau^2 + 250.39\tau$	$X_{CO} = -352.59\tau^2 + 304.23\tau$
2.0	$X_{CO} = -230.25\tau^2 + 217.24\tau$	$X_{CO} = -233.17\tau^2 + 237.18\tau$
1.5	$X_{CO} = -229.11\tau^2 + 191.1\tau$	$X_{CO} = -90.283\tau^2 + 172.74\tau$
1.0	$X_{CO} = -198.12\tau^2 + 160.96\tau$	$X_{CO} = 38.938\tau^2 + 105.14\tau$

Table 6 shows r_{CO} and P_{H_2} extrapolated to $\tau = 0$ for constant P_{CO} (0.498 MPa) and different H_2/CO ratios calculated from the regressed equations. The regressed equations are provided in Table 7. The plot of $\ell n r_{CO}$ vs. $\ell n P_{H_2}$ (extrapolated to time 0), with data obtained from the series of H_2/CO ratio experiments in which the P_{CO} was kept constant and was P_{H_2} varied, is shown in Figure 6. This plot has a good correlation coefficient (R-squared value), and the H_2 concentration dependency (b) was 0.4885, which we have rounded to 0.5.

Table 6. Rates of r_{CO} at Different Partial Pressures and H_2/CO Ratios Obtained from Regressed Equations

H_2/CO	P_{CO} (atm)	P_{H_2} (atm)	$y_{CO,in}$	$\left(\dfrac{dX_{CO}}{d\tau}\right)_{\tau=0}$	$r_{CO} = \left(y_{CO,in} \times 22.414\right) \times \left(\dfrac{dX_{CO}}{d\tau}\right)_{\tau=0}$
2.4	4.93	11.83	0.25	250.39	2.8
2.0	4.93	9.86	0.25	217.24	2.42
1.5	4.93	7.4	0.25	191.1	2.13
1.0	4.93	4.93	0.25	160.96	1.8
2.4	3.29	7.88	0.167	304.23	2.26
2.0	3.94	7.88	0.2	237.18	2.12
1.5	5.26	7.88	0.267	172.74	2.06
1.0	7.89	7.88	0.4	105.14	1.92

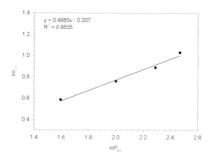

Figure 6. Plog of $\ell n r_{CO}$ vs. $\ell n P_{H_2}$ (atm) for rates measured at 493 K, 1.99 MPa (19.7 atm) pressure and a P_{CO} of 0.498 MPa (4.93 atm).

Table 7. P_{H_2} and r_{CO} Calculated from Regressed Equations at Different Conversion Levels (0.498 MPa (4.93 atm))

H_2/CO	X_{CO}, %	Space time (τ_1) ($g_{cat}.h.SL^{-1}$)	P_{H_2} (atm)	y_{CO}, in	Regressed Equation[a]	(mol/h/g_{cat})
2.4	5.0	0.0204	11.513	0.25	$X_{CO}=-284.82\ \tau^2 + 250.39\tau$	2.663
	20.0	0.0889	10.568	0.25		2.228
	30.0	0.1431	9.948	0.25		1.884
2.0	5.0	0.0236	9.531	0.25	$X_{CO}=-230.25\ \tau^2 + 217.24\tau$	2.302
	20.0	0.1034	8.554	0.25		1.892
	30.0	0.1680	7.910	0.25		1.560
1.5	5.0	0.0270	7.091	0.25	$X_{CO}=-198.12\ \tau^2 + 160.96\tau$	1.993
	20.0	0.1227	6.153	0.25		1.504
	30.0	0.2097	5.508	0.25		1.060
2.0	5.0	0.0324	4.576	0.25		1.652
	20.0	0.1531	3.664	0.25		1.119
	30.0	0.2896	3.181	0.25		0.515

a. From Figure 5.

Similarly, at constant H_2 partial pressure, the CO conversion as a function of space time is shown in Figure 7, while the regressed equations are summarized in Table 5. The r_{CO} and P_{CO} at constant P_{H_2} (0.796 MPa) calculated from the regressed equations are provided in Table 6. The regressed equations are given in Table 8. The plot of ℓnr_{CO} vs ℓnP_{CO}, with data obtained from a series of H_2/CO ratio experiments in which P_{H_2} was kept constant and P_{CO} was varied, is displayed in Figure 8. This plot shows a good correlation coefficient (R-squared value), and the CO concentration dependency (a) is -0.2, suggesting inhibition of the rate by CO adsorption. The data for holding P_{CO} and P_{H_2} constant are summarized in Tables 7 and 8, respectively.

Thus the kinetic expression becomes:

$$r_{CO} = \frac{kP_{CO}^{-0.2}P_{H_2}^{0.5}}{(1 + m\frac{P_{H_2O}}{P_{H_2}})} \tag{7}$$

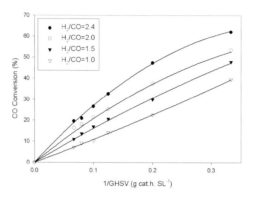

Figure 7. CO conversion as a function of space time at constant H_2 partial pressure 0.796 MPa (7.88 atm) over 25 wt.% Co/Al$_2$O$_3$ catalyst (T = 493 K, 1.99 MPa (19.7 atm) total pressure).

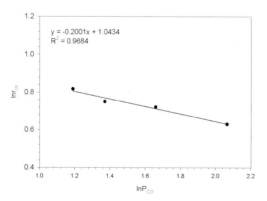

Figure 8. Plot of ℓnr_{CO} vs. ℓnP_{CO} at 493 K, 1.99 MPa (19.7 atm) total pressure and a P_{H_2} of 0.796 MPa (7.88 atm).

Table 8. P_{H_2} and r_{CO} Calculated from Regressed Equations at Different Conversion Levels (0.796 MPa (7.88 atm)).

H_2/CO	% CO Conv.	Space Time (τ_1) ($g_{cat}.h.SL^{-1}$)	P_{CO} (atm)	CO fraction in inlet gas	Regressed Equation[a]	(mol/h/g_{cat})
2.4	5.0	0.0168	3.134	0.167	$X_{CO}=-352.59\ \tau^2 + 304.23\tau$	2.179
	20.0	0.0717	2.669	0.167		1.890
	30.0	0.1136	2.351	0.167		1.670
2.0	5.0	0.0215	3.811	0.2	$X_{CO}=-233.17\ \tau^2 + 237.18\tau$	2.027
	20.0	0.0928	3.352	0.2		1.7630
	30.0	0.1480	2.996	0.2		1.500
1.5	5.0	0.0294	5.098	0.267	$X_{CO}=-90.283\ \tau^2 + 172.74\tau$	1.995
	20.0	0.1238	4.557	0.267		1.792
	30.0	0.1932	4.146	0.267		1.642
2.0	5.0	0.0467	7.719	0.4	$X_{CO}=38.938\ \tau^2 + 105.14\tau$	1.941
	20.0	0.1784	7.182	0.4		2.124
	30.0	0.2603	6.799	0.4		2.238

a. From Figure 7.

3.3.2. Determination of k and m

Equation (7) was linearized to the following form:

$$\frac{P_{CO}^{-0.2} P_{H_2}^{0.5}}{r_{CO}} = \frac{1}{k} + \frac{m}{k} \times \frac{P_{H_2O}}{P_{H_2}} \tag{8}$$

The values of the rate constant k and water effect parameter m were calculated from the slope and intercept of the plot of $P_{CO}^{-0.2} \bullet P_{H_2}^{0.5} / r_{CO}$ vs P_{H_2O} / P_{H_2} over the entire range of experimental data as shown in Figure 9. The k and m values found from the above plot were 1.16 (moles/h/g cat) and 0.93, respectively.

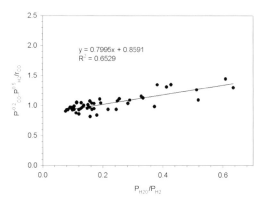

Figure 9. Plot of $P_{CO}^{-0.2} \bullet P_{H_2}^{0.5} / r_{CO}$ vs. P_{H_2O} / P_{H_2} for the determination of m and k over the entire range of experimental data.

Thus, the final form of the empirical kinetic rate expression is:

$$r_{CO} = \frac{1.16\, P_{CO}^{-0.2} P_{H_2}^{0.5}}{(1 + 0.93 \frac{P_{H_2O}}{P_{H_2}})}\,(mole\,/\,h\,/\,g_{cat}) \qquad (9)$$

4. Discussion

The power law kinetic equation could be a simplified form of a mechanistic scheme. A summary of some of the reported reaction orders for the partial pressure of hydrogen and carbon monoxide which have been obtained from power law fits by different groups are listed in Table 9. The partial pressure dependencies vary rather widely. The power law fits were obtained for different cobalt catalysts prepared using different supports and methods. The data in Table 9 show that there is not one best power law equation that would provide a good fit for all cobalt catalysts. Brotz [10], Yang et al. [12] and Pannell et al. [13] defined the Fischer-Tropsch rate as the moles of hydrogen plus carbon monoxide converted per time per mass of catalyst (r_{CO+H_2}). Wang [54] and Zennaro et al. [21] reported the rate as the conversion of carbon monoxide (r_{CO}). In the absence of the water gas shift reaction, the relation between the rates r_{CO+H_2} and r_{CO} differ only by a constant. The reaction order of 0.5 for H_2 obtained in our study is very similar to those reported in previous kinetic studies of FT synthesis on supported cobalt catalysts [13,15,21]. The order for CO in our study was found to be -0.2 which is in agreement with Zennaro et al. [21] and Ribeiro et al. [20]. The latter workers reviewed several kinetic studies and corrected the data to 473 K P_{tot} to 10 atm, and H_2/CO to 2.0 [36]. They obtained a simple power law $R_{CO} = kP_{H_2}^{0.7} P_{CO}^{-0.2}$, very close to the empirical relation reported here.

Table 9. Comparison of Reaction Orders for the Various Power Law Fits

Source	Reaction Order for CO (a)	Reaction Order for H_2(b)
Brotz [10]	-1.0	2.0
Yang et al. [12]	-0.5	1.0
Pannell et al. [13]	-0.33	0.55
Wang [15]	-0.5	0.68
Ribeiro et al. [20]	-0.0.2	0.7
Zennaro et al. [21]	-0.24	0.74
This Study	-0.2	0.5

The form of equation 4 was chosen on the assumption of inhibition by water. However, a large negative value of m will make the rate become negative and, while steam reforming is theoretically possible, it does not occur at a reasonable rate at this low temperature with FT catalysts. Thus, a power law equation of the following form was fitted with the data:

$$r_{CO} = -kP_{CO}^a P_{H_2}^b P_{H_2O}^c \tag{10}$$

The powers for a and b were taken as -0.2 and 0.5, respectively. The data fit was not very good ($R^2 = 0.486$) and slope yielded a value for c of -2.9. Using the values of added water partial pressure in the above power rate law (equation 10) to calculate the effect water addition should have on the rate did not reproduce the experimental data shown in Figure 3. Thus, further efforts to use equation 10 were abandoned.

The value for m together with the values for a and b were used to calculate the effect that the amounts of water added (shown in Figure 3) should have on the rates. The rates calculated using equation 9 fit the data very well in showing essentially no effect of water addition up to about 20 volume% and then a decrease in conversion for the 25 and 30% water addition data (Figure 10). Even though the value of *m* appears to be relatively large and should impact the rate over most of the CO conversion range, the values of

P_{H_2O} / P_{H_2} must off-set the impact of *m* at the lower conversion levels. One reason for this may be the effect of the increase of the WGS reaction with the addition of increasing amounts of water. This effect of water addition is shown in Figure 11 where CO_2 is nearly constant with time-on-stream, as it should be for nearly constant conversion levels, but shows increasing fractions of CO_2 with increasing water addition. One cannot rule out the possibility that this additional hydrogen may have a secondary impact on the rate constants that have been obtained.

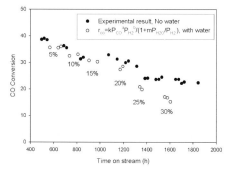

Figure 10. Comparison of the experimental data (closed circles) and the values calculated using equation [9] with m = 0.93 (open circles).

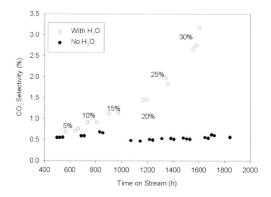

Figure 11. Production of carbon dioxide during periods when water was not added (closed circles) and when the vol% of the feed waqs added water (open circles).

The equations used to derive the reaction orders shown in Table 9, reference [10] and our equation 9 have been used to calculate the CO conversion after adjusting the rate constant of each equation to give a common conversion at a contact time of 0.3 arbitrary units (Figure 12). Even at 20% CO conversion the results for the various equations begin to deviate and this deviation increases with increasing conversion. Thus, with this catalyst, the correction for the reversible effects of water is needed.

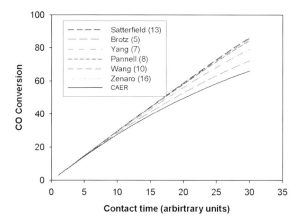

Figure 12. Calculated CO conversion using equations for references in Table 9, reference [10] and our equation [9] after adjusting k for each equation to give a common conversion at an arbitrary contact time of 0.03.

As with many attempts to calculate a constant from a linear plot such as, for example, the Linear Free Energy Relationship (LFER) plot frequently applied by organic chemists, the few end points largely define the value. This is the case for the plot used to calculate m and c. The data obtained for three conversion levels (up to 10 vol% water addition, up to 20 vol% water addition and all data) have been used to calculate values for k and m (Table 10). The rate constant calculated for the three cases is a constant ($k = 1.115 \pm 0.045$) while the values of m differ. The value of m for less than 10 vol% water addition is small (0.21), the value for less than 20 vol% water addition is intermediate (0.65), while a value of 0.93 is obtained when all of the data are used. While it is apparent that a water term is needed for the kinetics to describe the effect of water during FTS with a cobalt catalyst, defining the exact equation will be a demanding task.

Table 10. Values for the Rate Constant and m Calculated using Various Values for P_{H_2O} / P_{H_2}

Values for P_{H_2O} / P_{H_2}	k	m
< 0.2	1.0718	0.21
< 0.3	1.1249	0.65
All data	1.1640	0.93

5. Conclusions

The addition of water at higher levels in FT synthesis decreased the CO conversion but the activity recovered after water addition was terminated. A rate expression has been obtained for a 25 wt.% Co/Al_2O_3 catalyst operated in a 1 liter continuous stirred tank reactor (CSTR) at 493K, 19.7 atm. (1.99 MPa), over a range of reactant partial pressures. The data of this study are fitted by a simple power law expression of the form:

$$r_{CO} = kP_{CO}^{-0.2} P_{H_2}^{0.5} / (1 + mP_{H_2O} / P_{H_2})$$

where k =1.16 (moles/h/g cat) and m = 0.93 for the conditions used in this study.

6. Acknowledgments

This work was supported by US DOE contract number DE-AC22-94PC94055 and the Commonwealth of Kentucky.

- ## References

1. A.M. Hilmen, D. Schanke, K.F. Hanssen and A. Holmen, Appl. Catal. 186 (1999) 169.
2. D. Schanke, A.M. Hilmen, E. Bergene, K. Kinnari, E. Rytter, E. Adnanes, and A. Holmen, Catal. Lett. 34 (1995) 269.
3. S. Storsaeter, O. Bort, E. A. Blekkam and A. Holmen, J. Catal. 231 (2005) 405.
4. P.J. van Berge, J. van de Loosdrecht, S. Barradas and A. M. van der Kraan, Symp.on Syngas Conversion to Fuels and Chemicals, Div. Petrol. Chem., 217th Natl. Mtg. of the ACS, Anaheim, CA, March 21-25, 1999, p. 84.
5. P.J. van Berge, J. van de Loosdrech, S. Barradas and A.M. van der Kraan, Catal. Today 58 (2000) 321.
6. J. Li, G. Jacobs, T.K. Das and B.H. Davis, Appl. Catal. 233 (2002) 255.
7. J. Li, X. Zhan, Y. Zhang, G. Jacobs, T. K. Das and B.H. Davis, Appl. Catal. 228 (2002) 203 .
8. G. Jacobs, T.K. Das, P.M. Patterson, J. Li, L. Sanchez and B.H. Davis, Appl. Catal., A: Gen. 247 (2003) 335.
9. G. Jacobs, P. M. Patterson, T. K. Das, M. Luo and B. H. Davis, Appl. Catal., A: Gen. 270 (2004) 65.
10. W. Brotz, Z. Eletrochem, 5 (1949) 301.
11. R.B. Anderson in: Catalysis P.H. Emmett, (Ed) Vol. 4, Reinhold, New York, 1956, p 247.
12. C.H. Yang, F.E. Massoth, A.G. Oblad, Adv. Chem. Ser. 178 (1979) 35.
13. R.B. Pannel, C.L. Kibby and T.P. Kobylinski, Proceed. of 7th Intl. Cong. on Catalysis, Tokyo, (1980) 447.
14. A.O.I. Rautavuoma and H.S. van der Baan, Appl. Catal. 1 (1981) 247.
15. J. Wang, Physical, chemical and catalytic properties of borided cobalt Fischer-Tropsch catalysis, Ph.D. Thesis, Brigham Young University, Provo, UT, 1987.
16. B. Sarup and B.W. Wojciechowski, Can. J. Chem. Eng. 74 (1989) 62.
17. B.W. Wojciechowski, Catal. Rev. Sci. Eng. 30 (1988) 629.
18. I.C. Yates and C.N. Satterfield, Energy & Fuels 5 (1991) 168.
19. E. Iglesia, S.C. Reyes and S.L. Soled in "Computer-Aided Design of Catalysts", Chapter 7, p. 199 (R. E. Becker and C.J. Pereira, Eds.) Marcel Dekker, New York, 1993.
20. F.H. Ribeiro, A.E. Schach von Wittenau, C.H. Bartholomew and G.A. Somorjai, Catal. Rev.-Sci. Eng. 39 (1997) 49.
21. R. Zennaro, M. Tagliabue and C. H. Bartholomew, Catal. Today 58 (2000) 309.
22. B. Jager and R. Espinoza, Catal. Today, 23 (1995) 17.
23. H. Schulz, M. Claeys and S. Harms, Stud. Surf. Sci. Catal. 107 (1997) 193.
24. H. Schulz, E. van Steen and M. Claeys, Stud. Surf. Sci. Catal. 81 (1994) 455.
25. C. J. Kim, U.S. Patent 5,227,407 (1993) to Exxon Res. Eng. Co.
26. E. Iglesia, Appl. Catal. 161 (1997) 59.

27. J. Li, G. Jacobs, T.K. Das, Y. Zhang and B.H. Davis, Appl. Catal. 236 (2002) 67.
28. S. Krishnamoorthy, M. Tu, M. P. Ojeda, D. Pinna and E. Iglesia, J. Catal. 211 (2002) 433.
29. T.K. Das, W.A. Conner, J. Li, G. Jacobs, M.E. Dry and B. H. Davis, Energy & Fuels 19 (2005) 1430.
30. B. Jongsomjit, J. Panpranot and J. G. Goodwin, Jr., J. Catal. 204 (2001) 98.
31. B. Jongsomjit, C. Sakdamnuson, J. G. Goodwin, Jr., Catal. Lett. 94 (2004) 209.
32. A. Sirijaruphan, A. Horvath, J. G. Goodwin, Jr. and R. Oukaci, Catal. Lett. 91 (2003) 89.
33. R.L. Espinoza, J.L. Visagie, P.J. van Berge, F.H. Bolder, U.S. Patent 5,733,839 (1998).
34. G. Jacobs, T. K. Das, Y. Q. Zhang, J. Li, G. Racoillet and B. H. Davis, Appl. Catal. A: Gen. 233 (2002) 262.
35. W-J. Wang and Y-W. Chen, Appl. Catal. 77 (1991) 223.
36. G.P. van der Laan and A.A.C.M. Beenackers, Catal. Rev.-Sci. Eng., 41 (1999) 255.

Fischer-Tropsch Synthesis, Catalysts and Catalysis
B.H. Davis and M.L. Occelli (Editors)
Published by Elsevier B.V.

Early entrance coproduction plant – the pathway to the commercial CTL (coal-to-liquids) fuels production

John Shen[a],* and Edward Schmetz[a], Gary J Stiegel[b], John C Winslow[b], Robert M Kornosky[b], and Diane R Madden[b], Suresh C Jain[c]

[a]US Department of Energy (DOE), Germantown, MD 20874
[b]US DOE, National Energy Technology Laboratory (NETL), P.O. Box 10940, Pittsburgh, PA 15236
[c]US DOE, NETL, P.O. Box 880, Morgantown WV 26505

In coproduction mode, ultra clean CTL (coal-to-liquids) fuel is coproduced with electric power, chemicals, and steam by incorporating a Fischer-Tropsch (F-T) sub-system into an IGCC (integrated gasification combined cycle) complex. The near-commercial readiness of the IGCC technology suggests this mode could offer the earliest opportunity for the commercial production of CTL fuels. In 1999, the U.S. Department of Energy (DOE) initiated a solicitation seeking proposals to evaluate the feasibility of EECP (early entrance coproduction plant) projects which are aimed at demonstrating the advanced CTL technologies in integrated mode at pre-commercial or commercial scale units. This paper will review the highlights of the two EECP projects awarded by DOE. Successful implementations of EECP projects can reduce technical risks and are considered essential for the undertaking of commercial projects based on advanced CTL technologies.

1. Introduction

Ultra clean transportation fuels can be made from coal via four sequential steps including coal gasification, syngas cleanup, F-T (Fischer-Tropsch) synthesis, and F-T product workup. These fuels are known as CTL (coal-to-liquids) fuels to differentiate them from GTL (gas-to-liquids) and

BTL (biomass-to-liquids) fuels. Significant progress has been made since the late 1970s to improve CTL technologies in the first three steps (1, 2, 3). Based on these advanced technologies, it is estimated that CTL fuels could be competitive with crude at USD $36-42 per barrel depending on the configuration of the plant, economic assumptions and if carbon is contained (4).

One of the hurdles for the commercial CTL fuels production is technical risk, since the operation of advanced CTL technologies in integrated mode has not been commercially demonstrated. In response to this concern, DOE (the U.S. Department of Energy) issued an EECP (early entrance coproduction plant) solicitation in 1999 seeking proposals to address the issues related to the implementation of commercial demonstration efforts. In coproduction mode, a F-T subsystem is incorporated into an IGCC (integrated gasification combined cycle) plant to co-produce electric power, CTL fuels, chemicals, and steam. With the IGCC technology already commercially proven, this strategy is seen to offer the earliest opportunity for the commercial production of CTL fuels. A schematic diagram of the coproduction plant is shown in Figure 1. More discussions on the coproduction operations can be found elsewhere (3). This paper will discuss the highlights of the two cost-sharing EECP contracts awarded by DOE.

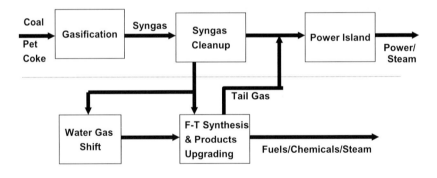

Figure 1. CTL (coal-to-liquids) fuels production via coal gasification – the coproduction mode.

1.1. EECP Solicitation by DOE

The objective of the EECP solicitation was to seek proposals to perform feasibility studies for CTL coproduction projects at pre-commercial or commercial scale units. The work scope includes three phases. Phase I is the project conception definition and planning for further R&D and testing needed to fill the data gaps. Phase II is to perform experimental work identified under Phase I. Phase III is preliminary engineering design for the project. The proposers were encouraged to form teams with members of complementary expertise to reduce capital cost and share the risks. Two EECP contracts were awarded: the first one to Waste Management and Processors, Inc. (WMPI) and the other to Texaco Energy Systems LLC (Texaco).

1.2. EECP Contract with WMPI

The current EECP contract participants include WMPI, Shell Global Solutions, Sasol, ChevronTexaco, Uhde GmbH, and Nexant, Inc. Shell and Uhde are the two new members added after the contract modification in May 2003. Shell will provide the gasification technology which replaces the Texaco gasification technology in the original plan. Uhde is to serve as the general engineering contractor.

The EECP project will be located adjacent to an existing power plant owned by WMPI in Gilberton, Pennsylvania. It will process 4,700 tons per day of eastern Pennsylvania anthracite coal waste (culm) to produce 3,732 barrels per day (b/d) of upgraded CTL diesel, 1,281 b/d of stabilized CTL naphtha, and 39 MW of electric power for export. Other products will include steam and sulfur. The gross plant efficiency is estimated to be 45% (5). A schematic flow diagram of the EECP is shown in Figure 2.

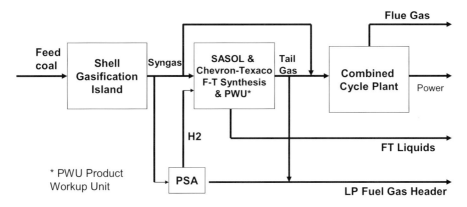

Figure 2. EECP project with WMPI – overall process configuration.

The anthracite coal waste feed will first be cleaned to reduce the ash content to 20%. The ash is high in silica and alumina contents and has high fusion temperature (> 1538°C). The properties of anthracite coal feed are listed in Table 1. Other feedstocks for EECP under considerations include mixtures of culm and other fuels including petroleum coke.

Table 1

Anthracite Coal Culm Characteristics

Proximate analysis, wt%		Ultimate Analysis, wt% dry	
Moisture	1.92	Carbon	72.64
Volatile Matter	7.21	Hydrogen	2.32
Fixed Carbon	71.25	Nitrogen	0.87
Ash	19.62	Sulfur	0.38
		Chloride	--
		Oxygen	3.89
		Ash	20.00
Gross Heating Value, Btu/lb (dry basis)	11,119		

In the gasification section, the replacement of Texaco with Shell gasifier is considered a plus for the project since the latter is operated at a higher temperature and is more suitable for gasifying high-ash culm feedstock. Two 50% gasifiers will be used instead of a single large gasifier to accommodate the high ash content in feed and to improve the plant RAM (reliability, availability, and maintenance). The coal slag removal from the gasifier will be aided by the addition of fluxant to the culm feed stream. In the syngas cleanup section, some of the experiences learned from the Puertollano IGCC plant in Spain will be incorporated. In F-T synthesis, the SSPD (Sasol Slurry Phase Distillate) technology will be employed, using iron catalysts. The syngas from coal gasifier will first undergo sour water-gas shift to increase the H_2/CO (hydrogen/carbon monoxide) ratio to 1.5 and then will be cleaned by the Rectisol process before being fed to the F-T reactor. The tail gas from the F-T reactor will be recycled to the feed stream to improve overall syngas conversion. The F-T reactor will be shop fabricated and transported to the site by water and land. It will have a diameter of 5.49 m (18-feet), as constrained by the transportation regulations. These regulations dictate that equipment up to 6.10m (20-feet) in width, 5.79 m (19-feet) in height, 91.44 m (300-feet) in length and 300 tons net can be transported to the Gilberton site. In F-T product workup section, hydrogen separated from the syngas-fed PSA (pressure swing adsorber) unit will be used. Oxygenates from the F-T reactor will be separated out in the water treatment step and then sent to the gasifier. The product slate from the plant will include naphtha, distillate, fuel gas, and sulfur.

Under Phase I, Uhde has begun the EECP design modification based on Shell gasification. Results from the earlier design based on Texaco will be used wherever applicable. This modified design will have enough details to supersede the preliminary engineering design planned for Phase III. The deliverable will be a LSTK (lump-sum-turnkey)-Offer EECP design package ready for EPC (engineering, procurement, and construction) bidding. Under Phase II, work will include feed characterization and Shell gasification process modeling based on anthracite culm feed and other feed mixtures. These results will enable Shell/Uhde to proceed with the gasifier design without the need for further pilot plant testing. Under Phase III, work will include project financing plan and EECP test plan based on the results of the LSTK-Offer design. On the environment side, work has continued on the preparation of NEPA (National Environmental Policy Act) documents and the applications with the PADEP (Pennsylvania Department of Environmental Protection) for various emissions permits. In March 2005, the PADEP approved the air permits for the project.

Progress on this project has been slowed by the inability to secure a LSTK-Offer EECP design contract between DOE/WMPI and Uhde. Intellectual property rights remain the issue. The project completion date has been extended to the end of December 2006 with a no-cost extension because of the delay. Separately, WMPI has been continuing to work with other parties for the off-take agreements of products from the project (6). This is a key requirement for project financing to obtain needed private sector funding for the project. More detailed discussions for this project can be found elsewhere (7).

1.3. EECP Contract with Texaco

The participants for the EECP contract included Texaco, Rentech, GE, Praxair, and Kellogg Brown & Roots. Texaco was the gasification technology provider, Rentech the F-T technology provider, GE the gas turbine technology provider, Praxair the air separation technology provider, and Kellogg Brown & Roots the engineering/technical services provider. The contract was completed in 2003.

1.3.1. Phase I Work

Phase I of the contract was completed in 2001. Petroleum coke was selected as the feedstock because of its low cost. The study includes two project host sites: one at a petroleum refinery (Motiva refinery at Port Arthur, Texas) and the other at an IGCC complex (Tampa Electric Company's Pole Power Generation Station near Tampa, Florida). For each site, two cases were developed with different process schemes and product slates. The case of "finished wax products at refinery site" was selected for the conceptual plant design because of its highest investment return. The refinery site has the advantages over the IGCC site because of its lower coke transportation cost and greater infrastructure compatibility. Finished wax products are favored over the transportation fuels because of the higher sale prices. The project will process petroleum coke from the refinery at a rate of 1,235 short tons per day. The product slate includes 457 barrels per day (b/d) of finished F-T wax, 125 b/d of diesel, 35 b/d of naphtha, and 55 MW of electricity for sale. Other salable products include steam, sulfur, nitrogen, and oxygen. The properties of a typical Motiva Port Arthur petroleum coke are listed in Table 2.

Table 2. Petroleum Coke (Motiva Port Arthur) Characteristics

Ultimate analysis, wt% dry basis	
Carbon	88.61
Hydrogen	2.80
Nitrogen	1.06
Sulfur	7.30
Oxygen	0.00
Ash	0.23
Moisture as Received, wt%	8.37
Chloride Content, ppm by weight, dry basis	20
Gross Heating Value, Btu/lb, dry basis	14,848

A schematic flow diagram of the EECP project is shown in Figure 3. Texaco's internal analysis of its gasification database indicates coal and petroleum coke are quite similar and hence, they will perform essentially the same in the Texaco Gasification Process. The syngas from the gasifier has a H_2/CO ratio of 0.76 and is fed to the downstream steps with a split of 25% to F-T reactor and 75% to power block. The former is cleaned to remove both H_2S (hydrogen sulfide) and CO_2 (carbon dioxide), and the latter cleaned to remove the bulk of H_2S with minimal CO_2 removal. A zinc oxide guard bed is placed before the F-T reactor to remove the trace impurities left in the cleaned syngas. The primary F-T wax products are upgraded using Bechtel's Hy-Finishing™ technology. Purchased makeup hydrogen is used in the product workup unit. In the power island, a GE PG6101 (FA) 60 Hz heavy-duty gas turbine generator is used in integrated mode with a two-pressure level HRSG (heat recovery steam regenerator) unit and a non-condensing steam turbine generator. The feed stream to the gas turbine consists of (a) syngas from the gasifier after the H_2S cleanup, (b) tail gas from F-T reactor, and (c) offgas from the syngas cleanup step.

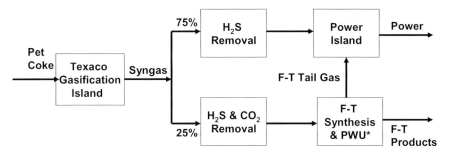

* PWU: Products Workup Unit

Figure 3. EECP project with Texaco – overall process configuration.

The F-T synthesis step uses the slurry-phase reactor technology developed by Rentech utilizing iron catalysts. The syngas feed to this reactor has a H_2/CO ratio of 0.76 which is the same as that from the gasifier exit. This arrangement can improve process efficiency and facilitate the coproduction operations, but will place more demands on the reactor/catalyst system performance (3). The F-T reactor has a diameter of 2.44 m (8-feet) which is considered adequate to provide scaleup data for commercial reactor design. It will be operated in once-through mode with tail gas fed to the power block. The Rentech F-T technology is still under development. The two identified main technical barriers are reactor design and catalyst/wax separation. In reactor design, the concerns include catalyst performance, hydrodynamics, reactor scaleup, and design of reactor internals for heat removal. In catalyst/wax separation, a two-stage system is likely to be needed and more development work for both stages is needed.

Other concerns from the Phase I study include the environmental issues of spent F-T catalyst disposal and the use of F-T water. In addition tail gas from F-T reactor is composed of uncondensed reactor products and unreacted F-T feed gas. Its thermal energy is lower than the syngas. There is no operating experience with the burning of this gas alone in a commercial gas turbine.

Results from a preliminary economic analysis indicate that the EECP project with the case of "finished wax at refinery site" can yield an 11.9% internal rate of return (IRR) with zero cost for the petroleum coke feedstock. For a project of this nature, however, the risk characteristics would require a 15% to 15.5% IRR in order to secure project financing. Therefore, government assistance would be needed to proceed with the project implementation. It should be noted that both the design and economic analysis are closely tied to the Motiva refinery and thus the results are site specific. Other results from the Phase I work include an overall EECP facility availability of 80.8% based on a risk assessment and an estimated gross plant efficiency of 66.5% achieved through the close integration within EECP and also between EECP and host site. More details of the Phase I work can be found elsewhere (3, 8).

1.3.2. Phase II Work

Based on the encouraging results from Phase I, Texaco decided to proceed with Phase II work which was completed in 2003. Under this task, Texaco performed experimental work as described in the Research, Development, and Testing (RD&T) plan prepared under Phase I. Its objective was to mitigate the identified risks in the conception design completed under Phase I. Results from the Phase II work are highlighted below. More details of the Phase II work can be found elsewhere (9).

Petroleum coke slurry: Petroleum coke from the Motiva refinery can be used to prepare slurries that meet the solids concentration, pumpability, and other characteristics required for Texaco proprietary gasification process.

F-T reactor design/operations: Experimental work was performed to mitigate the technical risks identified in Phase I, using the bench-scale units at Rentech (bubble column reactor) and at the University of Kentucky Center for Applied Energy Research (autoclave reactor). Results confirm the F-T reactor performance data at design conditions. The catalyst addition/withdrawal system is effective to maintain steady-state F-T reactor operation. In adding fresh catalysts to the F-T reactor, it is feasible to add them without a pre-activation step. The presence of CO_2 in the feed gas up to 5 volume % is not detrimental to the catalyst performance. Reuse of spent F-T catalysts after either regeneration or rejuvenation is deemed not economical.

F-T mathematical model and reactor scaleup: Texaco's proprietary F-T model was successfully used to fit the experimental data from Rentech bench unit which has a diameter of 3.8 cm (1.5-inch). It was also successfully used to fit the experimental data from LaPorte AFDU (alternative fuels development unit) which has a diameter of 0.56 m (22-inch). Based on these results, Texaco feels this model would be an adequate tool for the design of the EECP F-T reactor with a diameter of 2.44 m (8-feet). It should be noted that the LaPorte AFDU operation was carried out outside the EECP project.

F-T wax/catalyst separation: A 2-stage process was selected to effect the F-T wax/catalyst separation. The primary step uses Rentech's Dynamic Settler technology to reduce the catalyst concentration to less than 0.1 weight % in wax. This technology was selected based on testing performed outside the EECP project. The secondary step uses a LCI Sceptor® Micro-filtration system which is a cross-flow filter. The final clean wax is to have a catalyst content of 10 ppm or less by weight. The wax samples used in the secondary separation testing were prepared from the wax produced from the LaPorte AFDU operation mentioned earlier. Additional R&D and testing will be needed to obtain the long-term performance data for the secondary separation step.

Gas turbine testing: In commercial coproduction plant, one of the operation modes is to send all the syngas from gasifier to F-T reactor in once-through mode and then feed the tail gas to gas turbine. By Texaco's estimate, this tail gas will have a heating value of 89 to 94 Btu/standard cubic feet (SCF) which is much lower than the 300 Btu/SCF syngas used in commercial combustion. Under this task, GE has conducted the combustion of simulated low-Btu gases in its commercial 6FA gas turbine. Results show it is feasible to combust all the gas mixtures including the one with heating value as low as 75 Btu/SCF.

Environmental: Aqueous-phase products from the F-T reactor can be fed into the gasifier as coal-slurry medium. Spent F-T catalyst can be used as fluxant in the gasifier or disposed as non-hazardous landfill if required.

1.3.3. Phase III Work

Texaco decided not to proceed with Phase III work because the Motiva refinery site at Port Arthur, Texas became unavailable after the merger of Texaco and Chevron.

1.4. CTL Commercial Demonstration Project with WMPI

In 2003 WMPI was selected for award negotiation for coal-based coproduction applications under Round I of the DOE CCPI (clean coal power initiative) solicitation. This will be a follow-up of the WMPI EECP contract discussed earlier. Under the contract, WMPI will proceed to construct and operate the EECP plant in Gilberton as discussed above. The first step will be the procurement of an engineering, procurement, and construction (EPC) package based on the LSTK-Offer design by Uhde under the EECP contract. The total project cost is $620 million with a DOE share of $100 million. This project will be eligible for tax credits from the state of Pennsylvania under the Coal Waste Removal and Ultra Clean Fuels Tax Credit Act (Penna S.B. 650). The contract is still under negotiation because of the private financing issues. A recent article indicates that prospect for project financing could be improved with the provisions under the newly enacted Energy Policy Act of 2005 (6).

2. Conclusions

Results from the two DOE-supported EECP projects have been highlighted. The WMPI EECP project, if moved forward to construction and operation under the support of DOE/CCPI program, will represent the first effort to demonstrate the advanced CTL technologies in integrated mode at commercial scale units. The success from this demonstration could pave the way for more commercial CTL fuel production in the U.S. The Texaco EECP project indicates that commercial demonstration of advanced CTL technologies could also be viable at pre-commercial scale units using petroleum coke as feedstock under project specific circumstances.

- ## References

1. S.J. Stewart, G.J. Stiegel, and J.G. Wimer, "Gasification markets and technologies – present and future –an industry perspective", DOE/FE-0447 report, July 2002
2. J.N. Schlather and B. Turk, "Field testing of a warm-gas desulfurization process using a pilot-scale transport reactor system with coal-based syngas", Paper presented at the Pittsburgh Coal Conference, 12-15 September, 2005, Pittsburgh, PA. www.engr.pitt.edu/pcc/
3. J. Shen, E. Schmetz, G.J. Kawalkin, G.J. Stiegel, R.P. Noceti, J.C. Winslow, R.M. Kornorksy, D. Krastman, V.K. Venkataraman, D.J. Driscoll, D.C. Cicero, W.F. Haslebacher, B.C.B. Hsieh, S.C. Jain, and J.B. Tennant, "Commercial deployment of F-T synthesis: the coproduction option", Topics in Catalysis Vol. 26, Nos. 1-4, December 2003
4. E. Schmetz, C.L. Miller, J. Winslow, and D. Gray, Should there be a role for clean liquid transportation fuels from domestic coal in the nation's energy

future?", Paper presented at the Pittsburgh Coal Conference, 12-15 September, 2005, Pittsburgh, PA. www.engr.pitt.edu/pcc/

5. Project facts: Gilberton coal-to-clean fuels and power co-production project, www.netl.doe.gov

6. P. Bortner, Rich project receives boost, 3 August, 2005, www.ultracleanfuels.com

7. WMPI PTY., LLC, Early entrance co-production plant – decentralized gasification cogneration transportation fuels and steam from available feedstocks, Quarterly technical progress report submitted under DOE cooperative agreement DE-FC26-00NT40693, January to March 2002; October to December 2005. www.netl.doe.gov

8. Texaco Energy Systems Inc., Early entrance coproduction plant – Phase I preliminary concept report, submitted under DOE cooperative agreement DE-FC26-99FT40658, 17 May, 2002. www.netl.doe.gov

9. Texaco Energy Systems Inc., Early entrance coproduction plant – Phase II final report", submitted under DOE cooperative agreement DE-FC26-99FT40658, 26 January, 2004 www.netl.doe.gov

Fischer-Tropsch Synthesis, Catalysts and Catalysis
B.H. Davis and M.L. Occelli (Editors)

QA and optimization issues during development of the Statoil FT-catalyst

Erling Rytter, Dag Schanke, Sigrid Eri, Hanne Wigum, Torild Hulsund Skagseth and Edvard Bergene

Statoi Research Centre, Arkitekt Ebbells vei 10, N-7005, Trondheim, Norway

1. Background

Statoil has been involved in Fischer-Tropsch based GTL technology development since the mid 1980's (Rytter et al.,1990). In order to maximize distillate production, a low temperature, cobalt catalyst based Fischer-Tropsch technology has been selected. A slurry bubble column reactor offers the best performance in terms of economy of scale, throughput and yield, but presents several technical challenges. A highly active and selective cobalt catalyst is needed and must be adapted to suit the requirements of the slurry reactor. Separation of wax from the slurry is another critical aspect of this technology. Statoil has developed a supported cobalt catalyst and a continuous filtration technique that forms the heart of the Fischer-Tropsch process.

The present paper will describe a number of issues related to optimization and quality assurance during scale-up and production of the Statoil Co/Re/alumina catalyst.

2. Basic catalyst technology and considerations

The basic catalyst preparation method may consist of aqueous incipient wetness impregnation of a Co/Re solution on alumina giving a nominal (after reduction) composition in the range 12-30 wt% Co, 0,2-1,5 wt% Re and optional additional promoter(s) (Eri et al., 1987). Impregnation is followed by drying, calcination and reduction. In spite of this inherently simple procedure, a number of factors must be considered to obtain the optimal balance between:

- Catalyst cost, ref. recent years increase in Co price
- Activity pr. g catalyst and reactor volume
- Selectivity to C_{5+} products and avoidance of byproducts
- Long-term chemical stability of the catalyst and
- Physical integrity in the slurry process.

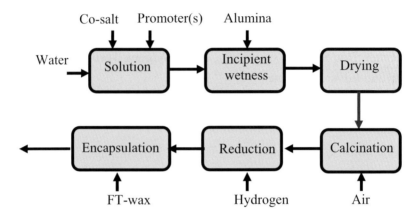

Figure 1. Basics steps in the catalyst production.

To optimize the performance factors, a number of parameters need to be evaluated in the overall production scheme, see Fig. 1. The actual preparation procedure may deviate somewhat from the figure. Although this catalyst production scheme is inherently simple, there are surprisingly many variables that need to be optimized. Some of these variables are:

- Selection of metal precursors, purity and loading.
- Pore characteristics, particle-density, -shape, -size and -size distribution, purity, crystalline phase, modification and attrition resistance of the alumina support (Eri et al., 2000; Rytter et al., 2002a).
- Impregnation technique, (liquid) volumes, number of impregnation steps and drying profiles.
- Calcination profile and hold temperature.
- Reduction procedures, e.g. gas composition and space velocity as well as pressure and temperature.

The catalyst performance varies widely with these parameters and some examples will be given. Note also that equipment selection is critical, like operation of a batch or continuous process, as well as tray, rotary or fluid-bed reactor for some of the steps. In fact, dedicated equipment may be needed to ensure efficient production and high quality. Further aspects to be considered are the operating conditions of the FT-reactor, especially the H_2/CO ratio and the conversion level. The latter is particularly important as it determines the water partial pressure which is closely related to the catalyst activity, selectivity and stability.

3. Alumina supports

A particularly important feature is the nature of the alumina support. The pore size distribution of a few selected γ–alumina samples are given in Fig.2. All are high surface area materials in the range $180 - 200$ m^2/g. Apart from the evident bimodal distribution for some samples with larger pores, these materials have high pore volumes and low bulk densities combined with high attrition resistance. A high pore volume is particularly beneficial for high-load impregnation, and a low density clearly ease a homogeneous distribution in the reactor slurry. As the catalyst support is a major factor in design of the FT-catalyst, some elaboration is justified.

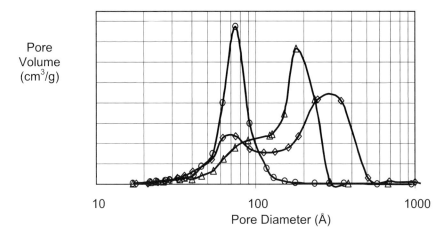

Figure 2. Examples of pore size distributions of γ-alumina.

We have previously described various important aspects of the alumina support including pore size distribution and alumina phase (Rytter et al., 2002b). A few of these effects will be described in further detail. Generally it is found that for comparable γ-aluminas with normal surface areas in the range 170-200 m^2/g, the C_{5+}-selectivity increases with pore volume and pore diameter as depicted in Fig. 3. This effect has also been verified for a larger selection of supports and the nature of it, e.g. in terms of surface diffusion, cobalt distribution or support reactivity, is being looked into.

330

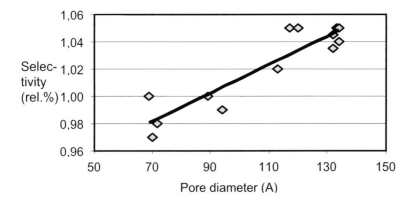

Figure 3. Relative selectivity (C_{5+}) for varying pore diameters of γ-aluminas. Fixed-bed tests at 210 °C and 20 bar with H_2/CO inlet ratio of 2.1.

However, as the density decreases, other properties like strength and morphology might be adversely affected; ref. Fig. 4. Here the upper curve depicts the behavior of a standard alumina material, whereas the next one shows that a stronger material can be made, even for a high pore volume support. The other lines illustrate how proper treatment may increase the attrition resistance further. One factor that may influence the strength is any inclusions or second phases as illustrated in Fig. 5, although the detailed correlations can be difficult to unravel.

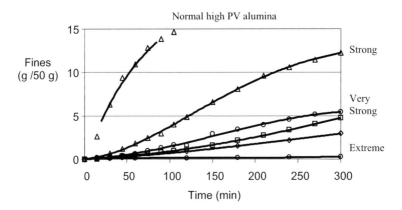

Figure 4. Variation in attrition resistance as measured by the ASTM air-jet method for different aluminas and modified aluminas.

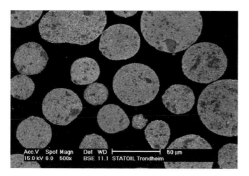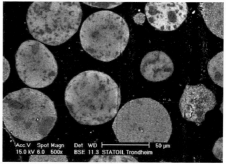

Figure 5. Cross-sections of alumina support (left) and catalyst particles (right).

4.   Modified alumina

Addition of modifiers can result in an extremely strong catalyst carrier (lowest line in Fig. 4). In this way it can be possible to strengthen the material independently of some other parameters, notably pore volume and/or diameter. However, still the total pool of alumina properties needs to be harmonized to the slurry reactor operation.

A particularly striking effect of the type of alumina is seen on the selectivity for alumina calcined at successively higher temperatures (Fig. 6), yielding low and very-low surface areas as more and more of the γ-alumina is transformed to other transition alumina forms and finally to α-alumina. The figure includes catalysts of different Co loadings as well as some that has been chemically modified. Pairs of catalysts with and without the Re promoter clearly point to a positive effect of Re on the selectivity. It is interesting that Pt has an adverse effect on this property.

To further illustrate the dramatic difference between γ-alumina and α-alumina, tests of varying particle size were performed in a fixed-bed reactor. It is well known that the selectivity decreases for larger particles due to diffusion limitations. The results indicate no such effect for particles below ca. 250 microns. A parallel selectivity performance is observed in the whole particle size range for both supports. This indicates that the selectivity increases because of a chemical surface effect of the α-alumina and is not due to changes in particle diffusion characteristics.

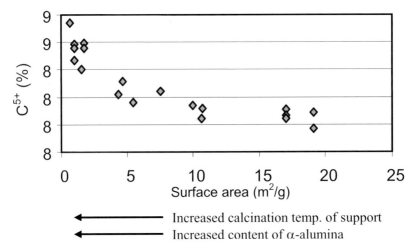

Increased calcination temp. of support
Increased content of α-alumina

Figure 6. Left: Selectivity (C_{5+}) as a function surface area for Co/(Re)/alumina catalysts..

5. Catalyst morphology

During scale-up of the catalyst production, one might experience deviations from ideal catalyst morphology with spherical particles following a normal distribution of particle size. Some artifacts are illustrated in Fig. 7, notably the appearance of a separate distribution of very fine particles and agglomerated particles, respectively. Proper handling and production will eliminate such effects.

Figure 7. SEM micrographs of alumina and catalyst particles showing non-ideal properties, and their rectification. Left: Fine surface particles. Right: Agglomerates.

6. Promoters

It is well known that Statoil has advocated the use of rhenium promoted cobalt on alumina catalysts (Eri et al., 1987; Rytter et al. 1990). Fig. 8 illustrates how one or more promoters may influence the catalyst selectivity. All the catalysts are on the same low surface area α-alumina type support (Eri et al., 2000), with 12 wt% cobalt loading. The filled diamond symbols represent an un-promoted catalyst. It is striking that promoters not only can increase the selectivity level, but actually can enhance the catalytic properties with time on stream. This observation indicates that a beneficial structural rearrangement takes place.

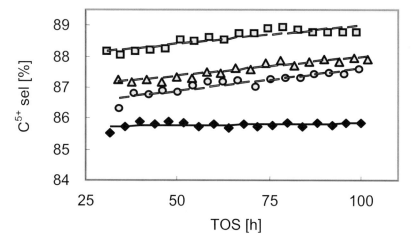

Figure 8. Selectivity (C_{5+}) for promoted catalysts. Fixed-bed tests at 210 °C and 20 bar with H_2/CO inlet ratio of 2.1.

7. Cobalt distribution

The insipient wetness technique allows for a reasonably even distribution of cobalt throughout the catalyst particle cross-section (Fig. 5, right). If one looks at the cobalt distribution within the pores, the situation becomes more complex. First it is noted that there is a significant reduction in typical crystal size as detected by XRD from 10-15 nm of the primary oxide (Co_3O_4) to 5-6 nm of the two-valent oxide (CoO) or Co metal particle, indicating that the oxide cracks during hydrogen reduction. Further, the oxide crystals are assembled in larger agglomerates in the size range 50-200 nm, see Fig. 9.

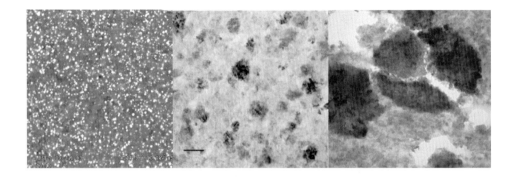

Figure 9. SEM and TEM images of successively larger magnification (left to right) showing cobalt oxide agglomerates as well as individual cobalt oxide particles.

An intriguing question is what the cobalt distribution looks like in a 3-dimentional pore structue. This is not easily unraveled as no good model exists for the pore geometry. However, 3D-TEM and HR-TEM resently has become available and can give us a vision of the pores themselves as well as the cobalt-alumina interphase, Fig. 10 (Midgley et al., 2005). Justification of these 3D plot cannot be given as a two-dimensional gray-scale image, but a video will better visualize the alumina crystals constituting the pores as well as the cobalt oxide agglomerates. It becomes evident that there is not a simple pore structure with the metal oxide impregnated to the walls of the pores. Rather, the alumina crystals (gray) are entangled into each other in what appears to be a caotic fashion and the cobalt agglomerates (light) clearly stretches over a number of alumina crystals and "pores". EELS spectra show that there is good contact between the oxide (green) and the support. This is further demonstrated by the high resolution image in Fig. 10, where one also can envisage an amorpheous section between cobalt oxide, which can be identified by the crystal plane distances, and the alumina. In other words, a rather comprehensive structural catalyst model appears to take form, but more work needs to be done on the reduced catalyst, preferably under operating conditions.

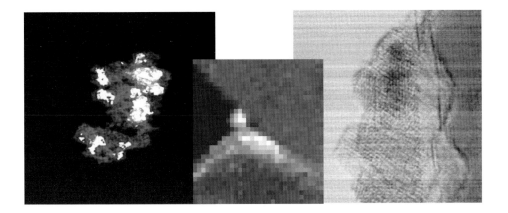

Figure 10. 3D-STEM, EELS and HR-TEM of cobalt on alumina catalysts in the oxide stage.

8. Scale-up of catalyst production

Scale-up of catalyst production from laboratory to full scale is a non-trivial task for any catalyst system. Starting from laboratory recipes for preparation of catalyst at typically <1 kg scale, a stepwise scale-up to a 1 ton/day order of magnitude production has been carried out in cooperation with a commercial catalyst supplier (Johnson Matthey Catalysts). The final catalyst is shipped to the semi-commercial FT-plant in Mossel Bay, SA, as active (reduced) catalyst protected by wax.

Several challenges have been encountered underway and important issues include raw materials selection, development and production of a suitable alumina support as well as selection and adaptation of conditions and equipment for impregnation, drying, calcination and reduction. Early production batches showed clearly inferior performance compared to laboratory samples. This challenge was resolved in collaboration with the catalyst manufacturer after detailed investigations into the basic mechanisms involved in the scale-up.

The performance of the catalyst critically depends on the distribution of Co through the alumina carrier, as well as of Co particle size and distribution, degree of reduction, and pore structure of the alumina. The relationship between these parameters is not trivial, e.g. a high dispersion catalyst is not necessarily a good catalyst in terms of activity, selectivity and stability. Another aspect is the influence of impurities in raw materials, and during catalyst production and

336

operation. An example is shown in Fig. 11 where a moderate amount of an impurity in raw materials will give an inferior catalyst. Clearly such effects have been the cause of some catalytic behavior claimed in the literature.

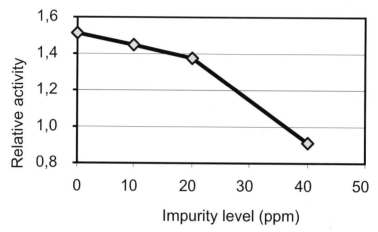

Figure 11. Variation in relative activity for a cobalt/rhenium/ alumina catalyst by adding traces of an impurity component.

9. Conclusion

The Co/Re/alumina system is highly suited for selective production of paraffinic wax in a slurry bubble column. Optimization of the catalyst formulation and quality assurance during catalyst production is vital.

References

Eri, S.; Goodwin, J.; Marcelin, G.; Riis, T., 1987, Catalyst for production of hydrocarbons , US 4.801.573.

Rytter, E.; Solbakken, Å.; Roterud, P. T., 1990, AIChE Spring National Meeting.

Eri, S., Kinnari, K. J.; Schanke, D.; Hilmen, A.-M., 2000, Fischer-Tropsch catalyst with low surface area alumina, its preparation and use thereof, WO 02/47816 A1.

Rytter, E.; Eri, S.; Schanke, D., 2002a, Catalyst and process for conversion of synthesis gas to essentially paraffinic hydrocarbonst, WO/GB03/04873, priority.

Rytter, E.; Schanke, D.; Eri, S.; Wigum, H.; Skagseth, T.H.; Sincadu, N., 2002b, ACS Petroleum Chemistry Division Preprints, 47(1).

Midgley, P.; Arslan, I. Univ. of Cambridge. Shannon, M. SuperSTEM lab., Daresbury. Walmsley, J. Sintef, Trondheim, 2005.

Fischer-Tropsch Synthesis, Catalysts and Catalysis
B.H. Davis and M.L. Occelli (Editors)
© 2007 Published by Elsevier B.V.

Magnetic Separation of Nanometer Size Iron Catalyst from Fischer-Tropsch Wax

R. R. Oder

EXPORTech Company, Inc., P.O. Box 588, New Kensington, PA 15068-0588, USA
724-337-4415 / FAX 724-337-4470, roder@magneticseparation.com

1. Abstract

This paper presents preliminary results using the Magnetic Micro-Particle Separator, (MM-PS, patent pending) which was conceived for high throughput isothermal and isobaric separation of nanometer (nm) sized iron catalyst particles from Fischer-Tropsch wax at 260 °C. Using magnetic fields up to 2,000 gauss, F-T wax with 0.3- 0.5 wt% solids was produced from 25 wt% solids F-T slurries at product rates up to 230 kg/min/m². The upper limit to the filtration rate is unknown at this time. The test flow sheet is given and preliminary results of a scale-up of 50:1 are presented.

2. Background

The novel separation technology (Magnetic Micro-Particle Separator, Patent Pending) evolved from a magnetic method for breaking solids-stabilized emulsions [1] which required implanting a ferromagnetic seed into the internal phase of the emulsion. The internal phase was coalesced in a magnetic field and drawn to collecting magnetic rods or wires where it was withdrawn from the separator under force of fluid flow. This is illustrated in Fig. 1 where the magnetic elements are permanent magnet rods. For the case of water in oil emulsions, iron ligno-sulfonate was used as the ferromagnetic seed. The technology was tested in recovery of organic acids from crud produced in caustic washing of crude oil but not pursued because of technical problems at the time with recovery of the magnetic additive. In the Fischer-Tropsch (F-T) application, however, there is no emulsion and the catalyst particles are magnetic so development of magnetic methods for separation of nanometer (nm) catalyst particles from F-T wax is expected to be straightforward.

338

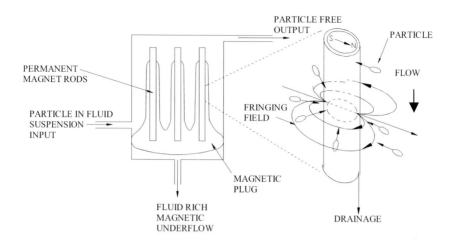

Fig. 1. Magnetostatic Coalescer

The significance of the magnetic technology lies in the fact that the force on magnetic particles, mσ∇H can be strong, where m is the particle mass, σ is the magnetic moment per unit mass, and ∇H is the gradient of the magnetic field at the location of the particle. For example, the force of attraction of a 10 μm diameter magnetic particle located on the surface of an alnico permanent magnet rod is about 100 times that of gravity, i.e., σ∇H/g ≈ 100, where g is the acceleration due to gravity. In the example, σ ≈ 50 emu/g, the alnico magnet is 1/8 inch (3.2 millimeters) in diameter and is magnetized transverse to its length so that ∇H ≈ 300 gauss/cm. In the case of mutual magnetic attraction of submicron size magnetized particles, the field gradient at the surfaces of the particles can be orders of magnitude greater than 300 gauss/cm, resulting in rapid coalescence and producing stable agglomerates.

The MM-PS is significantly different from the coalescence technology in that no rods or wires are present inside the separation chamber. The new approach takes advantage of rapid coalescence of the catalyst particles in an applied magnetic field and overcomes problems associated with plugging that can occur when strongly magnetic materials are present inside the separation chamber. Information has been presented elsewhere on characterization of the F-T catalyst/wax slurry and results of separation with the novel technology [2]. The discussion below presents information on scaling of the technology.

3. Apparatus and Fischer-Tropsch Catalyst Particle Slurries

The Magnetic Micro-Particle Separator (MM-PS) used in this work is shown in Fig. 2 as it was being assembled. The structures on the skid-mount are the electromagnet, power supply and chiller. The three tanks to the right in the photo are the feed, product, and underflow vessels.

Fig. 2. Magnetic Micro-Particle Separator Being Assembled

The apparatus, when completed, was tested in separation of nm iron catalyst particles from F-T wax at 260 °C at rates up to 59 BPD. The specific filtration rate for achieving 0.3-.5 wt.% wax from 20-25 wt.% feed slurry is greater than 230 kg/min/m^2. The upper limit is not known and the process was not optimized because of physical limitations in the apparatus and slurry supply. The process instrumentation and control diagram is shown in Fig. 3.

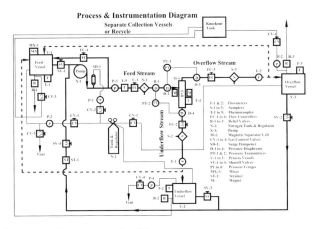

Fig. 3. Process and Instrumentation Diagram

The separation vessels employed in the work are shown in Fig. 4. The vessel lengths are 9, 12 and 16 inches. The volume was scaled up by more than a factor of 52 in the course of the work.

Fig. 4. MM-PS Separators

Catalyst particle images made using computer controlled transmission electron microscopy (TEM) showed that the catalyst particles are actually loose aggregates of very small particles ranging from a few nm to 60 nm in size. From the TEM images, it can be seen that many of the particles were agglomerated or chained together. Fig. 5 shows nm sized particles agglomerated in the MM-PS overflow. The sizes of individual particles range from 20 to 50 nm. Fig. 6 shows larger agglomerates of these particles observed in the MM-PS underflow.

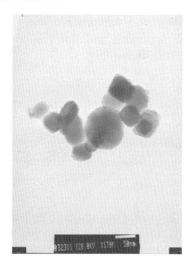 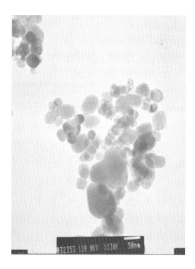

Fig. 5. Overflow Agglomerated Particles   Fig. 6. Underflow Agglomerated Particles

4. Results

Canister Size and Field Effects. The general effects of canister size and magnetic field strength can be seen in Fig. 7, which has been compiled using the results of runs under many different operating conditions. At low overflow rates, generally below 70 kg/min/m^2, changes in the operating variables, i.e., magnetic field strength, canister size, inlet and outlet port dimensions and configurations, inlet flow rate, and recycle ratio (the rate of underflow divided by the rate of overflow), etc., have little effect on overflow ash until an upper level in overflow rate is achieved where further increases in the flow rate make a precipitous increase in overflow ash as can be seen in the figure. The largest overflow rate achieved for the six-inch cell at 2,000 gauss for which the overflow ash was less than 0.5 wt.% was nominally 230 kg/min/m^2. This limit was imposed by the physical limitations in apparatus and supply of catalyst impregnated wax and not by the configuration employed with the six-inch diameter canister.

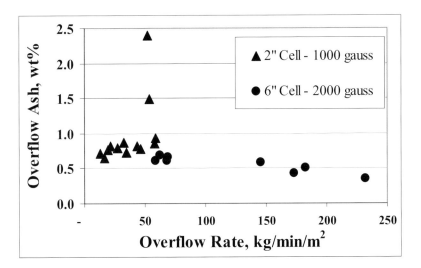

Fig. 7. Overflow Ash vs. Overflow Rate for Different Canister Diameters

Scale-Up. The maximum values of throughput achieved in the work reported here are shown versus canister diameter in the first two rows of Table 1. The range of values shown in the third row are projections based on scaling to 15 gpm product with underflow ten times product. The feed is 165 gpm.

342

Table 1. Scale-up to 15 gpm

		Ash		Product Rate		
ID inches	Feed wt%	Product wt%	Reduction %	gpm	gpm/Ft^2	kg/min/m^2
1.86	18.48	0.86	95	0.04	2.1	56
6.07	21.45	0.35	98	1.43	7.1	221
13 - 20				15	7 - 16	221-517

Projection of these data yields a filtration rate of 517 kg/min/m^2. Such a projection is probably an understatement, however, because a maximum rate was not achieved for the six-inch canister. Accordingly, scale up uses the measured throughput of 221 kg/min/m^2 which yields **very** conservative estimates. The range of 13-20 inches for the canister diameter for processing 15-gpm slurry product is obtained using the projected value and the measured rate of 221 kg/min/m^2, respectively.

The measured value of 221 for the specific filtration rate, kg/min/m^2, shown in Table 1, is 17 times greater than the Stokes settling rate for particles of density 1.75 g/cc in a fluid of 0.78 g/cc density similar to the operating conditions for the F-T separation, nominally 20 wt.% solids. This rate is estimated to be more than 400 times greater than Davis' estimate for the Sasol commercial unit and 95 times greater than that reported by Davis for one run with the University of Kentucky CAER one liter Fischer-Tropsch unit [3]. Table 2 shows an estimate of the size of MM-PS capable of preparing a low solids 15 gpm product. The estimate is based on using the conservative value of filtration, 221 kg/min/m^2. It is interesting to note that the cross-sectional area of the MM-PS could be increased by a factor of 10 and still occupy less than one percent of the volume of the Sasol reactor [3].

Table 2. Separator Footprint

Sasol Slurry Reactor	
Slurry Reactor Volume	288 m^3 (Liquid Fill)
Wax Production Rate	12,960 kg/hr
Magnetic Filter	
Filtration Rate @ 99.6% Catalyst Return	221 kg/min/m^2
Filter Cross-Sectional Area	0.98 m^2
Filter Volume	0.2 m^3
Percent of Slurry Reactor Volume	0.08%

5. Conclusions

Magnetic filtration shows potential for primary separation of nanometer iron catalyst from F-T wax and has achieved very high throughputs when compare to more conventional methods.

6. Acknowledgements

This work was carried out under US Department of Energy Small Business Innovation Research Contract DE-FG02-00ER83008. This support does not constitute an endorsement by the DOE of the views expressed here.

It is a pleasure to acknowledge the help of Russell E. Jamison, Dr. Cynthia A. Znati, Dr. Edward D. Brandner, and John J. St. Clair during the course of this work. The slurry of iron catalyst in Fischer-Tropsch wax from the LaPorte facility was supplied by Chevron/Texaco. The scanning electron microscope images of the catalyst particles were made by the RJ Lee Group of Monroeville, PA.

Introduction

Since the 1970's, environmental legislation regulating air quality has increasingly played an important role in determining automotive emission standards. Initially limited to the introduction of unleaded fuel and catalytic converters for gasoline vehicles, more recent regulation has focused on addressing emission aspects directly linked to fuel composition. With the increasing demand for diesel fuels, principally due to a combination of the inherently greater efficiency of high compression diesel engines (and associated lower CO_2 emissions per km travelled), attention has turned to the regulation of diesel properties most adversely affecting air quality; notably reduced sulphur, aromatics, fuel density and distillation endpoint, and increased cetane number as provided in European, American and Japanese specifications [1]. Of the various alternative fuels possible for use in diesel engines (including biodiesel, dimethylether, LPG, methanol and CNG), synthetic diesel from Fischer-Tropsch (F-T) based gas-to-liquids (GTL) processing is the most promising in terms of all of, potential to produce, quality and engine compatibility.

Although long recognized as a route for the conversion of coal and natural gas to liquids, the Fischer-Tropsch conversion is notoriously unselective, producing a wide product carbon-number distribution. Despite decades of research, efforts to improve the intrinsic selectivity of the F-T synthesis towards desired fractions has met with only limited success. With increasing global interest in F-T processing as a route to both associated and remote gas utilization, and the production of high quality, clean distillate fuels, attention is focused on routes to maximize overall middle-distillate selectivity. Consequently, it is generally accepted that the most effective route currently available for enhanced overall distillate selectivity in F-T based GTL processing, is an approach which makes use of high α-value F-T catalysts generating long-chain, wax, products in the F-T step, followed by cracking of the wax back into the distillate product range [2].

Hydrocracking is a well-established process for the conversion of heavy feedstock into lighter fractions. With origins based initially on coal liquefaction and later on the conversion of heavy gas oils, hydrocracking came into its own only during the 1960's when it was developed as the process of choice for the processing of refractory crude and cycle oils, the latter increasing with the then large expansion in refinery fluid catalytic cracking capacity [3]. Despite its now much wider and more versatile application, hydrocracking process and catalyst developments have almost exclusively been limited to the processing of crude and crude-derived fractions within traditional oil-refining complexes. Moreover, within the US market where this commercialisation drive took place, the desired product was principally gasoline. Consequently, process and catalysts have been optimised for such feedstocks, rich in heteroatoms (sulphur, nitrogen and metals) and generally highly unsaturated and aromatic. These feedstocks contrast greatly with that to be processed in a Fischer-Tropsch GTL environment, where the F-T wax comprises almost exclusively linear, paraffinic hydrocarbons free of sulphur, nitrogen and metals. Additionally, the only desired products of F-T wax hydrocracking are middle-distillate fuels.

Given that the capital cost of the GTL plant contributes a substantial component to the cost of production and that the refining step, including hydrocracking, comprises only 10 % of the overall plant capital investment [4], it is clear that the feedstock to the GTL hydrocracking step is a valuable (expensive) material. In light of the above, it is imperative to ensure optimal yield of the desired distillate product from hydrocracking and, consequently, it is pertinent to review and optimise the F-T wax hydrocracking process.

2.1 Existing and proposed F-T coal and gas to liquids capacity

Both coal and natural gas based F-T plants are operating commercially today. In South Africa, Sasol operated three coal-based plants [5] with a total product capacity of roughly 7 500 x 10^3 tpa in the early 2000's (Sasolburg and Secunda). The smaller, original plant in Sasolburg, dating from the early 1950's, was recently converted to natural gas feed. Two natural gas plants of approximately 1 200 x 10^3 tpa (Mossel Bay, South Africa) and 640 x 10^3 tpa (Bintulu, Malaysia) total product capacity have been in operation by PetroSA and Shell, respectively, for some ten years [6]. In the above plants, all of fixed- (Sasolburg and Bintulu), fluidised- (Secunda), entrained- (Mossel Bay) and slurry-bed (Sasolburg) reactor configurations are employed, as are coal gasification (Secunda), combined primary / autothermal reforming (Mossel Bay) and partial oxidation (Bintulu) syngas generation technologies.

Recently, some 12 plants based on natural gas have been proposed, totalling in the order of 33 000 x 10^3 tpa capacity [6]. Although several of the proposed plants may be considered competitive bids, it is reasonable to expect that at least three commercially viable projects will be implemented still this decade (more than doubling the current installed global capacity), viz. plants of 1 300 x 10^3 tpa in Nigeria (Sasol / Chevron Texaco [6]) and plants of 1 300 x 10^3 tpa (Sasol / QPC [6]) and 5 500 x 10^3 tpa (Shell / QPC [7]) in Qatar.

2.2 Refinery hydrocracking

The subject of refinery hydrocracking has been reviewed in detail elsewhere [3]. It is applied for the conversion of a range of 'heavy' fractions (typically rich in all of sulphur, nitrogen, metals and polyaromatics) to lighter products, mostly diesel, jet fuel and gasoline (the latter mostly in USA). A variety of catalysts have been developed, depending on feedstock and product demands. Hydrogen transfer is mediated via either noble metals (e.g. Pt, Pd) or combinations of various group VIA (Mo, W) and group VIIIA (Co, Ni) metals. An acidic carrier, typically comprising amorphous silica-alumina and zeolites, either alone or in combination, may provide an isomerisation and cracking function. When non-noble metals are used these are present in the form of metal-sulphides in the working catalyst. CoMo-type catalysts may be considered a typical non-noble metals selection, with Ni and W being introduced to provide increased hydrogenation and hydro-denitrification activity as required. Although other formulations are employed in special cases (e.g. residue upgrading, lube oil dewaxing, etc.), the above formulations are typical for the hydrocracking of vacuum gas oils (VGO) and FCC cycle oils.

In the case of maximum distillate yield processing, sulphided, non-noble metal formulations on silica-alumina, are preferred. Mild hydrocracking catalysts are similar but often employ an even milder acid function by carrier dilution or replacement with alumina. Zeolites (almost exclusively zeolite Y) are employed principally to achieve 'severe' hydrocracking, as is the case when high gasoline selectivity is desired and / or the feedstock is highly refractive. In this case, noble metals, especially platinum, are applied so as to increase the catalyst hydrogen transfer activity and so provide an appropriate cracking / hydrogenation balance for the intended application.

2.3 F-T wax hydrocracking

Considering published findings, it appears as if only very limited research and development has been committed to the special case of distillate production via F-T wax hydrocracking, e.g. [8, 2, 5, 9]. Results, for total wax conversion, reported by Sasol [5] and UOP [9], suggest that middle-distillate yields of the order of 80 wt-% may be achieved. These results are consistent with calculated values based on F-T wax with α-values of approximately 0.95 [2] (hydrocracking distillate extracted from data presented for the overall Shell Middle Distillate Synthesis process and direct F-T synthesis product composition). On the one hand, these results suggest a significant opportunity for improved middle-distillate selectivity in the wax hydrocracking process, especially considering that the non-selective products are poor quality gasoline / naphtha (15 wt-%) and $C_1 - C_4$ gas (5 wt-%) [5]. This is all the more pertinent when one considers the cost to produce the wax (inclusive of 90 % of the overall GTL plant capital investment prior to hydrocracking [4]), such that even modest improvements in the hydrocracking middle-distillate yield are likely to significantly impact on process economics. On the other hand, if indeed these distillate selectivities are essentially representative of the kinetically expected distribution, then the opportunity for improved middle-distillate production may be limited.

However, given the nature of the F-T wax feedstock and the ever more stringent distillate fuel specifications, targets for process optimisation, although strongly influenced by overall yield, are not limited only to issues of boiling range.

A further opportunity for process optimisation involves the judicious selection of reaction conditions (temperature, pressure and hydrogen / hydrocarbon ratio) so as to promote the transfer of primary middle-distillate product into the vapour phase with a view to reducing its residence time in the catalyst bed and, consequently, preventing distillate loss via secondary cracking. This approach to improved overall distillate selectivity has previously been referred to by Eilers *et al.* [2] and provides the incentive to optimise catalyst performance within the constraints of the operating window so defined.

2.4 Comparison of refinery and F-T wax hydrocracking

Whereas a wide choice of catalysts exists for hydrocracking, current commercial catalyst formulations have been developed and optimised for use in typical crude oil refinery settings. Consequently, the choice of metal function is driven not only by considerations of cost but also by technical constraints due to the conditions applicable.

As an example, noble metal catalysts are effective only when sulphur levels are below approximately 500 ppm. Hence, such catalysts are only applicable in two-stage hydrocracking processes with interstage removal of H_2S – they are not appropriate to single-stage hydrocrackers or two-stage units without intermediate H_2S removal [3]. Such limitations will not apply to F-T wax processing.

Indeed, in the case of F-T wax hydrocracking the feedstock contains negligible sulphur and nitrogen, and is almost completely saturated, comprising essentially linear paraffins (only in the case of Fe-derived F-T wax may low levels of olefins and oxygenates be present) [5]. Under these conditions both metal (in the case of noble-metals) and acid functions are un-poisoned by sulphur and nitrogen, respectively, and catalyst activity is likely to be significantly higher.

Also, whereas medium-pore zeolites, e.g. MFI-types (ZSM-5), have only limited application in refinery hydrocracking (e.g. selective de-waxing applications), due to their inability to process large, 'bulky' molecules, not only are these zeolites applicable to the linear paraffinic nature of F-T wax but may even be of advantage in limiting the extent of branching possible and so serve to maintain molecular linearity and its associated high cetane number properties.

352

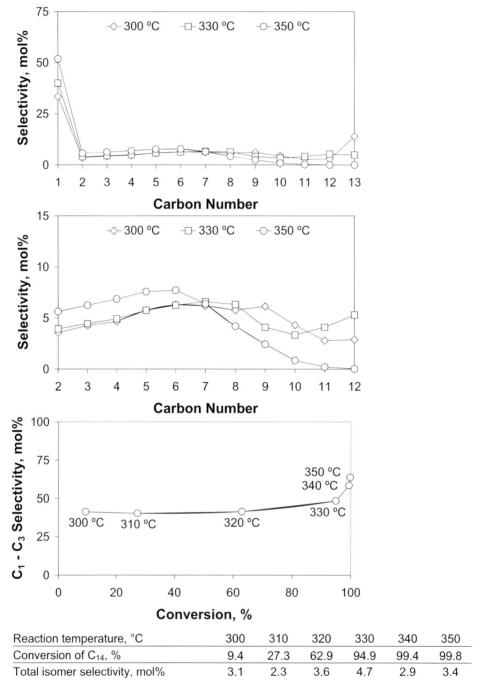

Figure 1: Effect of reaction temperature (p=80 bar, LHSV=0.2 1/h, H_2/n-C_{14}=116)

Reaction temperature, °C	300	310	320	330	340	350
Conversion of C_{14}, %	9.4	27.3	62.9	94.9	99.4	99.8
Total isomer selectivity, mol%	3.1	2.3	3.6	4.7	2.9	3.4

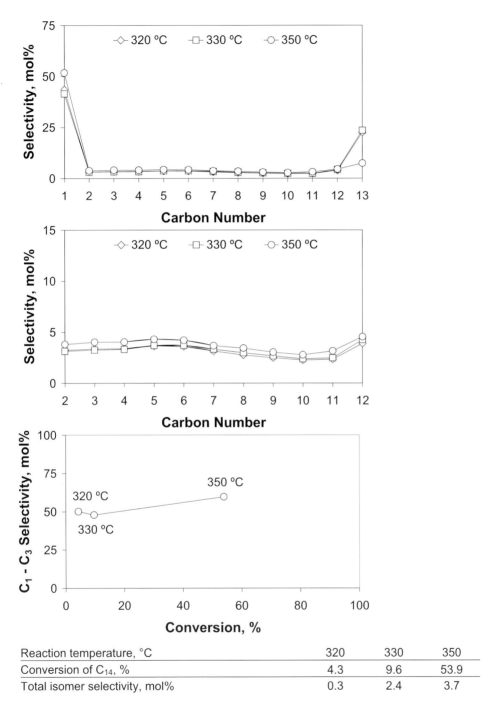

Reaction temperature, °C	320	330	350
Conversion of C_{14}, %	4.3	9.6	53.9
Total isomer selectivity, mol%	0.3	2.4	3.7

Figure 2: Effect of reaction temperature (p=80 bar, LHSV=1.3 1/h, H_2/n-C_{14}=10)

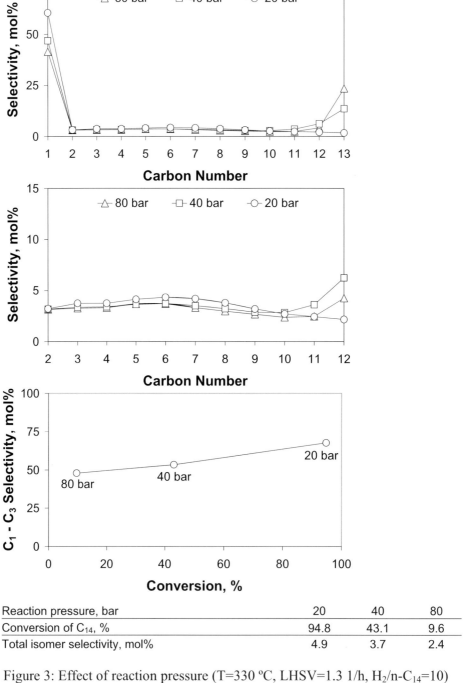

Reaction pressure, bar	20	40	80
Conversion of C_{14}, %	94.8	43.1	9.6
Total isomer selectivity, mol%	4.9	3.7	2.4

Figure 3: Effect of reaction pressure (T=330 ºC, LHSV=1.3 1/h, $H_2/n-C_{14}$=10)

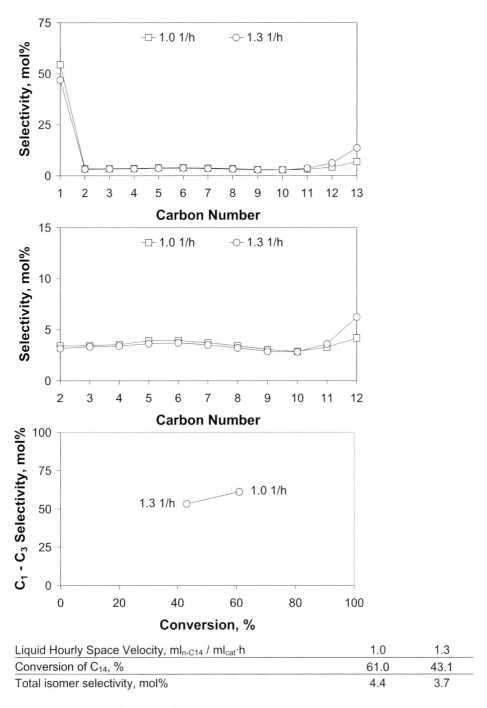

Figure 4: Effect of space velocity (T=330 °C, p=80 bar, H_2/n-C_{14}=116)

Liquid Hourly Space Velocity, ml_{n-C14} / ml_{cat}·h	1.0	1.3
Conversion of C_{14}, %	61.0	43.1
Total isomer selectivity, mol%	4.4	3.7

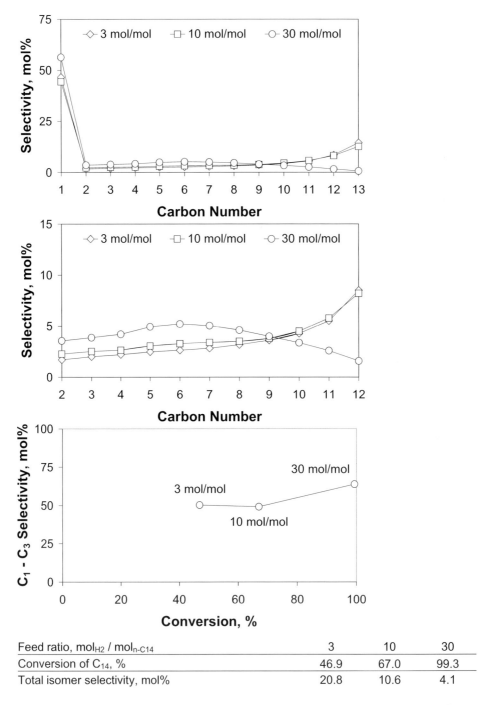

Feed ratio, mol_{H2} / mol_{n-C14}	3	10	30
Conversion of C_{14}, %	46.9	67.0	99.3
Total isomer selectivity, mol%	20.8	10.6	4.1

Figure 5: Effect of H_2 / n-C_{14} feed ratio (T=330 ºC, p=80 bar, LHSV=0.2 1/h)

Notable, also, is the high molar selectivity to the C_{13} fragment and, to a lesser extent, the C_{12} and C_{11} fragments at moderate conversion levels. At low and medium conversion levels, molar selectivities to the mid-carbon-number fragments (C_2 – ca. C_{10}) are essentially equal. With increasing conversion the molar carbon-number distribution shifts to lighter fractions (C_2 – C_7) but, like in case of methane selectivity, a more pronounced shift is obtained only on approaching 100% C_{14} conversion. The same overall pattern is observed for any increase in conversion resulting from a change in reaction conditions; be it increasing reaction temperature (Figures 1 and 2), decreasing reaction pressure (Figure 3), decreasing space velocity (Figure 4) and increasing H_2 / n-C_{14} feed ratio (Figure 5).

5. Discussion

Although the term hydrocracking is generally applied to processes involving a reduction in feedstock molecular mass in the presence of hydrogen transfer reactions, in practice three principal cracking pathways may be involved in the case of saturated hydrocarbons such as n-paraffins, viz. 'true hydrocracking', non-bond-specific hydrogenolysis and a variant of the latter, 'methanolysis'. These pathways are portrayed graphically in Figures 6 – 11 and 14 by means of their associated mechanisms and expected product carbon-number distributions. Note that in all these figures, selectivity is normalised to 100% across the full C_1 – C_{13} product distribution.

5.1 'True hydrocracking'

'True hydrocracking' (Figure 6) is mediated over dual-functional metal / acid catalysts and proceeds via adsorbed carbenium ion intermediates where cleavage is most likely on central C-C bonds and, from the fourth C-position, occurs with almost equal probability [5, 10, 11]. Moreover, due to the carbenium ion mechanism, the intermediate to cracked products is typically an isomerised carbenium ion with the result that products are generally highly branched.

As a consequence of the aforementioned, the theoretical product carbon-number distribution is as presented in Figure 7 where almost no methane is produced as are only low amounts of C_2 and C_3 (and likewise for the associated C_{11} – C_{13} fragments), while carbon chain fragments of intermediate length are formed with approximately equal selectivity. Only with severe conditions and / or increased conversion, respectively, where secondary cracking of the primary fragments is relevant, does the product carbon-number distribution shift towards lighter fragments.

'True hydrocracking' requires the presence of both strong acid and hydrogenation sites, the latter in order to activate the feed paraffins and minimise secondary cracking.

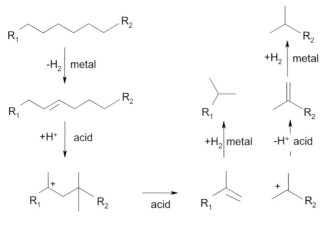

Figure 6: Reaction pathways of dual-functional, metal / acid catalysed 'true hydrocracking'

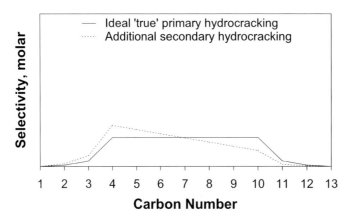

Figure 7: Theoretical carbon-number distribution of product from dual-functional, metal / acid catalysed 'true hydrocracking' of n-C_{14} (lines for additional secondary hydrocracking indicating only qualitative direction of shift)

5.2 Hydrogenolysis

Hydrogenolysis, on the other hand, proceeds via adsorbed hydrocarbon radical intermediates initially formed through abstraction of a hydrogen radical. These chemisorbed hydrogen-deficient hydrocarbon intermediates undergo C-C scission and the probability of such scission is almost identical, or non-selective, for all C-C bonds on the hydrocarbon chain (Figure 8).

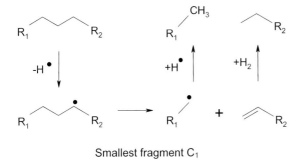

Smallest fragment C_1

Figure 8: Reaction pathways of mono-functional hydrogenolysis

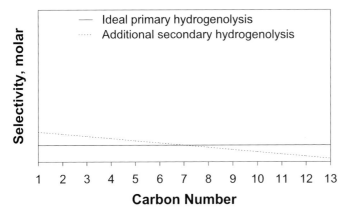

Figure 9: Theoretical carbon-number distribution of product from non-C-C-bond specific, mono-functional hydrogenolysis of n-C_{14} (lines for additional secondary hydrogenolysis indicating only qualitative direction of shift)

As a result, the hydrogenolysis pathway produces a product carbon-number distribution comprising essentially equal selectivities of hydrocarbon fragments from C_1 and higher [12], as shown in Figure 9. In contrast to 'true hydrocracking', the adsorbed radical intermediate mechanism results in low isomerisation activity and a product molecular structure, which is essentially unbranched. Again, only with more severe processing, does secondary hydrogenolysis of the primary fragments shift the overall product carbon-number distribution to lighter fractions.

The aforementioned non-carbon-number specific hydrogenolysis reaction is typically catalysed by base-metal oxides and sulphides and, to a lesser extent, by some metals such as Pt and Ir [13].

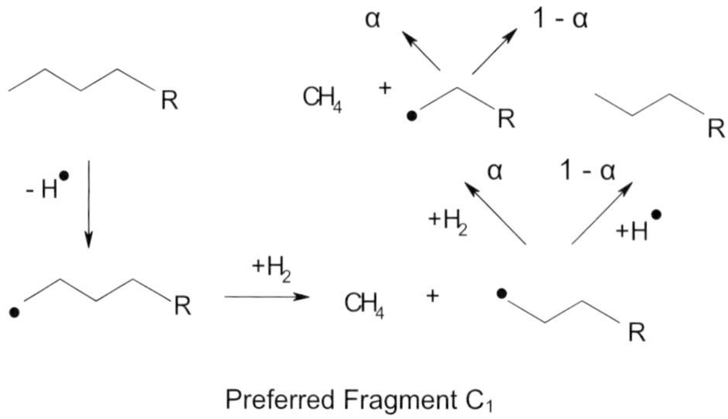

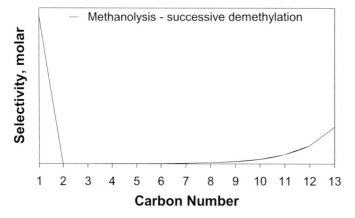

Preferred Fragment C_1

Figure 10: Reaction pathways of mono-functional hydrogenolytic demethylation ('methanolysis'). α = probability to remain adsorbed and be demethylated again

Figure 11: Theoretical carbon-number distribution of product from hydrogenolytic demethylation ('methanolysis') of n-C_{14} calculated with a probability for demethylation compared to desorption of $\alpha = 0.5$ in every successive step

5.3 'Methanolysis'

Additionally, certain metals, notably Ni [13] but also Co [12] show a tendency to promote successive hydrogenolytic demethylation ('methanolysis'). This reaction proceeds via a specific C-C bond cleavage mechanism (Figure 10) such that the terminal C-C bond of the adsorbed hydrocarbon radical cleaves with preference. While some of the higher fragments desorb, others remain adsorbed and undergo a subsequent demethylation step, etc.

Consequently, the overall product carbon-number distribution resulting from successive demethylation is dominated by the presence of methane and the corresponding higher primary fragments, as presented in Figure 11.

5.3.1 A 'methanolysis' model

A simple mathematical model was formulated for the successive methyl abstraction hydrogenolysis mechanism illustrated in Figure 10.

The model incorporates the following:
- The feed molecule, n-C_{14}, is used as a basis. It undergoes methyl abstraction to form methane and an n-C_{13} species
- Some of the n-C_{13} species desorb from the catalyst and leave the system (fraction 1-α), the remainder (fraction α) undergoing further methyl abstraction to form additional methane and, as the co-product, n-C_{12} species etc.
- The number of moles of methane that are produced is given by:

$$n_1 = \left[\sum_{i=2}^{13} n_i \cdot (14 - i) \right] + n_2$$

> i = carbon number of species
> n_i = mols of species i
> $+ n_2$ considers that the co-fragment obtained after methyl abstraction from C_2 is another methane molecule

- The C_1 / C_{13} molar ratio obtained from experimental results is used as a starting or convergence point for the modelling calculations to determine α from experimental results
- Fraction α of the species, the fraction that remains adsorbed and is subsequently further demethylated (Figure 10), was determined by convergence towards the observed C_1 / C_{13} molar ratio (using Microsoft Excel's SOLVER).

Results derived from the above kinetic model of the 'methanolysis' reaction are shown in Figures 11 – 14.

Note that the model does not include secondary methanolysis, i.e. readsorption and additional demethylation of the longer fragment once desorbed. This is for reasons such as the relative demethylation reactivities C_{12} / C_{13}, C_{11} / C_{13} etc. being unknown and, in particular, that products from the 'non-specific' hydrogenolysis mechanism were to be considered too. Therefore, the methanolysis model and the derived α-values, only hold for low to medium of conversion.

Figure 11 illustrates the expected molar carbon-number distribution from primary 'methanolysis' of n-C_{14} as calculated for an α-value of 0.5. The figure exhibits trends that are very similar in shape to what was obtained experimentally at low to medium conversion (Figures 1 – 5), namely very high molar selectivities of methane and high tail ends of the product carbon-number distribution.

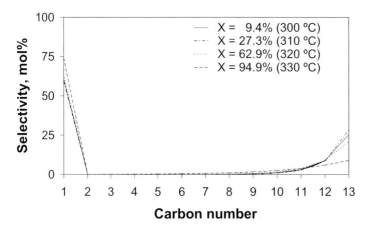

Figure 12: Carbon-number distribution of product from hydrogenolytic demethylation of n-C_{14}, derived from experimentally observed C_1 / C_{13} molar ratios by application of the 'methanolysis' model. Experimental data refers to Figure 1 and reaction conditions of p=80 bar, LHSV=0.2 1/h, H_2/n-C_{14}=116

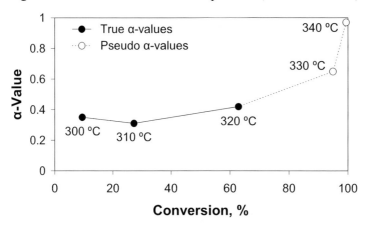

Figure 13: α-Values obtained by application of the 'methanolysis' model on experimentally observed C_1 / C_{13} ratios from hydrogenolytic conversion of n-C_{14} (experimental data from Figure 1, reaction conditions as in Figure 12)

Figure 12 represents the molar carbon-number distributions of the 'methanolysis' products, derived from the 'methanolysis' model, when applied to the results of the first of the temperature series (T_1 in Table 1, see also Figure 12) and based on the experimentally observed C_1 / C_{13} molar ratios. The respective α-values are presented in Figure 13 versus conversion.

As aforementioned, 'true' α-values are only obtained at low to medium conversion (that is the rather equal α-values around 0.4 in Figure 13) while α-values derived from high conversion C_1 / C_{13} molar ratios ('pseudo α-values') as well as the trends obtained in this high conversion range are only generally indicative of the inherent limit of the simple 'methanolysis' model when not considering secondary cracking.

5.4 Simultaneous 'methanolysis' and non-specific hydrogenolysis

'Methanolysis' and non-specific hydrogenolysis, occurring simultaneously over a catalyst, ideally produce a product carbon-number distribution comprising almost equal selectivities of hydrocarbon fragments in the medium carbon-number range (from C_2 to, say, C_9 or C_{10}) with additional high selectivity of methane and enhanced selectivity of higher fragments (ca. C_{10} or $C_{11} - C_{13}$), increasing with carbon number.

The product carbon number distribution from a reaction involving both mechanisms, for example with 50 mol% of ideal primary hydrogenolysis product (as illustrated in Figure 9) and 50 mol% of 'methanolysis' product (as illustrated in Figure 11), is presented in Figure 14.

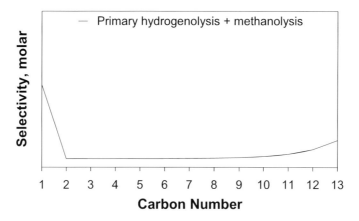

Figure 14: Theoretical carbon-number distribution of the combined product (1 : 1, molar) from primary, hydrogenolysis (Figure 9) and 'methanolysis' with α-value of 0.5 (Figure 11).

5.5   Carbon-number distribution over non-sulphided CoMo-catalyst

Returning to the findings for n-C_{14} conversion over the non-sulphided CoMo/silica-alumina catalyst of this study (Figures 1 – 5): the observed lack of isomerisation, together with the uniform distribution of carbon-number selectivities in the C_2 – C_{10} range (at low conversion), as well as the high methane and C_{13} selectivities, are findings which are not consistent with 'true' dual-functional hydrocracking reactions. Rather, the observed product distribution is typical of hydrogenolysis on the metal / metal-oxide component of the catalyst, including a significant contribution from 'methanolysis'.

Considering the catalyst composition employed in this study and recognizing that the catalyst is not present in the form of metal sulphides (as is the case in typical crude oil refinery hydrocracking applications for what it is designed), it may be reasonable to ascribe the high degree of 'methanolysis' observed to the presence of free Co metal, in much the same way as has been reported for free Co-sulphide islands in the case of sulphided CoMo-type hydroprocessing catalysts [14].

Although the above findings demonstrate the difficulties of applying typical maximum distillate type crude oil refinery hydrocracking catalysts to the case of normal paraffin (F-T wax) hydrocracking in the absence of sulphur, the high yields of linear paraffins observed in the mid- and high-carbon number range (C_4 – C_{13}), i.e. with minimal skeleton isomerisation, would be of interest to high cetane number distillate fuel production from F-T wax.

The challenge, therefore, may be to find ways to avoid the high methane and light gas yields, which correspond to fuel gas and poor quality gasoline (naphtha), in the case of wax feedstock. Note, however, that methane selectivities given in this paper are expressed on a molar, not a weight basis. Weight based selectivities are significantly lower, only exceeding 5 wt-% on exceeding 90% conversion.

6.   Conclusions

With increasingly stringent legislation in respect of transportation fuel specifications, a growing demand for high quality diesel fuel is likely to be met by synthetic production via proposed GTL plants based on F-T synthesis of heavy hydrocarbon wax followed by wax hydrocracking.

As this feedstock differs substantially from traditional crude oil refinery feedstock, existing hydrocracking catalysts and processes are not necessarily optimised for use in the GTL environment. Consequently, and recognizing the high value of wax feedstock to the GTL hydrocracking stage, selective wax hydrocracking presents a significant opportunity for improving overall GTL performance.

Base metal catalysts appear principally suitable for Fischer-Tropsch wax hydrocracking and have the advantage of producing less branched, high cetane number diesel. A drawback of using state-of-the-art 'diesel' hydrocracking catalysts, which are optimised for a sulphur-containing crude oil refinery environment, is the comparably high yield of light, gaseous compounds, in particular methane. Catalyst manufacturers have recognized this issue and are therefore developing dedicated catalysts for wax upgrading.

7. Acknowledgements

The authors gratefully acknowledge financial and technical support from Albemarle Catalysts Company B.V., PetroSA (Pty) Ltd, the South African National Research Foundation (NRF GUN 2053385), the South African Department of Trade and Industry THRIP Programme (PID 2445) and the University of Cape Town (URC Fund 457021).

8. References

[1] UOP LLC Publications (1998) http://www.uop.com/solutions and innovation/Issues%20&%20Solutions/UOPDieselFuel.pdf.
[2] J. Eilers, S.A. Posthuma and S.T. Sie, Catal. Letters, 7 (1990) 253.
[3] J. Scherzer and A.J. Gruia, Hydrocracking Science and Technology, Dekker, New York, N.Y., USA, 1996.
[4] M.E. Dry, J. Chem. Tech. Biotech., 77 (2001) 43.
[5] M.E. Dry, Fischer-Tropsch Synthesis – Industrial, in I.T. Horvath (ed.), Encyclopedia of Catalysis, vol. 3, 347, Wiley, New York, N.Y., USA, 2003.
[6] The Catalyst Review Newsletter, Catalyst Group – Resources, Spring House, Pa., USA, 5 Dec. 2002.
[7] P. Watts and N. Fabricius, 3rd GTL Commercialisation Conference, Doha, Qatar, 20 Oct. 2003, http://www.shell.com/static/qatar/downloads/nfabricius_speech.pdf
[8] S.T. Sie, M.M.G. Senden and H.M.H. van Wechem, Catal. Today, 8 (1991) 371.
[9] P.P. Shah, G.C. Sturtevant, J.H. Gregor, M.J. Humbach, F.G. Padrta and K.Z. Steigleder, Fischer-Tropsch Wax Characterization and Upgrading, Final Report for the U.S. Department of Energy, DOE/PC/80017-T1 (DE88014638), 1988.
[10] J. Weitkamp, P.A. Jacobs and J.A. Martens, Appl. Catal., 8 (1983) 123.
[11] J.A. Martens, P.A. Jacobs and J. Weitkamp, Appl. Catal., 20 (1986) 239.
[12] J.H. Sinfelt, Adv. Catal., 23 (1973) 91.
[13] B.C. Gates, J.R. Katzer and G.C.A. Schuit, Chemistry of Catalytic Processes, Chemical Engineering Series, McGraw-Hill, New York, N.Y., USA, 1979.
[14] H. Topsøe, B.S. Clausen and F.E. Massoth, Hydrotreating Catalysis, in J.R. Anderson and M. Boudart (eds.), Catalysis – Science and Technology, vol. 11, Springer, Berlin, 1996.

Fischer-Tropsch Synthesis, Catalysts and Catalysis
B.H. Davis and M.L. Occelli (Editors)
© 2007 Elsevier B.V. All rights reserved.

Methanol Synthesis in Inert or Catalytic Supercritical Fluid

Prasert Reubroycharoen[a], Noritatsu Tsubaki[b]

[a]*Department of Chemical Technology, Faculty of Science, Chulalongkorun University, Bangkok 10330, Thailand*
b*Department of Applied Chemistry, School of Engineering, University of Toyama, Toyama 930-8555, Japan*

1. Introduction

Methanol is one of the primary chemicals for industries, vehicle fuel, and fuel cell. It is commercially produced under 80-100 bar, 523-573 K with Cu/ZnO catalyst from CO_2-containing syngas, known as ICI process, in gas-phase reaction. Under the reaction condition, a one-pass conversion of the process even using H_2-rich syngas ($H_2/CO = 5$) is limited to 15-25% total carbon conversion due to the thermodynamic limitation of the highly exothermic reaction [1-4]. The recycle of unreacted syngas is necessary to enhance the syngas conversion, leading to the higher production cost. The recycling process can be omitted if the one-pass conversion is high enough. Consequently, developing a high one-pass conversion process will lower the cost of production. Air can be used instead of pure O_2, in the reforming process to produce syngas resulting in the lower cost of methanol production if recycling process is omitted in methanol synthesis, without N_2 accumulation. As described, methanol synthesis is an exothermic reaction. The excessive temperature leads to a serious catalyst deactivation when the heat of reaction is not rapidly removed while the reaction temperature is maintained constant. The reaction heat as well as the accumulated methanol must be removed rapidly from the reactor to prevent the decrease in the conversion of methanol synthesis and the catalyst deactivation. Liquid-phase methanol process (LPMeOH^{TM}), on the other hand, has been developed to overcome the gas phase. Methanol is synthesized in a mineral oil with fine catalyst particles. As the temperature is well controlled by the rapid heat removal of the liquid medium, removing the heat from the catalyst surface to the medium; therefore, the higher syngas conversion is achieved [5]. The syngas transportation into the active sites of liquid-phase reaction is, however, much slower compared to gas-phase reaction. It is known that a supercritical fluid (SCF) has a unique characteristic in its molecular diffusion and solubility parameter [6]. In addition, reaction conducted at supercritical condition enhances the solubility of reactants and products and eliminates interphase transport limitations [3]. The reaction in SCF is now widely used because it has many advantages such as high diffusion efficiency, extraction ability and solvent power. For example, Fischer-Tropsch synthesis with SC-phase n-hexane or n-pentane results in high conversion, efficient heat removal, olefin selectivity and long catalyst life time due to the in-situ wax extraction [7]. Similarly, the increase in the efficiency of methanol synthesis is realized via a coexisting mixture of SC phase C_{10}-C_{13} alkanes, which acted as an inert solvent by facilitating heat and product removal from the catalyst bed [8].

According to a novel methanol synthesis route, methanol is synthesized from CO_2-containing syngas using a conventional Cu-based oxide catalyst in alcohol as a catalytic solvent, at lower temperature such as 423K-443K with high one-pass conversion [9-12]. This process consists of three reaction steps: water-gas shift reaction (1), esterification (2), and hydrogenolysis of ester to form methanol and an alcoholic solvent (3).

$$CO + H_2O = CO_2 + H_2 \tag{1}$$

$$CO_2 + H_2 + ROH = HCOOR + H_2O \tag{2}$$

$$HCOOR + 2H_2 = CH_3OH + ROH \tag{3}$$

$$CO + 2H_2 = CH_3OH \tag{4}$$

Considering the above reactions, methanol is practically produced via esterification in which esterification is the rate-determining step depending solely on the nature of the alcohol. ROH acts as a catalytic solvent. Among different alcoholic solvents, 2-butanol was the most effective solvent due to its compromising effect between electronic distribution and spatial structure.

In this research, methanol synthesis at SC condition via the novel synthesis route is established. A higher CO conversion is expected due to the removal of accumulated methanol and reaction heat at the catalyst bed, assisted by the accompanying SCF, even if the maximum theoretical CO conversion from high temperature is limited. The SC alcohol acts as a catalytic solvent, as a result reactions (2)-(3) will undergo more efficiently.

2. Experimental

2.1. Catalyst preparation

The Cu/ZnO catalyst was prepared by the conventional co-precipitation method in an aqueous solution. 300-ml copper and zinc nitrates (each 0.13 M, Cu/Zn in molar ratio = 1) and 300-ml sodium carbonate (0.47 M) used as a precipitant were simultaneously added to 300 ml water under rapid stirring at 338 K and pH range of 8.3-8.5 and aged overnight. The precipitate was filtered and washed several times with distillated water. The drying process was conducted at 383 K for 24 h followed by calcination in the air at 623 K for 1 h. The reduction of oxide solid was taken at 473 K for 13 h by flowing 5% H_2 in N_2 and passivated by 2% O_2 in Ar. The homemade catalyst, molar ratio Cu/Zn of 1, is denoted as Cu/ZnO (A). The BET surface area of catalyst was 60 m^2/g and the Cu specific surface area, determined by N_2O adsorption method, was 30.1 m^2/g [13]. The Cu/MgO, Cu/MnO$_2$ and Cu/CeO$_2$ were prepared following the same procedures as for Cu/ZnO catalyst where magnesium nitrate, manganese nitrate or cerium nitrate was used instead of zinc nitrate.

In some experiments, the commercial ICI catalyst (ICI 51-2) was selected as a reference catalyst denoted as Cu/ZnO (B). The composition of Cu/Zn/Al was 62:35:3 wt% determined by EDX with a BET surface area of 20 m^2/g.

2.2. Supercritical methanol synthesis

The SC methanol synthesis was performed in a conventional fixed-bed reactor where a vaporizer and an ice-cooled high-pressure trap were set upstream and downstream of the reactor, respectively, as shown in Fig. 1. The catalyst was loaded into the reactor, and then reduced at 493 K by 10% H_2 diluted in He for 1 h. After the reduction, the reactors were adjusted to the reaction condition. The standard reaction conditions were as follows: partial pressure of syngas (CO/CO$_2$/H$_2$/Ar:32/5/60/3) = 10 bar, catalyst weight of 0.50 g (20-40 mesh), W/Fsyngas of 10 g of cat. h. mol^{-1}. Ar was an inner standard. Various alcohols were selected as the SC catalytic fluid as shown in Table 1. In order to clarify the catalytic effect of SC alcohol, n-hexane was selected as an inert SCF. Helium was used as a balance material, in the reaction of the gas phase for comparison. The effluent gas was analyzed with a TCD gas chromatograph equipped with an activated charcoal column. The liquid product collected in the trap was analyzed by an FID gas chromatograph equipped with a Porapak-N column.

Table 1. temperature and pressure of various supercritical fluids

SCFs	T_c (K)	P_c (bar)
1-Propanol	536.78	51.2
2-Propanol	508.3	47.9
2-Butanol	536.05	42.0
iso-Butanol	506.21	39.9
n-Hexane	507.6	30.4

Total carbon conversions of methanol synthesis were calculated as follows:

$$Total\ carbon\ conversion = [CO\ conv. \times \frac{a}{a+b}] + [CO_2\ conv. \times \frac{b}{a+b}] \quad (1)$$

(a, b were the contents of CO, and CO_2 in feed gas, respectively)

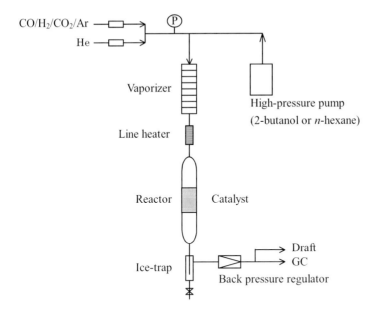

Figure 1. Reaction apparatus configuration.

3. Results and Discussion

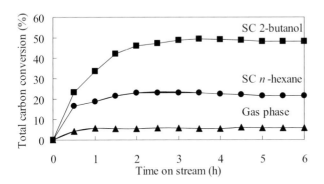

Figure 2. Total carbon conversion vs time-on-stream of SC methanol synthesis. Reaction conditions: T = 538 K, P (total) = 55 bar, Cu/ZnO (B) = 0.5 g, W/Fsyngas = 10 g.h.mol^{-1}, P$_{syngas}$ = 10 bar, (▲) P$_{He}$ = 45 bar, (●) P$_{C6}$ = 45 bar, (■) P$_{2-C4OH}$ = 45 bar.

The conversion in the SC phase was higher than that in the gas phase, as shown in Fig. 2. In addition, SC 2-butanol exhibited highest conversion.

The low conversion of gas phase resulted from the high reaction temperature and low partial pressure as the methanol synthesis was a volumetric reduction and exothermic reaction. Due to the exothermic reaction, the heat of reaction could not be easily removed from the catalyst bed, leading to lower conversion. SC *n*-hexane, an inert SC solvent, was introduced into the reaction resulting in the higher conversion of methanol synthesis. The addition of proper solvents such as *n*-hexane could improve the conversion due to the high heat capacity and molecular diffusion efficiency of SC *n*-hexane [14]. SC *n*-hexane removed methanol and reaction heat smoothly from the catalyst bed. Consequently, the presence of the SCF improved the conversion.

When SC 2-butanol was used in methanol synthesis, the conversion was as high as 48% which was higher than that in SC *n*-hexane. The conversion of methanol synthesis was improved due to a combination effect of the SCF and catalytic solvent. As an SCF, SC 2-butanol promoted the higher conversion by facilitating heat and product removal. Simultaneously, as a catalytic solvent, it promoted the higher conversion by accelerating a new reaction route. Even if a part of 2-butanol was reactively involved in a catalytic reaction as shown in equations (1) - (3), possibly changed to ester temporarily, most of the 2-butanol was stable and acted as a conventional SCF. It should be noted that the change of critical point of coexisting 2-butanol derived from the slight variation of alcoholic solvent and formed ester during the reaction should be negligible due to the large amount of coexisting 2-butanol.

Table 2. Selectivity and total carbon conversion of supercritical methanol synthesis with various kinds of the SCFs.

SCF	Total carbon conversion (%)	Selectivity (%)			
		MeOH	HCHO	HCOOR	CH$_4$
No	5.7	96.8	3.2	0.0	0.0
n-Hexane	21.4	98.1	0.0	0.0	1.9
2-Butanol	48.1	90.7	0.0	9.3	0.0

Reaction conditions: T = 538 K, P (total) = 55 bar, Cu/ZnO (B) = 0.5 g, W/ F_{syngas} = 10 g.h.mol^{-1}, P_{syngas} = 10 bar.

Methanol and formaldehyde were the main product and by-product of the gas-phase methanol synthesis, respectively, as shown in Table 2. For SC n-hexane, the selectivity of methanol was as high as 98.1% and very small amount of methane was found. In case of 2-butanol, the conversion was as high as 48.1% with 90.7% methanol selectivity and ester was the only by-product which formed from esterification in step (2).

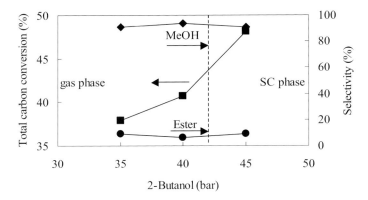

Figure 3. Relationship between total carbon conversion and partial pressure of 2-butanol. Reaction conditions: T = 538 K, P (total) = 55 bar, Cu/ZnO (B) = 0.5 g, W/F$_{syngas}$ = 10 g.h.mol^{-1}, P$_{syngas}$ = 10 bar, balance material = N$_2$.

Fig. 3 shows the conversion and selectivity change with partial pressure of 2-butanol. The conversion gradually increased in gas phase when the partial pressure of 2-butanol increased from 35 to 40 bar. Moreover, the conversion remarkably increased when 2-butanol changed to SC 2-butanol, whereas the methanol selectivity was nearly the same.

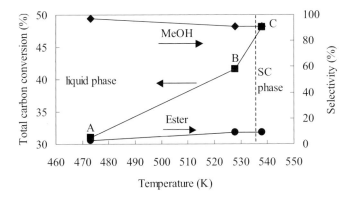

Figure 4. Relationship between total carbon conversion and reaction temperature. Reaction conditions: P (total) = 55 bar, Cu/ZnO (B) = 0.5 g, W/F$_{syngas}$ = 10 g.h.mol^{-1}, A: P$_{syngas}$ = 40 bar, B: P$_{syngas}$ = 18 bar, C: P$_{syngas}$ = 10 bar. Balance pressure: 2-butanol

As shown in Fig. 4, the conversion increased by increasing the reaction temperature when 2-butanol changed from liquid phase to SC phase. According

374

to Fig. 4, the calculated partial pressures [15] of syngas in states A, B, and C were 40, 18, and 10 bar, respectively. Even if the partial pressures in states A and B were higher than that in state C, the conversions were lower than that in state C. If all partial pressures were equalized, the same trend of conversion would still be achieved. This trend indicated that the conversion could be significantly improved by the SCF.

Table 3. Methanol synthesis performance in various SCFs.

SCFs	Conversion (%)			Selectivity (%)	
	CO	CO_2	Total carbon	MeOH	HCOOR
Helium (gas-phase)	15.5	22.5	16.4	99.3	0.7*
n-Hexane	43.5	60.0	45.6	100.0	0.0
1-Propanol	50.8	61.0	52.1	99.1	0.9
2-Propanol	77.4	22.5	70.2	88.1	11.9
2-Butanol	70.0	28.5	64.6	98.9	1.1
iso-Butanol	21.3	62.2	26.7	97.3	2.7

Reaction conditions: T = 543 K, P (total) = 62 bar, Cu/ZnO (A) = 0.5 g, W/F_{syngas}= 10 g.h.mol^{-1}, P_{syngas} = 10 bar, time = 6 h. * Formaldehyde

The effects of SCFs on the catalytic activity of methanol synthesis are shown in Table 3. It is obvious that the gas phase (He) exhibited the lowest total carbon conversion (about 16.4%). As methanol synthesis is a highly exothermic reaction, the low conversion in gas phase possibly resulted from the heat of reaction which cannot be removed rapidly from the catalyst bed. The total carbon conversion increased to 45.6% when SC-C_6 was introduced into the reaction. This illustrates that the SCF enhanced the reaction activity of methanol synthesis. The rate of reaction was possibly enhanced while the experiment was operated in the mixture critical region because of a favorable pressure dependence of the reaction rate constant as well as the unusual volumetric behavior of heavy solutes solubilized in an SCF solvent [16].

The higher conversion of methanol synthesis was obtained by applying more effective heat removal to keep reaction temperature as low as possible [17]. The removal of methanol product during the reaction also shifts the equilibrium to higher conversion. Consequently, the higher total carbon conversion in SC-C_6 can be illustrated in terms of both more effective heat transfer and high molecular diffusion efficiency. The more effective heat transfer resulted from the higher thermal conductivity of the SC phase which was as approximately 5 to 6 times as the gas phase [18]. The heat transfer in the SC phase was more effective than that in the gas phase [19-20].

The concept of "beating the equilibrium" by using a high-boiling inert solvent to remove the methanol from the catalyst bed during the reaction was proposed by Berty *et al.* [21]. This resulted in the shift of an equilibrium and high conversion. The same concept could be used to explain the reason of high conversion when SCF was introduced into the reaction. Furthermore, SCF should work more effectively due to the higher mass transfer efficiency in SC phase than that in liquid phase.

Interestingly, SC 2-propanol exhibited the highest total carbon conversion of 70%. SC *n*-hexane, as an inert SCF, promoted the reaction by improving heat and product removal from the catalyst bed. SC 2-propanol, as a catalytic solvent, promoted the reaction not only by the SCF advantage, but also by the catalytic effect as behaved in the low temperature methanol synthesis [9-12] via the new reaction route. This showed that the total carbon conversion of methanol synthesis could be significantly improved by the combination of SCF advantage and catalytic effect when alcohol was used as SCF. It should be noted that even if a part of the alcoholic solvent, 2-propanol, was reactively involved in a catalytic reaction as shown in equations (1) - (3), possibly changed to ester temporarily, most of the alcoholic solvent was very stable and acted as a conventional SCF when great quantity of the coexisting alcoholic solvent was used. The change of critical point derived from slight variation of the solvent composition can be omitted.

Among alcoholic solvents, 2-propanol exhibited the highest total carbon conversion whereas *i*-butanol showed the lowest total carbon conversion. In addition to SCF effect, this indicated that the conversion of methanol synthesis solely depended on the structure of alcohols. As the total carbon conversion decreased when the carbon number of alcohol molecule increased; therefore, the conversion of 2-butanol was lower than that of 2-propanol. Furthermore, the high total carbon conversion of 2-propanol compared to 1-propanol indicated that 2-propanol was more active due to its strong electronic density. Esterification is initiated by nucleophilic attack from oxygen in alcohol to carbon in carboxylic acid. Branched alcohol has high electron density on oxygen atom, which is favorable to esterification but large spatial obstacle of the branched alcohol is not promotional to esterification. As 2-butanol exhibited higher total conversion than *i*-butanol which has higher electron density on its oxygen atom, it is clear that the spatial obstacle also had an effect on the conversion, the greater spatial obstacle, the lower methanol synthesis activity. As electronic and spatial factors were well balanced, 2-propanol showed the highest activity, corresponding to the results of methanol synthesis at low temperature [9-12].

The gas-phase (He) and SC n-hexane exhibited high methanol selectivity as shown in Table 3. The methanol selectivity of the SC alcohols was high as well, however, the HCOOR selectivity of 2-propanol was slightly high (approximately 12%). It should be noted that HCOOR is easily converted to methanol and the alcoholic solvent under high H_2 partial pressure.

As mentioned, a part of the alcohol solvent was invloved in the catalytic reaction whereas the most parts of the alcohol solvent acted as SCF, but were not involved in the catalytic reaction. The coexisting SCFs were studied by varying the molar ratio of 2-propanol as a catalytic cosolvent in SC n-hexane. The role of the solvent was based on the assmuption that 2- propanol proceeded the methanol synthesis via reaction (2) – (3), while SC n-hexane acted as the sovent media effectively transported 2-propanol into the catalyst surface, and then removed the methanol product and reaction heat form the catalyst bed.

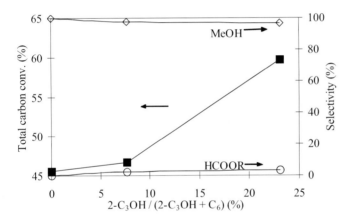

Figure 5. The effect of co-existing SCF on the methanol synthesis. Reaction conditions: T = 543 K, P (total) = 62 bar, Cu/ZnO (A) = 0.5 g, W/Fsyngas= 10 g.h.mol⁻¹, P_{syngas} = 10 bar, time = 6 h.

The effect of coexisting SC solvent on the methanol synthesis was shown in Fig. 5. The total carbon conversion siginificantly increased from 45 to 60% when the amount of 2-propanol increased from 0 to 23 mol% whereas the selectivity of methanol slightly decreased. This indicated that the coexisting 2-propanol in SC n-hexane enhanced the reation activity of methanol synthesis. As the small amount of 2-propanol was used; therefore, the synthesis was considered as an economic process. A small amount of cosolvent added to SCF possibly had a significant impact on the economics of SC separation processes [22].

Table 4 compared the reaction activity of different Cu-based oxide catalysts. Cu/ZnO catalyst exhibited the highest total carbon conversion (74.1%). Cu/MgO catalyst as well showed high total carbon conversion (69.8%) while the conversion of Cu/MnO$_2$ catalyst and Cu/CeO$_2$ catalyst were 23.8% and 18.0%, respectively. The low activity of Cu/MnO$_2$ catalyst could be explained by its low BET surface area. For, Cu/CeO$_2$ catalyst, this catalyst gave very high HCOOR selectivity and low activity, leading to believe that hydrogenolosis could not perform as well as other catalysts. Thus, the low activity of Cu/CeO$_2$ was due to the low hydrogenolysis activity. It should be noticed that the CO$_2$ conversion of Cu/MgO catalyst (68.5%) was higher than that of Cu/ZnO catalyst (36.5%). The higher CO$_2$ conversion could be explained by the higher adsorption of CO$_2$ on the surface of MgO, known as a base catalyst [23]. The conversion of Cu/MnO$_2$ catalyst and Cu/CeO$_2$ catalyst was low.

Table 4. Reaction performances of various catalysts on the methanol synthesis.

Catalyst	Conversion (%)			Selectivity (%)		BET surface area (m^2/g)	Pore volume (cm^3/g)	Pore size (nm)
	CO	CO$_2$	Total carbon	CH$_3$OH	HCOOR			
Cu/ZnO (A)	79.7	36.5	74.1	90.0	10.0	63.8	0.24	15.1
Cu/MgO	70.0	68.5	69.8	95.6	4.4	82.5	0.19	9.1
Cu/MnO$_2$	21.0	48.6	23.8	95.1	4.9	40.4	0.15	14.7
Cu/CeO$_2$	13.0	51.5	18.0	6.8	93.2	74.2	0.20	10.8

Reaction conditions: T = 543 K, P (total) = 62 bar, W = 0.5 g, W/F$_{syngas}$= 10 g.h.mol^{-1}, P$_{syngas}$ = 10 bar, time = 6 h, SCF = 2-propanol, Cu/M molar ratio of 1/1.

4. Conclusion

The low total carbon conversion of the gas-phase reaction resulted from the highly exothermic reaction from which heat and product cannot be removed rapidly from the catalyst bed. The heat and product removal from the catalyst bed, which would improve the reaction activity, was achieved when SC *n*-hexane was introduced into the reaction. The highest total carbon conversion was obtained by using the alcohols as SC catalytic fluids. SC alcohol improved the conversion by promoting the reaction not only by the SCF advantage, but also by the catalytic effect as behaved in the low temperature methanol synthesis. The methanol synthesis could significantly be improved by the combination of SCF advantage and catalytic effect when alcohol was used as an

SCF. Among alcoholic solvents, 2-propanol exhibited the highest total carbon conversion whereas *i*-butanol showed the lowest total carbon conversion indicating that the conversion of methanol synthesis solely depends on the structure of alcohols. When electronic and spatial factors were balanced, 2-propanol exhibited the highest activity.

References

1. S. Lee, Methanol Synthesis Technology, CRC Press, Boco Raton, FL, 1990.
2. I. Wender, Fuel Process. Technol. 48 (1996) 189.
3. G.C. Chinchen, P.T. Denny, J.R. Jennings, M.S. Spencer, K.C. Waugh, Appl. Catal. 36 (1988) 1.
4. M. Marchionna, M. Lami, A.M. Raspolli Galletti, CHEMTECH (1997) 27.
5. Air Products Liquid Phase Conversion Co., L.P. Commercial-Scale Demonstration of the Liquid Phase Methanol (LPMEOH™) Process Final Report. U.S. DOE Contract No. DE-FC22-92PC90543. June (2003).
6. P.G. Debenedetti, R.C. Reid, AIChE J. 32 (1986) 2034.
7. L. Fan, K. Yokata, K. Fujimoto, AIChE J. 38 (1992) 1639.
8. T. Jiang, Y. Niu, B. Zhong, Fuel. Proc. Tech. 73 (2001) 175.
9. N. Tsubaki, M. Ito, K. Fujimoto, J. Catal. 197 (2001) 224.
10. P. Reubroycharoen, Y. Yoneyama, N. Tsubaki, Fuel 82 (2003) 2255.
11. P. Reubroycharoen, Y. Yoneyama, N. Tsubaki, Catal. Comm. 4 (2003) 461.
12. P. Reubroycharoen, Y. Yoneyama, V. Tharapong, N. Tsubaki, Catal. Today 89 (2004) 447.
13. J.W. Evans, M.S. Wainwright, A.J. Bridgewater, D.J. Young, Appl. Catal. 7 (1983) 75.
14. J. Liu, Z. Qin, J. Wang, Ind. Eng. Chem. Res. 40 (2001) 3801.
15. R. H. Perry, D. W. Green, Perry's Chemical Engineers' Handbook, McGraw-Hill, Australia, 1997.
16. B. Subramaniam, M. A. McHugh, Ind. Eng. Chem. Process Des. Dev., 25 (1986) 1.
17. G. A. Mills, Fuel, 78 (1994) 1243.
18. N. V. Tsedenberg, Thermal Conductivity of Gases and Liquids; MIT Press: Cambridge, MA, 1965.
19. K. Yokota, K. Fujimoto, Ind. Eng. Chem. Res. 30 (1991) 97.
20. L. Fan, K. Fujimoto, Appl. Catal. 186 (1999) 343.
21. J. M. Berty, C. Krishman, J. R. Jr. Elliot, CHEMTECH, 20 October (1990), 624
22. S. Kim, K. P. Johnston, AIChE J., 33 (1987) 1603.
23. K. Tanabe, W. F. Hölderich, Appl. Catal. A, 181 (1999) 399.

Fischer-Tropsch Synthesis, Catalysts and Catalysis
B.H. Davis and M.L. Occelli (Editors)

Fischer-Tropsch based GTL Technology: a New Process?

L.P. Dancuart, A.P. Steynberg

Sasol Technology Research & Development, PO Box 1, Sasolburg 1947, South Africa

1. Abstract

The 21st century is witnessing the establishment of a new global business based on natural gas processing. The Gas-to-Liquids (GTL) industry is entering a new phase of expansion based on the use of the Fischer-Tropsch (FT) synthesis. While for many this might look like new technology, most of the fundamentals are not so new.

Decades ago, the pioneers of this industry were able to foresee with ingenuity and provide with science the foundation that is used by today's engineers and scientists. They also predicted the unique benefits that could be expected from the use of synthetic fuels. Moreover, based on the unique composition of the primary products, they anticipated their importance as chemicals and other non-fuel products.

The development of the FT-based GTL technology is intimately related to the initial efforts to apply it using coal as feedstock. Its evolution followed a logical process that was delayed by years of abundant, low cost petroleum and a lack of stringent fuel specifications aimed at protecting the environment.

This work highlights some of these concepts, giving recognition to the FT technology pioneers.

2. Introduction

Historically the Fischer-Tropsch process is indeed the second technology used to produce synthetic hydrocarbons. Friedrich Bergius in

Rheinau-Mannheim, Germany developed the high pressure coal hydrogenation process from 1910 to 1925. The objective at the time was to ensure that Germany was capable of continuous production of hydrocarbons based on their indigenous coal reserves (1). Coal hydrogenation was the most important source of synthetic hydrocarbons for Germany during the Second World War. Coal hydrogenation is no longer considered to be a desirable approach for the production of synthetic hydrocarbons. This is due to the significantly higher environmental impact and the much higher capital cost when compared to natural gas conversion. Also, the highly aromatic products obtained from coal hydrogenation are not compatible with modern fuel specification trends.

In spite of the high capital costs, the indirect conversion of coal to synthetic hydrocarbons using FT technology is enjoying renewed interest. This is due to the relatively low cost of coal compared to natural gas in certain regions, and the strategic incentive of improved energy supply security. Moreover, advances in carbon capture and storage are adding impetus to this trend, particularly in the USA and in China.

The development of the FT industry may be divided in five stages (2):

1902-1928	Discovery	<ul><li>Successful research on synthesis gas conversion.</li><li>The first patents are filed in Germany.</li></ul>
1929-1949	First era of cobalt catalyst	<ul><li>Commercial developments in Germany and other countries, with coal derived synthesis gas.</li></ul>
1950-1990	The iron catalyst era	<ul><li>The Sasol commercial plants in South Africa, using coal derived synthesis gas.</li></ul>
1990-2004	The initiation of the FT GTL commercial era	<ul><li>Accelerated FT research and development using both iron and cobalt catalysts.</li><li>Two commercial plants: PetroSA (Mossel Bay, South Africa) and Shell GTL (Bintulu, Malaysia).</li></ul>
2004-	Commercial expansion	<ul><li>Two large projects confirmed: ORYX GTL (Sasol) and Pearl GTL (Shell) in Qatar.</li><li>Many more concepts proposed: Sasol Chevron (six), ExxonMobil, Syntroleum and Statoil/PetroSA.</li></ul>

Recently, interest in GTL technology using cobalt catalyst has revived due to the progress that has been made in reducing capital costs for large scale plants together with increases in the prevailing crude oil prices and the expectation of an upward price trend. In addition there is now greater awareness of the significant environmental benefits associated with the synthetic diesel fuel product. Another important factor is the existence today of vast quantities of known reserves of remotely located natural gas that have no suitable alternative uses.

2.1 Discovery

Frans Fischer (1877-1947) and Hans Tropsch (1889-1935) discovered the reaction that carries their names around 1923 at the Kaiser Wilhelm Coal Research Institute in Mülheim, Germany. This process converts synthesis gas, a mixture of hydrogen and carbon monoxide, into a complex set of hydrocarbons that can be 'refined' using petroleum processing technology.

The primary FT synthesis reaction can be summarised as follow (3):

$$n\, CO + 2n\, H_2 \rightarrow (\text{-}CH_2\text{-})_n + n\, H_2O$$

Two other important reactions also occur to different degrees of completion: (i) water gas shift, and (ii) syngas to alcohols.

$$CO + H_2O \leftrightarrow CO_2 + H_2 \qquad\qquad \text{(i)}$$

$$n\, CO + 2n\, H_2 \rightarrow C_nH_{2n+1}(OH) \qquad \text{(ii)}$$

The relative importance of these reactions to each other depends on many factors including the FT catalyst, the type of reactor and the composition of the synthesis gas, in particular its hydrogen to carbon monoxide ratio.

While petroleum refining and the use of its derivatives was known at the time, its availability and known proven reserves were limited. As a consequence, the work of these scientists was closely followed by many of the fuel corporations, some of them still active today. The perceived need for alternative sources for hydrocarbon liquids motivated the creation of the Hydrocarbon Synthesis Corporation around 1938-39. The objective was to produce a broad range of hydrocarbons, including fuels and lubricants. Standard Oil Development (now Exxon) took 680 shares of the new corporation, Shell and Kellogg 425 each and I.G. Farben 170 (4). This venture was negatively affected by the Second World War.

The interest in synthetic hydrocarbons dropped significantly after the war. Although demand for liquid hydrocarbons continued to increase, the growth in petroleum supply continued to outstrip the demand due to major discoveries which lead to lower prices. In spite of this, and for some years, a few research teams continued working in the area, primarily based on the German demonstrated technology (5). It was during this time that the expression 'Gas to Liquids' appears in the technical literature associated with FT synthesis using natural gas as feedstock (6, 7).

2.2 First era of cobalt catalyst (1930-1950)

A total of 21 plants were in operation producing synthetic fuels at the end of the Second World War; nine of them used FT technology, and twelve were based on coal liquefaction. Most of these FT units were operated by IG Farben and Rührchemie (8). During the war the FT plants were able to supply 9% of the German demand for hydrocarbons. The low octane of the FT naphthas was the critical factor that resulted in a larger production from coal liquefaction. The contribution of the different sources of hydrocarbons to the German fuel demand is shown in Table 1.

Table 1. Hydrocarbon Sources in Germany (1940-1945)

Source	tpa	bpd	%
Coal Liquefaction	3918000	94760	60.1%
Petroleum	1920000	46440	29.5%
FT Synthesis Fuels	591000	14290	9.1%
Brown Coal Tars	50000	1210	0.8%
Bituminous Coal Tars	36000	870	0.6%
Total	6515000	157570	100.0%

The FT fuels were produced in ten plants, nine of them located in Germany, see Table 2. These plants made gasoline, diesel fuel, waxes and lubricants. Some included process integration with liquids from coal gasification

Table 2. Commercial FT Plants in Germany 1939-1945

Plant	Location	tpa	bpd
Braun Hohle Benzin	Ruhland, Saxony	170000	4110
Essener Benzin	Berghamen, Ruhr	80000	1930
Rührchemie	Holten, Ruhr	72000	1740
Rheinpreuseen	Moers, Ruhr	70000	1690
Krupp Benzin	Wanne-Eickel, Ruhr	60000	1450
Hoesch Benzin	Dortmund, Ruhr	47000	1140
Gewerkschaft Viktor	Castrop-Rauxel, Ruhr	40000	970
Schatfootsch Benzin	Leschowitz, Silesia	40000	970
Wintershall	Luetzkendorf, Saxony	12000	290
Total Germany		591000	14290
Kuhlmann	Harnes, France	11000	270
Total		602000	14560

Processing in these early units was both simple and complex. Their basic configuration was apparently simple but there were a large number of synthesis reactors: 65 primary reactors and 35 secondary units were required to produce about 1500 bpsd using FT catalysts based in cobalt and thorium. The typical configuration is shown in Figure 1 (7).

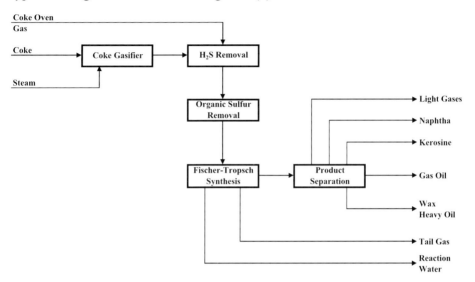

Figure 1 – Typical Process Scheme of an early German FT plant

The FT synthesis was conducted at temperatures between 180-200°C, slightly below the range considered to be of commercial interest at present. A process limitation related to the low operating pressure; most of the reactors were run at atmospheric pressure. A few plants included reactors designed for operation at pressures of up to 10 bar. The synthesis gas in these plants was desulfurised to ca 5 ppm sulfur to protect the FT catalyst (7). As a consequence, and as it is the case at present, all the FT products had very low sulfur contents.

Table 3 Typical Mass Balance of German FT Plants (1940-1945)

Products	% mass
Gas (LPG)	10.0%
Gasoline	52.2%
Diesel Fuel	15.8%
Fuel Oil	10.7%
Waxes	11.3%
Total	100.0%

The low reactor operating pressure resulted in a high selectivity towards light products and low reactant conversion levels. The FT products were essentially paraffinic, including some olefins and smaller quantities of oxygenates. The former, primarily alpha-olefins, were recovered as feedstock for the production of base oils through polymerization or detergents via sulfonation. The oxygenates, primarily heavy alcohols, were recovered for the production of lubricants through esterification reactions. Table 3 shows a typical mass balance (8).

The typical quality of the diesel from these plants is presented in Table 4 based on a sample of fuel obtained from the Kuhlmann plant located in Harnes, France. This was analyzed after the war (9). As indicated, this fuel was obtained by distillation of the FT products, whose composition includes mostly linear paraffins. This translates directly in a high Cetane number, as well as in poor cold temperature characteristics.

Table 4. Characteristics of an early FT Diesel (1945)

Density (20°C)		0.768
ASTM D-158 Distillation	PI, °C	193
	T10, °C	218
	T50, °C	248
	T90, °C	291
	PF, °C	311
Flash Point	°C	78
Cloud Point	°C	0
Sulphur content	ppm	400
Corrosivity		pass
Cetane Number		80
Bromine Number		6.9

This fuel was obtained by distillation of the FT products, whose composition includes mostly linear paraffins. This results in a high Cetane number, as well as poor cold temperature characteristics. The 400 ppm sulphur content appears abnormally high for a FT product. This might be related to analytical errors and/or product contamination or it might also be that sulphur was deliberately added to improve the lubricity of the synthetic diesel. These FT diesels were used to improve the quality of other fuels, including the highly aromatic products derived from the liquids from coal devolatilisation, whose Cetane numbers were between 10 and 15 (8). It is interesting to note that this straight run synthetic diesel produced using a cobalt FT catalyst, contained olefins as indicated by the high Bromine number.

2.3 The Iron Catalyst Era

The first post-war FT plant was operated in the early 1950s by a consortium lead by Texaco operated the 4 500 bpd Carthage Hydrocol plant in Brownsville, Texas. Natural gas was used as feedstock, making this the first GTL plant. The primary product for this iron catalyst based HTFT process was the production of 82 octane gasoline (10). The process scheme for this plant is shown in Figure 2.

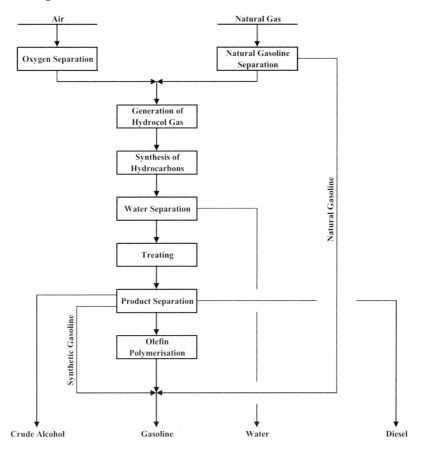

Figure 2 – Process Scheme of the Carthage Hydrocol plant in Brownsville, Texas

Their gasoline was a blend of synthetic naphtha with natural gasoline and an oligomerised naphtha from the olefin polymerization unit (11). Unfortunately, this plant suffered from many operating problems during their first years and, once solved, the consortium was forced to close down the plant due to significant natural gas price increases.

Roughly at the same time the operation of Sasol started in Sasolburg, South Africa. Due to many circumstances including geo-political and economical factors Sasol developed four FT processes: two low temperature systems (LTFT) and two high temperature (HTFT). All have been demonstrated commercially and all are in operation in South Africa. The LTFT Arge reactors are an optimization of the Rührchemie design. The HTFT circulating bed units were derived from an original Kellogg concept tested only on a small scale in the USA – these reactors experienced many problems when commissioned, forcing Sasol to make significant changes that resulted in the Synthol circulating fluidized bed (CFB) reactor technology. Both LTFT and HTFT reactor designs have now evolved into more advanced fluid bed type reactor designs that are capable of very high production capacities.

These process schemes include three primary processing steps (i) synthesis gas production, (ii) production of synthetic hydrocarbons, and (iii) upgrading of the synthetic hydrocarbons to commercial products. Additionally, all of them include oxygen plants and all the utilities normally included in petroleum refineries and chemical plants.

Table 5 – Commercial FT Processes of Sasol

FT Process	Low Temperature (LTFT)	High Temperature (HTFT)
Temperature	220-260°C	320-350°C
Catalyst	Fe / Co (a)	Fe
Traditional Reactor Type	Arge Tubular	Synthol Circulating Bed
Advanced Reactor Type	Sasol Slurry Phase Distillate (Sasol SPD) Slurry (Three-phase)	Sasol Advanced Synthol (SAS) Fluidized Bed

(a) Base catalyst proposed for the ORYX GTL plant (Qatar)

The original process scheme used in Sasolburg is shown in Figure 3. This integrated the Arge LTFT and Synthol HTFT synthesis units (12). The light olefins from both FT processes were oligomerised together in the Cat Poly units to produce diesel and gasoline. The reaction water processing and subsequent recovery of oxygenated products were also combined. The Arge units were used primarily for the production of waxes while the Synthol products were mainly fuels. The complex also produced chemicals: alcohols, ketones, ammonia etc. The Synthol units were replaced by a slurry phase LTFT reactor in 1993 (13). The process scheme changed once more in 2004 when coal gasification was replaced by natural gas reforming thus making the Sasol

One Site a GTL plant although the commercial products target the wax and chemicals rather than synthetic fuels.

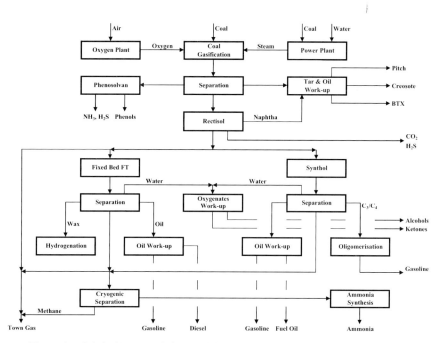

Figure 3 – Original Process Scheme of the Sasol 1 Plant - Sasolburg, South Africa

This first Sasol LTFT plant had a complex processing scheme. This original concept, shown in Figure 4, showed a high degree of integration to maximize the benefits of the unique characteristics of the FT products (14).

This process scheme included eight separate units. While its objective was to produce gasoline and diesel, it also included the production of LPG and hydrogenated waxes. The quality of the gasoline was improved by a Hot Refining step, a 400°C fixed catalyst bed treatment with two objectives: (i) conversion of oxygenates to other hydrocarbon species and (ii) shifting the double bond in olefins to the centre of the molecule. Both changes improved the fuel stability and the octane number of the synthetic naphthas. The diesel was a blend of the straight run FT products and the cracked stock from the Paraformer, a thermal cracker unit – therefore still containing some olefins and oxygenates.

388

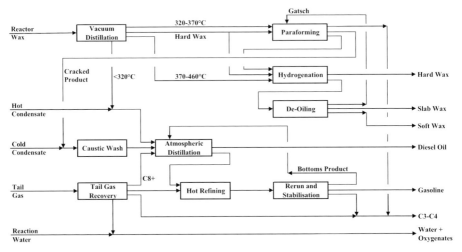

Figure 4 - Process Work-up for the first Sasol LTFT plant in Sasolburg (1957)

The process flow diagram for the Sasol 2 plant in Secunda, South Africa is presented in Figure 5 (12). The original Synthol HTFT reactors installed at Secunda were an improved and scaled-up version of the ones used in Sasolburg.

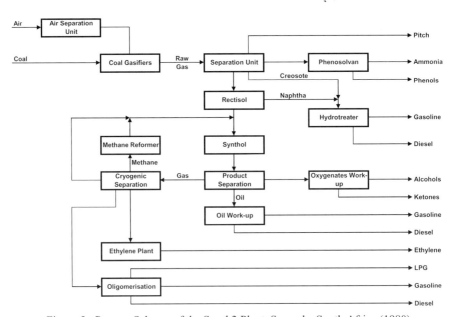

Figure 5 - Process Scheme of the Sasol 2 Plant, Secunda, South Africa (1980)

The Secunda process scheme was conceived to maximize gasoline production – therefore, it includes hydroprocessing and catalytic reformers similar to those used in petroleum refineries. Due to the scale of operation, it includes facilities for the recovered of ethylene, alcohols, ketones, phenols, ammonia and other chemical products. Its twin plant at the same location, Sasol 3, has a very similar configuration. At present all the original Synthol reactors have been replaced by the more efficient Sasol Advanced Synthol (SAS) reactors, with capacities of up to 20 000 bpd per train.

In 1983 Dr Mark Dry, working for Sasol, proposed a process scheme for maximum diesel production in 1983 (15). While this scheme - shown in Figure 6 - was conceived for a coal based LTFT plant, it is also applicable when natural gas is selected as feedstock thus replacing the coal gasifiers with natural gas reformers. The major difference relative to most LTFT GTL concepts is the inclusion of an oligomerisation unit for the conversion of the light hydrocarbons to liquid fuels which might lead to some 7% additional distillates production compared to equivalent schemes that exclude this process.

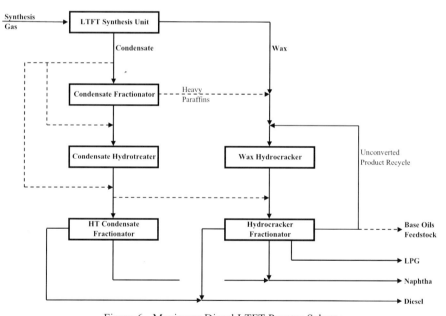

Figure 6 - Maximum Diesel LTFT Process Scheme

The diesel from this concept is made up of three distillates. Almost a third would come from fractionation of the LTFT Condensate, and is primarily a paraffinic product with smaller olefins and oxygenates contents. The LTFT Wax would be hydrocracked at relatively mild conditions selectively to a diesel, recycling to extinction unconverted species to lead some two thirds of the final product. The balance would be made by oligomerising C3-C6 LTFT olefins to a low Cetane number, highly branched distillate. These diesels have very low contents of aromatics and naphthenes, as well as negligible levels of sulphur and nitrogen components. Therefore, they are very attractive fuels from an environmental perspective.

At present all the commercial Sasol FT units are operating with Fe-based catalysts, developed and manufactured in own facilities. The LTFT Co-based catalyst developed by Sasol Technology for the Sasol Slurry Phase Distillate™ (SPD™) process is being manufactured at a commercial plant in the Netherlands.

The Fluidised Bed Reactors

Fluidised bed reactors have an inherent advantage with higher heat transfer coefficients which is important due to the large amounts of heat that must be removed from the FT reactors to control their temperature. Fluid bed (also called fluidized bed) reactors may be two-phase (gas-catalyst) or three-phase (gas with catalyst suspended in a hydrocarbon liquid slurry). The three phase reactor is also known as a slurry phase reactor. It has been calculated that the heat transfer coefficient for the cooling surfaces in a slurry phase reactor are five times higher than those for fixed bed reactors (16-18). The magnitude of the heat transfer coefficient for a two phase fluidized bed reactor is similar to that for a slurry phase reactor.

The HTFT two-phase fluid bed concept was the basis used for the Hydrocol process, the world's first GTL commercial plant that was in operation in Brownsville, Texas in the 1950s. This reactor technology has now been perfected by Sasol and is used in the world's largest FT application at Secunda, South Africa.

The largest capacity proposed to date for a fixed bed reactor has been about 8 000 bpsd of product while fluid bed reactors producing 20 000 bpsd are in commercial operation in South Africa. This is an important advantage for large scale applications.

Most of the more recently proposed GTL concepts are based on the use of the LTFT three phase fluid bed reactor. This development can also overcome the fixed bed reactor limitations on the temperature control of the very exothermic FT reaction. Additionally, a higher catalyst utilisation efficiency is possible and, as mentioned previously, larger reactor capacities can be attained, compared with tubular fixed bed units. The first commercial scale LTFT slurry reactor, a 2 500 bpd unit, was commercialised by Sasol in 1993 and is in continuous operation using an iron FT catalyst.

The tubular reactor developed from the original German concept has the tendency of developing hot-spots in the catalyst bed. This is overcome in a slurry system because of the churning nature of the slurry and controlled slurry mixing. As a consequence, the slurry phase is sufficiently well mixed to give isothermal operation, giving a different perspective to temperature control in the highly exothermic FT system (13). Moreover, the average synthesis temperature can be higher than that used in a tubular reactor, without the risk of catalyst degradation.

During the liquid phase FT program completed at the demonstration scale slurry column at the La Porte Alternative Fuels Development unit it was found that the system was essentially isothermal, with a temperature difference smaller than 2°C along the height of the reactor. ExxonMobil has reported that their AGC-21 FT reactor can be operated with an "*essentially flat temperature profile*". All these publications confirmed the earlier results observed at the Rheinpreussen-Koppers demonstration scale plant: "the temperature gradient in the (slurry FT) reactor never amounted to more than ±1°C".

Dr Dry from Sasol proposed in 1982 to use slurry phase reactors to ultimately produce mainly diesel with naphtha as a significant co-product by using a scheme in which the reactor wax is hydrocracked (19, 20). It was further proposed that this naphtha is a good feedstock for thermal cracking to produce ethylene. Gulf Oil in 1985 proposed the use of a slurry reactor with a modern precipitated cobalt catalyst to produce mainly diesel as a final product (21). The advent of still more active cobalt catalysts has now resulted in the ability to consider gas velocities for the LTFT reactors that are in line with those used for the HTFT fluid bed reactors.

In the 1980's the US Department of Energy (DOE) and the South African Council for Scientific and Industrial Research (CSIR) were actively promoting the use of slurry phase reactors for FT synthesis. In the case of the DOE, a pilot plant design was prepared in 1989 and the La Porte pilot plant mentioned above was operated: first as a methanol reactor; then using iron catalyst and finally using cobalt catalyst. In the case of the CSIR, useful hydrodynamic information was obtained from a large scale non-reactive column using hot wax as the liquid phase. This program was completed in 1989.

In his text book, W-D Deckwer begins with a description of 'the simple bubble column' and then proceeds to describe various modified bubble column designs and bubble columns with directional liquid circulation, both internal and external (22). This modified bubble columns included:
- a cascade of distributors within a single column that divide the column into a number of stages.
- a packed bed.
- a static mixer in the slurry bed.

- use of vertical baffles to segregate the bed.

It is also significant that all these options for modified bubble columns are illustrated with liquid up-flow. There are several subsequent patents that use these concepts, in some cases together with additional features and in others with questionable validity.

The most commonly proposed approach considers catalyst particles in the range from about 30 to 300 μm for a reactor system in which they are uniformly fluidized by an upward liquid velocity. There are several patented techniques to achieve this result (23-34). This is somewhat surprising since this general design approach had been proposed as early as 1948 (32). There are some proprietary features which provide important advantages, for example, the use of staged internal circulation devices, also known as downcomers (26). Although the basic concepts for the design of modern slurry phase reactors are well known, there are clearly some pitfalls when it comes to practical implementation. It will be preferable to use designs that have been implemented successfully on a commercial scale or at least proven on a demonstration scale (e.g. in reactors with diameters of around 1 m or more).

A 1956 patent assigned to the Koppers Company probably comes closest to describing most of the features that are now proposed for use with modern slurry bubble column reactors (33). This patent describes the following features:

- enhanced internal slurry circulation
- internal cooling
- multiple banks of internal coolers using vertical tubes with water as cooling medium
- staged reactors with product knock-out between stages
- oil reflux to uniformly wash the walls of the reactor freeboard zone to avoid catalyst deposits on these walls
- external liquid recirculation to generate an upward liquid velocity through the slurry bed
- high shear mixing zone to generate small gas bubbles with a high interfacial area to enhance mass transfer
- high aspect ratio and relatively high gas velocity for that time

This patent describes a 1 m diameter and 18 m high reactor with a gas flow of more than 4000 m^3/h at a reaction pressure of 20 atm; this corresponds to about 13 cm/s gas velocity assuming standard conditions for the stated flow. It seems that internal slurry circulation using downcomers is now a more popular approach than external circulation but the principles remain the same. Modern commercial reactors would obviously also be larger and use significantly higher gas velocities. Many companies propose to use variations of this type of reactor, e.g. Statoil, ExxonMobil, Syntroleum, IFP/ENI/Agip and

ConocoPhilips. These reactors are given various names e.g. 'slurry phase', 'slurry bubble column' and 'liquid phase'.

With the modern approach, which is accompanied by high liquid wax yields from the reactor, it is important to avoid particle degradation so as to facilitate the catalyst/liquid separation. Sasol has solved problems related to the catalyst degradation of its supported cobalt catalyst.

The most important product by volume from the proposed GTL plants are distillates. The best use of the synthetic LTFT diesels will be to upgrade, by blending, the quality of conventional diesel fuels. Based on their superior quality, the synthetic distillates are the ideal, high value, blending component for upgrading of lower-quality stock derived from catalytic and thermal cracking operations, for example cycle oils (35, 36). In 1948 JA Tilton from the Esso Standard Oil Company suggested that "*there is a possibility that the Fischer-Tropsch synthesis process may eventually be a source of premium quality Diesel fuels*" (37). It is interesting to note that it was also anticipated that blends of the use of these high Cetane number diesels with lesser fuels were usable to meet a 50 Cetane specification, concluding that these high Cetane number fuels "might be used either alone or as blending agents". The blend material from conventional processing includes distillate cuts from FCC and Coker units. The blended final product will be low in sulphur and aromatics, a fuel compatible with demanding environmental legislation. Hence the products could enter a market where the LTFT diesel characteristics are valued as blend material to meet local requirements. It is also possible to use this fuel as a neat fuel in applications where its premium characteristics are desired.

2.4 The Initiation of the FT GTL Commercial Era

Shell has been involved in syngas chemistry for many years, giving special attention to the options for the conversion of natural gas into more easily transportable liquid hydrocarbons. The first result of this effort has been the Shell Middle Distillate Synthesis (SMDS) plant commissioned in Malaysia in 1993. This plant makes use of cobalt FT catalyst and tubular reactors in the Heavy Paraffin Synthesis unit (HPS). A simplified flow scheme of this plant is presented in Figure 7.

The plant includes hydrocracking of the LTFT products over a dual functional catalyst in the Heavy Paraffins Conversion (HPC) unit. The products of the SMDS Bintulu plant include naphtha, kerosene, diesel and some fuel gas. The HPC unit is operated typically at 30-50 bar total pressure and at a temperature of about 300-350°C, actually performing four functions:

- hydrogenation of the olefins present in the FT products,
- removal of the oxygenates, mainly primary alcohols,
- hydroisomerisation, and

- hydrocracking of the n-paraffins to iso-paraffins.

The reported distillation ranges for the kerosene and gasoil products were 150-250°C and 250-360°C, respectively (35, 36, 38).

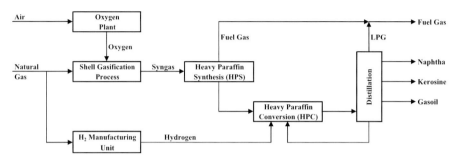

Figure 7 – Simplified Flow Scheme of the Shell SMDS plant in Malaysia

A different GTL concept, based on the HTFT Synthol technology developed by Sasol, is in commercial operation at the PetroSA GTL plant in Mossel Bay, South Africa since 1993 (39). These reactors are an improved version of those originally designed for the Secunda plants. This plant takes advantage of the process integration of natural gas liquids with the synthetic hydrocarbons and is presented in Figure 8.

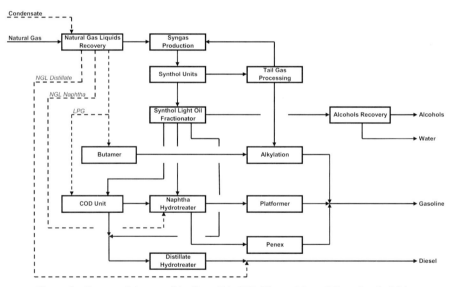

Figure 8 – Process Scheme of the PetroSA GTL Plant - Mossel Bay, South Africa

The process scheme is designed primarily for the production of gasoline (39) with smaller volumes of distillate fuels. As expected, the hydrogen required for the hydroconversion units is obtained from the synthesis gas. Two important integration features of this plant are (i) the simultaneous hydrotreatment of the naphtha contained in natural gas liquids with the synthetic naphtha before the catalytic reformer, and (ii) the blending of natural gas condensate and synthetic distillates in the final products.

2.5 Commercial Expansion

There was a time when it was thought that future large scale GTL plants will not be able to target higher value hydrocarbon products due to limitations imposed by the size of the markets. For example, Shell stated in 1995 that their future SMDS projects will be based primarily on transportation fuels (36). However, for at least three important products namely ethylene, propylene and lubricant base oils (also known as waxy raffinate), the markets are large enough to sustain large scale co-production of these products. In the recently announced Shell plant in Qatar, that is eventually expected to produce about 140 000 bpsd of hydrocarbon products, the production of lubricant base oils and paraffins is included in the initial product slate.

Sasol has proposed two approaches for the large scale production of higher value hydrocarbons via FT technology. Firstly, there is the option to modify the product upgrading schemes for the primary products from supported cobalt catalyst to produce olefins and lubricant base oils in addition to diesel fuel (40). Secondly, Sasol believes that its proprietary iron based FT processes can be used to further enhance or expand the production of higher value hydrocarbon products with fuels as significant by-products. Being highly olefinic, the nature of the products from iron catalysts is well suited to extraction and processing for the production of commodity chemicals.

The Sasol Chevron global alliance announced in 2004 the next phase of the projects proposed for Qatar: (i) an expansion of the 34 000 bpsd Oryx GTL plant to a total capacity of 100 000 bpsd by 2009, and (ii) a new 130 000 bpsd upstream/downstream integrated project by 2010. These projects will include production of FT-derived Base Oils (41). Additional prospects are under evaluation in Australia and Algeria.

3. State of Development of the LTFT GTL Technology

Most of the recent interest for future LTFT GTL plants has been for the use of supported cobalt catalyst in slurry phase reactors. There are good reasons for this approach not least of which is the relative simplicity of the process that lends itself to successful application at remote locations. Work is in progress to construct plants at Ras Laffan, Qatar (ORYX GTL) and Escravos, Nigeria (EGTL). This concept is used for the Sasol Slurry Phase Distillate™ (Sasol SPD™) process to produce mainly diesel fuel with by-product naphtha. The Shell Middle Distillate Synthesis (SMDS™) using supported cobalt catalyst in fixed bed reactors will also be applied on a large scale at the Pearl GTL plant in Ras Laffan, Qatar. Although these processes are ideally suited for the production of diesel, both Sasol and Shell have recognized the opportunity to produce chemicals in a synergistic manner.

Both HTFT iron catalysts and cobalt catalysts are suitable for natural gas conversion. Only Sasol offers technologies based on both cobalt catalyst and HTFT iron catalyst.

At present there are ten major companies actively working on the development of GTL technology. Eight of them have been able to get to the demonstration scale phase, with production capacities at the scale of hundreds of barrels per day. The current development status of the GTL FT technology of these companies is shown below.

		Research Scale	Pilot Plant Scale	Demonstration Scale	Basic Engineering	Detailed Engineering	Construction	Commercial Operation
Slurry Bed	Sasol	Y	Y	Y	Y	Y	ORYX GTL	Sasolburg
	Sasol Chevron					EGTL		
	ExxonMobil	Y	Y	Baton Rouge				
	Rentech	Y	Y	Denver				
	BP	Y	Y	Nikiski				
	ConocoPhillips	Y	Y	Ponca City				
	IFP/Agip/Eni	Y	Y	Sannazaro				
	JNOC	Y	Y	Yufutsu				
	Statoil	Y	Y	Mossel Bay				
	Syntroleum	Y	Y	Catoosa				
	Shell	Y	La Porte					
Fixed Bed	Sasol	Y	Y	Y	Y	Y	Y	Sasolburg
	Shell	Y	Y	Y	Y	Pearl GTL	Y	Bintulu
HTFT	Sasol	Y	Y	Y	Y	Y	Y	Secunda
	PetroSA							Mossel Bay

The substitution of a natural gas reforming technology for a coal gasification technology for the synthesis gas production step used by Sasol commercially for many years is relatively straightforward. Iron catalysts are better suited to the use of coal derived synthesis. Removal of catalyst poisons is much easier when using natural gas which is particularly beneficial when using the more expensive but longer lasting cobalt catalysts.

4. Conclusions

- The FT GTL process of today is well related to the commercial concepts of yesterday. The term GTL was used by R.C. Alden in 1946. The first commercial FT GTL plant was in operation in the 1950s in the USA.
- The cobalt FT catalyst, key for performance in the early years, is being revived as the catalyst of choice for the LTFT GTL projects under consideration.
- The Sasol plants in South Africa have been a fundamental element in the development of the LTFT synthetic fuels industry.
- The synergy between the products from FT and Natural Gas Separation plants has been proposed since the late 1940s as a sensible commercial opportunity.
- The LTFT Slurry reactor system is a flexible unit that can be operated over a broad range of process conditions.
- The primary LTFT products can be converted into commercial products using proven technology. process configurations are very much related to those proposed in the 1980s.
- The LTFT GTL diesel can be used to improve the quality of poor diesels like those derived from cracking operations.

This chapter is taken from a paper presented at the 227th National Meeting of the American Chemical Society (Anaheim, California – March 2004).

References
(1)   A.N. Stranges, "Germany's Synthetic Fuel Industry 1927-45", Paper 80a presented at the AIChE National Spring Meeting, March-April 2003, New Orleans
(2)   C.H. Bartholomew, "History of Cobalt Catalyst Design for FTS", Paper 83b presented at the AIChE National Spring Meeting, March-April 2003, New Orleans
(3)   A.P. Steynberg and M.E. Dry (editors), "Fischer-Tropsch Technology", Elsevier, Studies in Surface Science and Catalysis, v.152, (2004)
(4)   Y.Q. Zhang and B.H. Davis, "Indirect Coal Liquefaction – Where do we stand?", paper 26-6 presented at the Fifteenth Annual International Pittsburgh Coal Conference, September 1998, University of Pittsburgh, Proceedings, pp.1739-1803 (1998)

398

(5)   H.R. Batchhelder, "Synthetic Fuels" in Advances in Petroleum Chemistry and Refining, edited by J.J. McKetta, vol.5 chp.1 – Interscience Publishers, John Wiley & Sons (1962)

(6)   R.C. Alden, "The Conversion of Natural Gas to Liquid Fuels", California Oil World, Oct.1946, No. 2 pp. 5-16

(7)   R.C. Alden, "Conversion of Natural Gas to Liquid Fuels", Oil Gas J, 09.Nov.46 pp.79-85

(8)   US Naval Technical Mission in Europe, "The Synthesis of Hydrocarbons and Chemicals from CO and H2", Technical Report No. 248-45 – September 1945

(9)   C.C. Ward, F.G. Schwartz and N.G. Adams, "Composition of Fischer-Tropsch Diesel Fuel (cobalt catalyst)", Ind.Eng.Chem., 43 (5) pp.1117-1119, May 1951

(10)   FH Bruner, "Synthetic Gasoline from Natural Gas", Ing Eng Chem, 41 (11) pp.2511-2515 (1949)

(11)   P.C. Keith, "Gasoline from Natural Gas", Oil Gas J, 15 Jun 1946, pp.102-112

(12)   M.E. Dry, "The Fischer-Tropsch Synthesis", Cat Sci & Tech, ed. by JR Anderson, pp.159-255 (1981)

(13)   B. Jager, R.C. Kelfkens and A.P. Steynberg, "A Slurry Bed Reactor for Low Temperature Fischer-Tropsch", Natural Gas Conversion II, Elsevier Science B.V. (1994)

(14)   J.C. Hoogendoorn and J.H. Salomon, "Sasol: World's Largest Oil-from-Coal Plant" - Part 3, British Chem. Eng., July 1957, p.368-373

(15)   M Dry, "The Sasol Fischer-Tropsch Processes", App. Ind Cat, v2 Ch.5, ed. by BE Leech, pp.167-213 (1983)

(16)   J.J.C. Geerlings, J.H. Wilson, G.J. Kramer, H.P. Kuipers, A. Hoek, and H.M. Huisman, "Fischer-Tropsch technology - from active site to commercial process", Applied Catalysis A: General 186 (1999) 27.

(17)   J.W.A. de Swart, PhD Thesis, University of Amsterdam, (1996).

(18)   J.W.A. de Swart, R. Krishna, and S.T. Sie, Studies in Surface Science and Catalysis, 107 (1997) 213.

(19)   M.E. Dry, "Sasol's Fischer-Tropsch Experience," Hydroc.Processing, pp. 121-124, August 1982

(20)   M.E. Dry, "The Sasol route to fuels," Chemtech, pp. 744-750 (1982)

(21)   H. Beuther, T.P. Kobylinski, C.E. Kibby, and R.B. Pannell, South African Patent No. ZA 855317 (1985).

(22)   W-D. Deckwer, "Reaktionstechnik in Blasensaeulen," Otto Salle Verlag GmbH & Co, Frankfurt am Main Verlag Sauerlaender AG, Aarau, Switzerland (1985); W.-D. Deckwer, "Bubble Column Reactors", (translated by Valeri Cottrell) R.W. Field (ed.), John Wiley and Sons, New York, 1992.

(23)   C. Maretto, V. Piccolo, J.-C. Viguie and G. Ferschneider, "Fischer-Tropsch Process", U.S. Patent 6,348,510, 19 February 2002.

(24)   W.C. Behrmann, C.H. Mauldin and L.E. Pedrick, "Hydrocarbon synthesis reactor employing vertical downcomer with gas disengaging means", U.S. Patent RE37,229, 12 June 2001.

(25)   M. Chang, "Enhanced gas separation for bubble column draft tubes" U.S. Patent 5,332,552, 26 July 1994.

(26)   A.P. Steynberg, H.G. Nel and R.W. Silverman, "Process for producing liquid and optionally, gaseous products from gaseous reactants", U.S. Patent 6,201,031, 13 May 2001.

(27)   D. Casanave, P. Galtier and J-C. Viltard, "Process and Apparatus for operation of a slurry bubble column with application to the Fischer-Tropsch synthesis", U.S. Patent 5,961,933, 5 October 1999.

(28)   W.C. Behrmann, C.H. Mauldin and L.E. Pedrick, "Hydrocarbon synthesis reactor employing vertical downcomer with gas disengaging means", U.S. Patent 5,382,748, 17 January 1995.

(29)   W.C. Behrmann and C.J. Mart, "Gas and solids reducing slurry downcomer", U.S. Patent 5,866,621, 2 February 1999.

(30)   S.C. Leviness, "Multizone downcomer for slurry hydrocarbon syntheses process", U.S. Patent 5,962,537, 5 October 1999.

(31)   J-M. Schweitzer, P. Galtier, F. Hugues and C. Maretto, "Method for producing hydrocarbons from syngas in three-phase reactor", U.S. Patent Application 2003/0109590, 12 June 2003

(32)   H.V. Atwell, "Method of effecting catalytic conversions", U.S. Patent, 2,438,029, 16 March 1948.

(33)   H. Lethäuser, W.L. Linder and E. Sattler, "Apparatus for the production of hydrocarbons", U.S. Patent 2,775,512, 25 December 1956.

(34)   A.P. Vogel, A.P. Steynberg and P.J. van Berge, "Process for producing liquid and, optionally gaseous products from gaseous reactants", U.S. Patent 6,462,098, 8 October 2002.

(35)   P.J.A. Tijm, H. van Wechem and M. Senden, "The Shell Middle Distillate Synthesis Project – New Opportunities for Marketing Natural Gas", Alternate Energy '93 (Colorado Springs, April 27-30 1993), p.9

(36)   P.J.A. Tijm, J.M. Marriott, H. Hasenack, M.M.G. Senden, and Th. Van Herwijnen, "The Markets for Shell Middle Distillate Synthesis Products", Alternate Energy '95, Vancouver, Canada, May 2-4, 1995.

(37)   J.A. Tilton, W.M. Smith and W.G. Hockberger, "Production of High Cetane Number Diesel Fuels by Hydrogenation", Ind.Eng.Chem, 40, 7, 1269-1273 (1948).

(38)   M.M.G. Senden, S.T. Sie, M.E.M. Post and J. Ansorge. "Engineering Aspects of the Conversion of Natural Gas into Middle Distillates", HJ de Lasa et al (eds.), Chemical Reactor Technology for Environmentally Safe Reactors and Products, pp.227-247 (1992)

(39)   K. Terblanche, "The Mossgas Challenge", Hydroc.Engineering, v2 n2, pp.2-4 (1997)

(40)   A.P. Steynberg, W.U. Nel and M.A. Desmet, "Large Scale Production of High Value Hydrocarbons using Fischer-Tropsch Technology", paper to be presented at the 7th Natural Gas Conversion Symposium, Dalian, China, June 2004

(41)   Press Release from Sasol Limited Pty (Ltd), "Record $6 Billion GTL Initiative", 24 March 2004.

Fischer-Tropsch Synthesis, Catalysts and Catalysis
B.H. Davis and M.L. Occelli (Editors)

Concepts for Reduction in CO_2 Emissions in GTL Facilities

Dennis J. O'Rear and Fred Goede

ChevronTexaco Corporation, 100 Chevron Way, P.O. Box 1627, Richmond, CA 94952
Sasol Limited, Baker Square West, 33 Baker Street, Rosebank 2196, South Africa

1. Introduction

Life Cycle Assessments (LCA) of Gas-To-Liquids (GTL) processes and conventional petroleum refining have found comparable Green House Gas (GHG) emissions[1,2]. In the GTL system, higher GHG emissions are associated with the production phase are offset by lower GHG emissions in the use phase. While about $1/3^{rd}$ of the carbon in the natural gas ends up as CO_2 during the GTL process step, portions of this CO_2 are concentrated in various streams in the process. LCA studies have indicated that conventional amine scrubbing coupled with sequestration can be used to capture most of this CO_2 thus giving GTL an advantage over conventional refinery based systems. However, the conventional amine scrubbing/sequestration approach suffers a high capital cost – approximately 30% of the costs to make the syngas[3] -- primarily associated with compressing the captured CO_2. This paper explores improvements over the conventional amine scrubbing/sequestration system. The two general approaches are improved absorption systems and reaction systems.

2. Absorption Systems

As noted, a significant cost for CO_2 sequestration in a GTL facility is associated with the compression of the captured CO_2. In the conventional amine scrubbing system, the CO_2 is freed from the amine by reducing its pressure and raising the temperature. The CO_2 could be desorbed at high temperature without pressure decrease. However when amines are operated in this Temperature-Swing mode, the high temperature desorption step causes the amines to decompose. Absorption systems based on water or caustic alkali metal solutions, do not suffer from this problem, and may be used in a Temperature-Swing mode.

Shown below is a GTL process and points where CO_2 can be captured.

Figure 1.
Locations for CO_2 capture in a GTL Facility

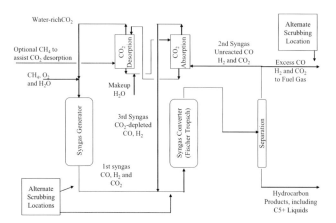

CO_2 primarily forms in the syngas generation step and to a smaller extent in the Fischer Tropsch (FT) step, especially if non-shifting FT catalysts are used. But being an inert, it builds up in the recycle gas system of the GTL process and can reach levels approaching 50%. This requires the use of a bleed stream for fuel. Figure 1 shows that the recovered CO_2 is recycled back to the syngas generator, where it is used to control the H_2/CO ratio. Recycling relatively pure CO_2, vs the unprocessed tail gas, has the advantage of reducing the CO and H_2 fed to the syngas generator, and increasing the capacity of the generator. The recovered CO_2 can also be sequestered in an underground or marine environment. Since it is produced at higher pressure, the compression costs are reduced. Since CO_2 builds up slowly in the recycle gas loop, all of it does not need to be removed.

The technical feasibility of this operation was evaluated as follows. A typical recycle gas stream composition is shown in Table 1 along with Henry's Law constants[4].

Table 1. FT Tail Gas Analysis (Mol %) and Henry's Law Constants

CH_4	13%	37,800
CO_2	35%	1,460
CO	25%	53,600
H_2	25%	68,300
Inerts (N_2)	2%	80,400

Four cases were studied by the Kremser Brown Method. This method provides only a theoretical estimate of the staging, scrubbing, and stripping requirements. It is also designed to handle multi component systems. Utility requirements are not addressed in this preliminary calculation.

Case 1   Scrubbing the tail gas. 50% of the CO2 was absorbed from the unreacted syngas using water at 20°C and 315 psia. Then, 99.5% of the CO2 was recovered in a water rich CO2 stream and recycled back to the syngas generator. No CH4 purge is used, so all of the stripping comes from steam generation at 315 psia (217°C).

Case 2   Scrubbing the tail gas. 50% of the CO2 was absorbed from the unreacted syngas using water at 20 C and 315 psia. Then, 99.5% of the CO2 was recovered using CH4 stripping at 100°C. No reboiler is used, so there should be very little water in the desorbed gas.

Case 3   Scrubbing the tail gas. Same flowsheet, except that CH4 stripping is done at 50°C.

Case 4   Scrubbing the fuel gas. 75% of the CO2 was absorbed from the unreacted syngas using sea water at 20°C and 315 psia. The scrubbed gases are used as fuel. Then, 99.5% of the CO2 was recovered in a water rich CO2 stream. The CO2 rich, desorbed gases are sequestered. No external purge is used, so all of the stripping comes from steam generation at 315 psia (217°C).

The composition of the scrubbed gas and the desorbed gases for the four cases were calculated with a 100 mol basis of gas to the absorber.

Table 2. Gas Compositions

	Case 1	Case 2	Case 3	Case 4
Scrubbed Gas				
Total Flow, kmol	81.59	81.59	81.59	72.34
Composition: Mol/100 mol to absorber and % removed				
CO_2	17.5 50	17.5 50	17.5 50	8.8 75
CH_4	12.7 2	12.7 2	12.7 2	12.6 3
CO	24.6 1	24.6 1	24.6 1	24.4 2
H_2	24.7 1	24.7 1	24.7 1	24.6 2
N_2	2.0 1	2.0 1	2.0 1	2.0 1
Desorbed Gas				
Total Flow, kmol/hr	23	69	139	36
Composition: Mol/100 mol to absorber and Mol % (H_2O free)				
CO_2	17 99	17 25	17 12	26 73
CH_4	0.26 1.1	0.26 0.4	0.26 0.2	0.40 1.1
CO	0.35 1.5	0.35 0.5	0.35 0.3	0.55 1.5
H_2	0.28 1.2	0.28 0.4	0.28 0.2	0.43 1.2
N_2	0.018 0.08	0.018 0.03	0.019 0.011	0.03 0.08
H_2O	4.56 -			8.49 -
CH_4 as Purge		51 73	121 87	

Table 3. Process Analysis

Absorber	Case 1	Case 2	Case 3	Case 4
Feed Gas Flowrate (kmol/hr)	2000	2000	2000	5000
Temp (°C)	20	20	20	20
Pressure (psia)	315	315	315	315
Theo. Stages	4	4	4	6
Theo. Water Requirement. (mol liq/mol gas)	35.35	35.35	35.35	54.83
Practical Water Requirement. (mol liq/ mol gas)	52.5	52.5	52.5	82.25
Stripper				
Desorbed Gas Flowrate (kmol/hr)	460	1380	2780	1800
Temp (°C)	217	100	50	217
Pressure (psia)	315	315	315	315
Theoretical Stages	3	3	3	3
Theo. Stripping Requirement (mol liq/mol gas)	153	51.5	25.38	153
Practical Stripping Requirement (mol liq/mol gas)	76.5	25.75	12.69	76.5

These results show that reasonable designs for adsorption and desorption columns can be used to provide high levels of CO_2 removal. Removal of CO_2 does not remove significant quantities of valuable components (H_2, CH_4 and CO). The product gas streams are high purity CO_2 (when a reboiler is used) or CO_2/CH_4 mixtures when CH_4 is used as a stripping gas. The recovered CO_2, either as a neat stream or as a CO_2/CH_4 mixture, can be fed to the syngas generator.

There are many alternatives. For example, if water is readily abundant, it can be used on a once-through basis to scrub CO_2 from the gas and sequester the mixture. Also, water can be produced from an underground aquifer, used to scrub CO_2, and reinjected into the same or different aquifer thereby eliminating the stripping and gas handling systems. Further details on these options are described elsewhere[5].

CO_2 can also be present in the natural gas feedstock. These scrubbing concepts can be extended to purify the feedstock prior at processing in the GTL facility, or for other uses (LNG, sales, combustion, etc). This can eliminate the need for both CO_2 compression and the CO_2 stripping section. Water is pressurized and circulated rather than gas. Ideally an aquifer is found that is slightly caustic. During discharge into the aquifer, the pressure of the water needs to be controlled to prevent damage to the formation. If aqueous scrubbing does not result in sufficient removal of the CO_2, a smaller amine scrubbing system can be used, with the CO_2 and acid gas discharge directed to the water or original gas stream.

Figure 2.

Use of aqueous scrubbing with elimination of CO_2 compression.

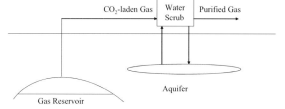

3. Reaction Systems

The other general approach is to convert the CO_2 into liquid products. Several concepts are discussed below.

3.1. Reduction with H_2 to CO.

The simplest approach is to react the CO_2 with H_2 and convert it into additional CO which can be processed in the FT reactor. However a source for the H_2 needs to be found. If it is generated by processing of natural gas, there will likely be no net benefit in GHG emissions. Alternatively, H_2 can be produced by processing the FT naphtha in a naphtha reformer to produce aromatics and a H_2 rich by-product gas stream. The aromatics can be used in gasoline or petrochemicals. The H_2 and the CO_2 (either in the entire tail gas stream or as a separate component) can then be processed in the stream reformer (or a separate shift reactor) to form additional CO. Water is separated. Further details of this concept are presented elsewhere[6].

3.2. Reduction with H_2 in a Shifting FT Reactor.

CO_2 will react in a FT reactor when shifting catalysts are used (Fe). Thus it is possible to consider a small FT reactor using a Fe catalyst and H_2 (recovered from a naphtha reforming operation) to convert the CO_2 into FT product that resemble those from the larger non-shifting slurry bed reactor[7].

The products from the two reactors can be combined and distilled in a single operation. This concept may have particular merit as GTL facilities expand and multiple trains are used.

3.3. Reduction with H_2 in a Dual Functional Reactor.

A reaction between H_2 and CO (or CO_2) over composite catalysts for methanol synthesis (Zn, Cr, Cu, etc) and methanol conversion catalysts (zeolites) was studied extensively during the 1970's. The product from this process resembles that from the MTG process and is a mixture of aromatics and isoparaffins. It has a high octane and can be used as a gasoline component. The advantage of this process is that it produces very little methane by-product. The disadvantage of this process is that it operates at higher pressures than typically used in a FT process. Because these products differ from the highly paraffinic FT products it may not be best to recover them in the same distillation facilities. But since they resemble the products from naphtha reforming, the distillation section of these two operations can be combined[8].

3.4. Dry Reforming with LPG.

Considerable research has been done in dry reforming of methane, e.g. the reaction between CO_2 and methane to form syngas with a stoichiometric ratio near unity. This reaction suffers from several problems that appear to have a common origin in the relatively low reaction rate. In a O_2-blown GTL facility it is critical to get high methane conversions (low methane slip) otherwise the methane rapidly builds up in the recycle gas loop where it must be purged. When attempts are made to operate the dry reforming reaction at high methane conversion, solid carbon forms. This carbon not only plugs equipment directly, but is well know to attack the metal surfaces of the processing equipment and cause metal dusting and erosion. For these reasons, the dry reforming reaction remains only of academic interest.

However the analogous reaction between higher hydrocarbons (propane, butane, LPG etc) and CO_2 appears to operate at much higher rates, and with little or no formation of methane by-product[9]. Unlike dry reforming of methane, the dry reforming of LPG is a commercial process used to generate CO for use in hydroformylation. A commercial process called CALCOR and licensed by Caloric. In a GTL location processing dry gas rather than associated gas, a small amount of LPG is often produced. Also LPG is produced in the FT process. In order to be sold into existing markets, the LPG must be distilled into specification propane and butane, and then shipped in pressurized containers. The cost of this operation often makes the value of LPG at the GTL site low or even negative. Thus other options besides sale of LPG are desired.

408

The equivalent dry-reforming reaction between LPG and CO_2 is an option to both eliminate LPG handling and reduce CO_2 emissions. Furthermore, the syngas produced by such a reaction has a H_2/CO ratio below 1. A source of such a syngas can be of value in a FT-GTL facility. A FT process consumes a syngas with a stoichiometric ratio of 2, but for practical operation, the ratio fed to the FT reactor needs to be below 2. With a ATR syngas generator, production of syngas with a ratio below 2 can only be achieved by recycling of CO_2. Insertion of CO_2 into the ATR creates potential problems of carbon deposition and metal dusting analogous to those found in dry reforming. To avoid these problems, steam is added to the feed, but this acts to increase the H_2/CO ratio. In contrast, if a low H_2/CO ratio syngas is available from dry reforming of LPG and CO_2, the ATR can be operated without CO_2 recycle and with reduced H_2O. This increases the capacity of the ATR unit. This concept is described in greater detail elsewhere[10].

A direct comparison of the dry reforming of methane and LPG is apparently not available in the open literature. Such a comparison would be worthwhile, as would be greater emphasis on the LPG reaction rather than the methane reaction.

4. Conclusions

The presence of high concentrations of CO_2 at elevated pressures in the GTL process offers some unique opportunities to reduce CO_2 emissions. These concepts can be extended to purification of CO_2-laden natural gas and to hydrogen or power production from coal or heavy oil.

5. Acknowledgement.

Daniel Chinn's provision of the absorption studies is gratefully appreciated. Thanks to others who contributed to this work: Curt Munson, Steve Zavell, and Charles Kibby.

- **References**

[1] Gas to Liquids Life Cycle Assessment Synthesis Report, August 2004, Five Winds International

[2] CO2 Abatement in Gas-To-Liquids Plant: Fischer-Tropsch Synthesis, IEA Greenhouse Gas R&D Program, Report Number PH3/15 November 2000.

[3] EP 0516441

[4] Arthur Kohl and Richard Nielsen, Gas Purification, Gulf Publishing Company, 1997, pp 417 465. This source also provides the temperature dependence of the constants.

[5] US6723756, US6720359, US6667347, US6620091

[6] US6693138, US6723756, US20030191197

[7] US20050113465

[8] US20050113463

[9] Hydrocarbon Processing, Vol. 64, May 1985, pp. 106-107 and Hydrocarbon Processing, Vol. 66, July 1987, pg. 52.

[10] US6774148

STUDIES IN SURFACE SCIENCE AND CATALYSIS

Advisory Editors:
B. Delmon, Université Catholique de Louvain, Louvain-la-Neuve, Belgium
J.T. Yates, University of Pittsburgh, Pittsburgh, PA, U.S.A.

412

416

418

419

420